Methods in Molecular Biology

For further volumes:
http://www.springer.com/series/7651

Sepsis

Diagnostic Methods and Protocols

Edited by

Nicasio Mancini

University Vita-Salute San Raffaele, Milan, Italy

Humana Press

Editor
Nicasio Mancini
University Vita-Salute San Raffaele
Milan, Italy

ISSN 1064-3745 ISSN 1940-6029 (electronic)
ISBN 978-1-4939-1775-4 ISBN 978-1-4939-1776-1 (eBook)
DOI 10.1007/978-1-4939-1776-1
Springer New York Heidelberg Dordrecht London

Library of Congress Control Number: 22014949861

Printed on acid-free paper

Humana Press is a brand of Springer
Springer is part of Springer Science+Business Media (www.springer.com)

Preface

Sepsis is a major clinical problem that takes an unbearable toll both on lives and economical resources. Clinical and microbiological diagnosis are therefore of pivotal importance in the management of septic patients, as it is widely recognized that an inappropriate treatment is associated, especially in the first hours, with a dramatic increase in mortality. To put it in two words: "Time matters!"

When Prof. John M. Walker contacted me proposing to edit this volume, I was completely aware, as a medical microbiologist, of the difficulty of the task. However, I accepted by proposing a somehow peculiar table of contents considering the standard format of an MiMB volume. Microbiological, clinical, and pathophysiological aspects of sepsis should have been included in the volume. This is the reason why in *Sepsis: Diagnostic Methods and Protocols*, the usual protocol format of an MiMB volume coexists with general overview chapters and with chapters discussing the real clinical impact of the diagnostic approaches.

I am profoundly indebted to all authors who contributed to this volume with their different expertise and to Prof. John Walker for inviting me to edit it.

Milan, Italy *Nicasio Mancini*

Acknowledgments

I am grateful to Prof. Massimo Clementi and Prof. Roberto Burioni for useful discussions and valuable suggestions in the planning of this volume. I also thank Dr. Silvia Carletti for the helpful assistance in revising its chapters.

Contents

Contributors

DAMIEN BARRAUD • *Department of Medical Intensive Care, Hôpital Central, Nancy Cedex, France*
FRANK BLOOS • *Department of Anesthesiology and Intensive Care Medicine, Jena University Hospital, Jena, Germany*
ROBERTO BURIONI • *Laboratory of Microbiology and Virology, University Vita-Salute San Raffaele, Milan, Italy*
JEAN-MARC CAVAILLON • *Unit Cytokines and Inflammation, Institut Pasteur, Paris, France*
MASSIMO CLEMENTI • *Laboratory of Microbiology and Virology, University Vita-Salute San Raffaele, Milan, Italy*
CLAUDIA DISQUÉ • *Molzym GmbH & Co. KG, Bremen, Germany*
SHIRAZ GEFEN-HALEVI • *Infectious Disease Unit/Microbiology Laboratory, Sheba Medical Center, Tel Hashomer, Israel*
MARIANNA GEORGITSI • *4th Department of Internal Medicine, University of Athens Medical School, Athens, Greece*
EVANGELOS J. GIAMARELLOS-BOURBOULIS • *4th Department of Internal Medicine, University of Athens, Medical School, Athens, Greece*
SÉBASTIEN GIBOT • *Department of Medical Intensive Care, Hôpital Central, Nancy Cedex, France*
GERD HABERHAUSEN • *Roche Diagnostics GmbH, Penzberg, Germany*
ACHIM HAECKER • *Roche Diagnostics GmbH, Penzberg, Germany*
MUSA Y. HINDIYEH • *Central Virology Laboratory, Public Health Services, Department of Epidemiology and Preventive Medicine, School of Public Health, Sackler Faculty of Medicine, Ministry of Health, Chaim Sheba Medical Center, Tel-Hashomer, Israel; Real Time Molecular Diagnosis Unit, Israel Central Virology Laboratory, Chaim Sheba Medical Center, Tel-Hashomer, Israel*
JAROSLAV HRABÁK • *Department of Microbiology, Faculty of Medicine and University Hospital in Plzen, Charles University in Prague, Plzen, Czech Republic*
KLAUS-PETER HUNFELD • *Institute of Laboratory Medicine, Microbiology and Infection Control, Northwest Medical Centre, Frankfurt, Germany*
FRANCESCA ISELLA • *Department of Anesthesia and Intensive Care, IRCCS San Raffaele Scientific Institute, Milan, Italy*
NATHAN KELLER • *Infectious Disease Unit/Microbiology Laboratory, Sheba Medical Center, Tel Hashomer, Israel*
HARALD H. KESSLER • *Institute of Hygiene, Microbiology and Environmental Medicine (IHME), Medical University of Graz, Graz, Austria*
PHILIPPE LAGACÉ-WIENS • *Department of Medical Microbiology and Infectious Diseases, Infectious Diseases, University of Manitoba, Winnipeg, MB, Canada*
GIOVANNI LANDONI • *Department of Anesthesia and Intensive Care, IRCCS San Raffaele Scientific Institute and University Vita-Salute San Raffaele, Milan, Italy*
EVA LEITNER • *Institute of Hygiene, Microbiology and Environmental Medicine (IHME), Medical University of Graz, Graz, Austria*

CHRISTOPHE LELUBRE • *Department of Intensive Care, Erasme Hospital, Université Libre de Bruxelles, Brussels, Belgium*
JÉRÉMIE LEMARIÉ • *Department of Medical Intensive Care, Hôpital Central, Nancy Cedex, France*
BENEDIKT LOHR • *Institute of Laboratory Medicine, Microbiology and Infection Control, Northwest Medical Centre, Frankfurt/Main, Germany*
MICHAEL G. LORENZ • *Molzym GmbH & Co. KG, Bremen, Germany*
YING-YI LUAN • *Department of Microbiology and Immunology, Burns Institute, First Hospital Affiliated to the Chinese PLA General Hospital, Beijing, People's Republic of China*
MADELEINE MAI • *Institute of Laboratory Medicine, Microbiology and Infection Control, Northwest Medical Centre, Frankfurt/Main, Germany*
MINNA MÄKI • *Orion Diagnostica Oy, Espoo, Finland*
NICASIO MANCINI • *Laboratory of Microbiology and Virology, University Vita-Salute San Raffaele, Milan, Italy*
DANIELA MANEG • *Institute of Laboratory Medicine, Microbiology and Infection Control, Northwest Medical Centre, Frankfurt/Main, Germany*
ELLA MENDELSON • *Central Virology Laboratory, Chaim Sheba Medical Center, Public Health Services, Ministry of Health, Tel-Hashomer, Israel; Department of Epidemiology and Preventive Medicine, School of Public Health, Tel-Aviv University, Tel-Aviv, Israel*
GIACOMO MONTI • *Department of Anesthesia and Intensive Care, IRCCS San Raffaele Scientific Institute, Milan, Italy*
YOSHITOMO MORINAGA • *Department of Laboratory Medicine, Nagasaki University Graduate School of Biomedical Sciences, Nagasaki, Japan*
HELGE MÜHL • *Molzym GmbH & Co. KG, Bremen, Germany*
IRIS MÜLLER • *Institute of Laboratory Medicine, Microbiology and Infection Control, Northwest Medical Centre, Frankfurt/Main, Germany*
GIACOMO PARDINI • *Alifax S.p.A., Polverara, PD, Italy*
MARIANNA PARLATO • *Unit Cytokines and Inflammation, Institut Pasteur, Paris, France*
SCOTT POWELL • *Nanosphere, Inc., Northbrook, IL, USA*
TERESA RAICH • *Nanosphere, Inc., Northbrook, IL, USA*
ZHI-YONG SHENG • *Department of Microbiology and Immunology, Burns Institute, First Hospital Affiliated to the Chinese PLA General Hospital, Beijing, People's Republic of China*
GILL SMOLLAN • *Infectious Disease Unit/Microbiology Laboratory, Sheba Medical Center, Tel Hashomer, Israel*
DAIANA TADDEO • *Department of Anesthesia and Intensive Care, San Raffaele Scientific Institute, Milan, Italy*
MARC VAN NUFFELEN • *Department of Intensive Care, Erasme Hospital, Université Libre de Bruxelles, Brussels, Belgium*
JEAN-LOUIS VINCENT • *Department of Intensive Care, Erasme Hospital, Université Libre de Bruxelles, Brussels, Belgium*
KATSUNORI YANAGIHARA • *Department of Laboratory Medicine, Nagasaki University Graduate School of Biomedical Sciences, Nagasaki, Japan*

YONG-MING YAO • *Department of Microbiology and Immunology, Burns Institute, First Hospital Affiliated to the Chinese PLA General Hospital, Beijing, People's Republic of China*
ALBERTO ZANGRILLO • *Department of Anesthesia and Intensive Care, IRCCS San Raffaele Scientific Institute and University Vita-Salute San Raffaele, Milan, Italy*
QING-HONG ZHANG • *Department of Microbiology and Immunology, Burns Institute, First Hospital Affiliated to the Chinese PLA General Hospital, Beijing, People's Republic of China*

Chapter 1

Microbiological Diagnosis of Sepsis: The Confounding Effects of a "Gold Standard"

Nicasio Mancini, Roberto Burioni, and Massimo Clementi

Abstract

The need of rapid and sensitive diagnostic techniques for sepsis is every day more compelling. Its morbidity and mortality loads are dramatically high, with one quarter of patients eventually dying. Several diagnostic progresses have been made in the last years using both molecular- and nonmolecular-based approaches, and they have to be broadly shared in the scientific community also under the technical point of view. The initial chapters of this book give a thorough overlook of the state of the art in the actual diagnosis of sepsis. The other chapters provide a broad range of protocols describing both already used and futuristic tools, covering both microbiological and nonmicrobiological aspects. The potential role of each described protocol is evidenced by a brief introduction on the specific topic of each chapter. A final chapter describing algorithms potentially useful in stratifying the risk of sepsis in each single patient and suggesting the future perspectives in the diagnosis of sepsis closes the book.

Key words Blood culture, Mass spectrometry, Molecular-based assays, Nonmolecular biomarkers

1 The Impact of Sepsis

Sepsis is still a major clinical challenge whose medical, but also economical, impact is still probably underestimated [1–3]. Recent data on its real incidence is lacking, due to several factors certainly influencing it and, therefore, to the difficulty of setting up reliable epidemiological studies [4, 5]. However, each practitioner certainly bears in mind the effects of sepsis (especially of severe sepsis, complicated by severe organ dysfunctions) on his or her patients, with mortality ranging from 20 to 45 % [1, 3, 5–7].

Clinical diagnosis is pivotal to allow prompt recognition of the ongoing clinical picture and, therefore, prompt therapeutical intervention. The recently revised "International Guidelines for Management of Severe Sepsis and Septic Shock" within the "Surviving Sepsis Campaign" clearly stress this need [8]. This is the reason why, differently of most of the manuals published in the *Methods and Protocols* series, this volume includes several clinically oriented chapters.

Nicasio Mancini (ed.), *Sepsis: Diagnostic Methods and Protocols*, Methods in Molecular Biology, vol. 1237, DOI 10.1007/978-1-4939-1776-1_1, © Springer Science+Business Media New York 2015

Indeed, the protocols reported in the more bench-oriented chapters have to be focused to their potential clinical usefulness; unfortunately, this is not a concept always followed by companies or research group developing diagnostic assays in this field.

As a matter of fact, it is every day more compelling the need of diagnostic assays capable of a sensitive and, importantly, rapid microbiological diagnosis of sepsis. Indeed, a late clinical, but also microbiological, diagnosis of sepsis is associated with several possible short-, medium-, and long-term drawbacks. As an example, several papers report that a clinically unrecognized sepsis is burdened by higher mortality [8, 9]. However, at the same time, a delayed or incorrect microbiological diagnosis may inevitably lengthen the use of empirically administered broad-spectrum antibiotics, which in the medium and long term may favor the selection of multidrug-resistant strains [10–12] with all the resulting medical, epidemiological, but also economical effects [13–15].

2 The Microbiological Diagnosis of Sepsis

The role of the microbiology lab is still very limited in the acute phases of sepsis, especially considering the abovementioned guidelines which strongly recommend prompt antibiotic therapy (possibly within 1 h of clinical suspect) [8]. As reported in Chapters 3 and 4 of this manual, clinicians have not sufficient time to wait for microbiological results, especially when only culture-based assays are used. Blood culture is still considered the "gold standard" in confirming the clinical suspect of sepsis, and most of the epidemiological studies performed to date are based on it. However, blood culture suffers from the usual drawbacks associated with culture-based assays, that is, the interfering effects of ongoing antibiotic therapy, the long time to positivity which often are not compatible with the real clinical needs, and the possible presence of fastidious pathogens not growing in available culture media [16]. The suboptimal sensitivity of blood culture is certainly one of the above-cited confounding factors influencing the results of epidemiological studies on sepsis. To date, the real advantage of blood cultures is related to the possibility of performing phenotypical antibiotic susceptibility testing on grown isolates.

Molecular-based diagnostic assays have been repeatedly considered as potentially adjuvant tools, allowing a faster and reliable diagnosis and a more targeted therapeutic approach [16–21]. Several molecular approaches have been suggested as a possible support to culture in the microbiological diagnosis of sepsis [16–18]. They may be classified in two main groups: (1) the assays performed on positive blood culture bottles, potentially useful in allowing a rapid identification at the genus or species level of grown pathogens and, more importantly, of the main multidrug-resistance genetic markers (i.e., extended spectrum β-lactamases, carbapenemases), but obviously influenced by the abovementioned drawbacks associated with culture, and (2) those performed directly on blood, which could theoretically represent a dramatic revolution in the microbiological diagnosis of sepsis. Several protocols related to examples belonging to both groups of assays are reported in the chapters of this manual.

A really promising approach, not covered in this manual due to the need of improving and validating the available protocols, is PCR mass spectrometry, that is, PCR amplification of a broad-range molecular target followed by analysis of base composition by mass spectrometry (MS) of genus- and species-specific sequences. The mostly used technology is PCR coupled with electrospray ionization mass spectrometry (PCR-ESI), which combines the potential sensitivity of PCR with the extreme rapidity of MS techniques thus allowing its use directly on blood samples.

3 The Nonmicrobiological Biomarkers of Sepsis

The need of a rapid diagnosis has also opened new perspectives in the research field of nonmicrobiological biomarkers of sepsis, that is, molecules rapidly produced by the immune response during the different phases of a septic episode. Most of the possible markers are still in the first steps of study, but it is easy to foresee that some among them will certainly make their way into the clinics. Although not substituting microbiological diagnosis, these markers will (and some of them already are, such as procalcitonin) certainly help the clinicians in making the right choice in the first crucial hours of a septic episode. Under this perspective, we are very grateful to Prof. Jean-Marc Cavaillon that makes an unprecedented overview on all biomarkers actually under investigation in Chapter 15 of this manual.

4 Concluding Remarks

This volume is the first manual specifically addressing both laboratoristically and clinically some of the most compelling diagnostic aspects related to sepsis. Several different strategies have already been applied to this field, and most of them are reviewed in this manual.

As foreseen by some of the protocols reported in this book, the possibility of confirming laboratoristically a clinical suspect of sepsis and, more importantly, of tailoring therapy in the first crucial hours is certainly closer. This will certainly dramatically change both the clinical flowchart and the laboratory organization in coping with this clinical emergency. We all have to prepare and to be ready for that moment.

References

1. Melamed A, Sorvillo FJ (2009) The burden of sepsis-associated mortality in the United States from 1999 to 2005: an analysis of multiple-cause-of-death data. Crit Care 13(1):R28
2. Martin GS et al (2003) The epidemiology of sepsis in the United States from 1979 through 2000. N Engl J Med 348(16):1546–1554
3. Ginde AA et al (2013) Impact of older age and nursing home residence on clinical outcomes of US emergency department visits for severe sepsis. J Crit Care 28:606–611. doi:10.1016/j.jcrc.2013.03.018
4. Lagu T et al (2012) What is the best method for estimating the burden of severe sepsis in the United States? J Crit Care 27(4):414, e1–9
5. Hall MJ et al (2011) Inpatient care for septicemia or sepsis: a challenge for patients and hospitals. NCHS Data Brief 62:1–8
6. Wang HE et al (2007) National estimates of severe sepsis in United States emergency departments. Crit Care Med 35(8):1928–1936
7. Dombrovskiy VY et al (2007) Rapid increase in hospitalization and mortality rates for severe sepsis in the United States: a trend analysis from 1993 to 2003. Crit Care Med 35(5):1244–1250
8. Dellinger RP et al (2013) Surviving sepsis campaign: international guidelines for management of severe sepsis and septic shock—2012. Crit Care Med 41(2):580–637
9. Conde KA et al (2013) Differences in sepsis treatment and outcomes between public and private hospitals in brazil: a multicenter observational study. PLoS One 8(6):e64790
10. Cornaglia G, Giamarellou H, Rossolini GM (2011) Metallo-beta-lactamases: a last frontier for beta-lactams? Lancet Infect Dis 11(5):381–393
11. Calfee DP (2012) Methicillin-resistant Staphylococcus aureus and vancomycin-resistant enterococci, and other Gram-positives in healthcare. Curr Opin Infect Dis 25(4):385–394
12. Yezli S, Li H (2012) Antibiotic resistance amongst healthcare-associated pathogens in China. Int J Antimicrob Agents 40(5):389–397
13. Burchardi H, Schneider H (2004) Economic aspects of severe sepsis: a review of intensive care unit costs, cost of illness and cost effectiveness of therapy. Pharmacoeconomics 22(12):793–813
14. Moerer O et al (2007) A German national prevalence study on the cost of intensive care: an evaluation from 51 intensive care units. Crit Care 11(3):R69
15. Chalupka AN, Talmor D (2012) The economics of sepsis. Crit Care Clin 28(1):57–76, vi
16. Mancini N et al (2010) The era of molecular and other non-culture-based methods in diagnosis of sepsis. Clin Microbiol Rev 23(1):235–251
17. Reinhart K et al (2012) New approaches to sepsis: molecular diagnostics and biomarkers. Clin Microbiol Rev 25(4):609–634
18. Murray PR, Masur H (2012) Current approaches to the diagnosis of bacterial and fungal bloodstream infections in the intensive care unit. Crit Care Med 40(12):3277–3282
19. Mancini N et al (2008) Molecular diagnosis of sepsis in neutropenic patients with haematological malignancies. J Med Microbiol 57(Pt 5):601–604
20. Mancini N et al (2009) Molecular diagnosis of polymicrobial sepsis. J Clin Microbiol 47(4):1274–1275
21. Mancini N et al (2012) Potential role of the detection of enterobacterial DNA in blood for the management of neonatal necrotizing enterocolitis. J Med Microbiol 61(Pt 10): 1465–1472

Chapter 2

Pathophysiological Aspects of Sepsis: An Overview

Yong-Ming Yao, Ying-Yi Luan, Qing-Hong Zhang, and Zhi-Yong Sheng

Abstract

Sepsis is defined as severe systemic inflammation in response to invading pathogens, or an uncontrolled hyperinflammatory response, as mediated by the release of various proinflammatory mediators. Although some patients may die rapidly from septic shock accompanied by an overwhelming systemic inflammatory response syndrome (SIRS) triggered by a highly virulent pathogen, most patients survive the initial phase of sepsis, showing multiple organ damage days or weeks later. These patients often demonstrate signs of immune suppression accompanied by enhanced inflammation. Sepsis is a result of a complex process; there is interaction of various pathways, such as inflammation, immunity, coagulation, as well as the neuroendocrine system. This treatise is an attempt to provide a summary of several key regulatory mechanisms and to present the currently recognized molecular pathways that are involved in the pathogenesis of sepsis.

Key words Sepsis, Inflammation, Immunity, Coagulation, Neuroendocrine

1 Introduction

Sepsis with subsequent multiple organ dysfunction syndrome (MODS) is a distinct systemic inflammatory response to concealed or known infection and is a leading cause of death in intensive care units. As a result of a concerted effort to disclose the underlying pathogenetic mechanisms, there have been accumulating evidences to suggest that the profound proinflammatory and anti-inflammatory responses that occur in sepsis are balanced by an array of counter-regulatory molecules involved in an effort to restore immunological equilibrium. Thus, sepsis has been theorized to develop through two stages: initially there is a release of large quantities of inflammatory mediators, including tumor necrosis factor (TNF)-α, interferon (IFN)-γ, and interleukin (IL)-2; with the progression of the disease condition, negative immune regulation would be elicited, including reduction of reproductive activity of lymphocytes, decreased proinflammatory cytokines and antigen presentation, as well as increased expression of immunosuppressive cytokines (IL-10) and other molecules. This T helper cell (Th) 2 cells-mediated phase

Nicasio Mancini (ed.), *Sepsis: Diagnostic Methods and Protocols*, Methods in Molecular Biology, vol. 1237,
DOI 10.1007/978-1-4939-1776-1_2, © Springer Science+Business Media New York 2015

might lead to an increase in susceptibility to infection, also caused by massive apoptosis of lymphocytes.

In addition to inflammation and immunity, coagulation and neuroendocrine systems are also considered to be important regulatory pathways in the pathogenesis of sepsis. Here, we will attempt to review what is presently known about the four key regulatory mechanisms of sepsis, particularly with focus on inflammation and cell immunity, and to discuss the interplay of these factors in the pathogenesis of sepsis. We suggest that the successful treatment of sepsis will require a better understanding of its pathophysiological mechanisms, allowing more rational and targeted prophylactic and therapeutic approaches.

2 Inflammatory Response During Sepsis

Traditionally, sepsis has been defined as a clinical syndrome with manifestation of fever, tachycardia, leukocytosis/leukopenia, and other symptoms and signs of infection. It might represent an appropriate but inadequate response against an overwhelming infection or uncontrolled inflammation [1, 2]. As a result of a concerted effort to disclose the underlying pathogenetic mechanisms, there have been accumulating evidences to suggest that sepsis is described as the systemic inflammatory response syndrome (SIRS) resulting from infection [3]. Insults, such as hemorrhagic shock, traumatic and severe tissue injury, thermal injury, and ischemia-reperfusion injury, can lead to SIRS.

2.1 Induction of Inflammation by Pathogens

Toll-like receptors (TLRs) expressed on the cell surface and intracellular compartment could specifically recognize molecules shared by a variety of microbial components called pathogen-associated molecular patterns (PAMPs). Then, through the recognition of PAMPs, a wide array of cytokines and chemokines may be released from the immune cells, thus igniting the inflammatory process. Induction of TLR signaling by low doses of bacteria/toxins has been implicated as an important event in the induction of a protective innate immune response. However, uncontrolled stimulation of TLRs can potentially lead to disproportionate inflammation and tissue injury [4], and it may occur during sepsis. It is known that TLR signaling is tightly regulated, and there are several negative regulators designated to prevent excessive TLR signaling. It has been found that an overexpression of Triad3A (one of the ubiquitin modifying enzyme) promotes substantial degradation of TLR4 and TLR9, but it does not affect TLR2 expression, and a decrease in induction of TLR4 or TLR9 signaling may not affect TLR2 signaling [5]. Nevertheless, it should be noted that the ubiquitin-proteasome pathway is not the only negative regulator. There is growing evidence that tripartite-motif protein (TRIM) 30α is a negative regulator of TLR in mediating nuclear factor (NF)-κB, and it may target TAB2 and TAB3 for degradation. Recent studies suggested that TLRs was discriminated among different pathogen-associated molecules and activated signaling cascades that led to immune response. Radioprotective105 (RP105) expression is a specific homologue of TLR4 and together with its helper molecule, myeloid differential protein (MD) 1, has a comforted association with TLR4/MD2, and this association inhibits lipopolysaccharide (LPS)-TLR4/MD2 complex formation [6, 7]. Myeloid differential factor (MyD) 88 is an essential factor located in all TLRs except TLR3, and the latter is composed of three main domains, namely, the N-terminal death domain, the intermediate domain, and the Toll/interleukin-1 receptor domain. Overexpression of MyD88 can inhibit LPS-induced NF-κB activation. Therefore, it has become evident that disorder in functions of immune cells is closely related with signal transducing signal pathway induced by TLRs.

2.2 Release of Inflammatory Mediators During Sepsis

The host response to sepsis can either be balanced or unbalanced. Many different components of this host response involved in sepsis may contribute to different outcomes. The interaction between pathogens and innate immune receptors triggers the release of a myriad of inflammatory mediators, among them are cytokines. Cytokines are small proteins that promote a wide variety of inflammatory reactions at tissue level and play an eminent role in the pathogenesis of bacterial infection and sepsis. In addition to the proinflammatory cytokines (TNF-α and IFN-γ), many other cytokines have been shown to be of importance in regulating the septic host response, such as IL-6, IL-35, and high-mobility group box-1 protein (HMGB1).

HMGB1 has been indicated in animal experimentation that the serum levels of HMGB1 are increased at late time points after endotoxin exposure [8], which can amplify and exacerbate the inflammatory response by triggering the release of these cytokines (Fig. 1). Up to now, a series of research on its extrachromosomal activity and inflammation-promoting activity have demonstrated that HMGB1 is also released by various activated immune cells. HMGB1 has been shown to be able to provoke inflammation, to regulate the migration of monocytes [9], to contribute maturation of many antigen-presenting cells (APC) via the receptor for advanced glycation end products (RAGE) in vitro [10], and to stimulate $CD4^{+}CD25^{+}$ regulatory T cell (Treg) activity via binding RAGE on the surface of Tregs and trigger a shift of Th1 to Th2 with suppression of T lymphocyte immune function [11]. The potential role of RAGE signaling in the inflammatory response accompanying sepsis has been documented in mice with abdominal sepsis that both RAGE deficient mice and wild-type mice treated with soluble RAGE were partially protected against lethality in this model of severe sepsis.

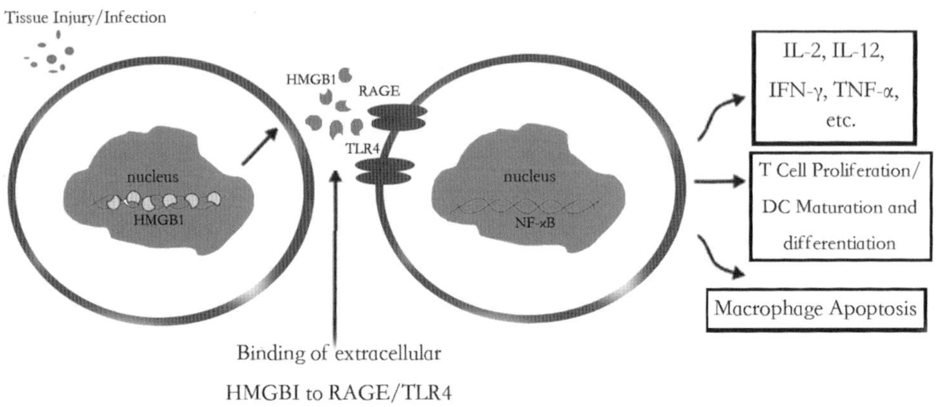

Fig. 1 Functions of HMGB1 in the extracellular environment. HMGB1 can be passively released from damaged and infected cells undergoing necrotic or pyroptotic cell death. Extracellular HMGB1 can bind to its receptors RAGE or TLR4 on effector cells to contribute to secretion of proinflammatory cytokines, macrophage apoptosis, T cell proliferation, and DC maturation and differentiation. *DC* dendritic cell, *HMGB1* high-mobility group box-1 protein, *IFN-γ* interferon-γ, *RAGE* receptor for advanced glycation end products, *TLR4* Toll-like receptor-4

3 Sepsis and Host Immune Response

To date, it is known that the complicated sepsis syndrome may lead to both widespread activation and dysfunction of the innate immune system [12]. Activation of host innate immunity may occur not only after a microbial invasion but also result from exposure to internal "danger" signals produced by cell injury, tissue ischemia, hypoxia, and necrosis. As the innate immune system is activated highly enough, the host response itself is able to render the patient to manifest SIRS or even shock and MODS/multiple organ failure (MOF). Although some patients survive the initial SIRS insult, these patients are still at risk of developing secondary or opportunistic infections, because of the frequent onset of a compensatory anti-inflammatory response syndrome (CARS). A growing body of evidence shows that innate immune cells from patients with severe sepsis are able to promote an upsurge in anti-inflammatory cytokines and reverse the Th2 type response, including regulatory T cells (Tregs) and regulatory dendritic cells (DCs), rendering the patient to enter a state of immune depression and CARS. Therefore, both innate immunity and inflammation must be taken into consideration in the development of severe sepsis.

3.1 Neutrophils and Innate Immune Response in Sepsis

Neutrophils are abundant in the blood but absent in normal tissues. They have the shortest life span among leukocytes, surviving only a few hours after leaving the bone marrow. In the early phase of sepsis, a considerable reserve pool of mature neutrophils within the bone marrow can be rapidly mobilized, resulting in a dramatic rise in circulating neutrophil number that is available for recruitment to the sites of infection [13]. Compared to mature neutrophils, immature neutrophils have a longer life span and resistance to spontaneous apoptosis and higher basal intracellular TNF-α/IL-10 ratio and are also capable of mediating important innate immune functions though less efficiently [3].

Mechanisms governing neutrophil function in sepsis are complex. A failure of neutrophil migration in lethal sepsis and a reduced survival rate have been demonstrated. During the process of sepsis, neutrophil migratory responses can be regulated by bacterial products, cytokines/chemokines, leukotrienes, and immunomodulatory hormones via induction of cytoskeletal changes, disruption of polymorphonuclear leukocyte (PMN)-endothelial cell interactions, and alterations in G-protein-coupled receptor expression or signaling [14]. It is known that the dysregulated PMN G-protein-coupled receptor and TLR expression and/or signaling can result in the alteration of leukotriene functions, further leading to proinflammatory and immunomodulatory gene suppression, as well as decreased production of reactive oxygen species in sepsis. In addition, recently an elevated levels of circulating form of C5aR (cC5aR) in serum and reduction of the C5a receptor on neutrophils were detected in septic shock [15]. Therefore, as an essential effector of the innate immune response, impaired recruitment and migration of neutrophils are correlated with a poor outcome in severe sepsis.

3.2 The T Lymphocyte-Mediated Immunity and Sepsis

In the development of sepsis, several lines of evidence have indicated that T lymphocytes play a central role in cell-mediated immune response, including a remarkable attenuation of reproductive activity and the predominant Th2 immune reaction (Fig. 2). The initiation of immunological reaction mediated by Th2 cells accompanied by apoptosis of a large number of lymphocytes might lead to an increase in susceptibility to infection [16]. Besides, γδ-T cells may play a potential role in the postburn survival and sepsis, cytokine formation by Th1 and Th2 cells, initiation of neutrophil-mediated tissue damage, and wound healing. Treg-mediated immunosuppressive effect is mainly dependent on the drift of Th1/Th2 caused by the activation of T cell receptor signal. Tregs can release a variety of immunosuppressive cytokines (IL-4, IL-10), which might markedly promote the Th2 immune reaction.

Previous studies clearly illustrated that multiple apoptotic pathways might be involved in sepsis. During sepsis, there is an increase in proapoptotic protein Bim and a decrease in the level of antiapoptotic molecules (Bcl-2 and Bcl-xL) in T cells [17]. Transgenic mice that selectively overexpress Bcl-2 and Bcl-xL in T lymphocytes display appreciable protection against lymphocyte

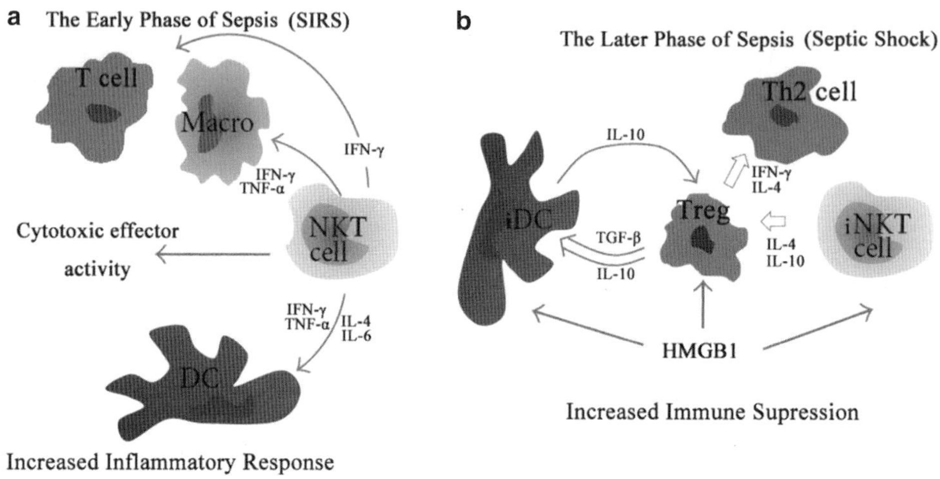

Fig. 2 The immunoregulatory effect of immune cells in different phases of sepsis. (**a**) In the early phase of sepsis, macrophages, DCs, T lymphocytes, regulatory T cells, and NKT cells play pivotal roles in the maintenance of peripheral homeostasis and regulation of inflammatory response by releasing various proinflammatory cytokines, including TNF-α and IFN-γ. NKT cells also possess cytotoxic effector activity via Fas/FasL and perforin pathway. (**b**) In the later phase of sepsis, production of inhibitory cytokines ensued, including IL-10 and IL-4, which limit the strength of immune cell activation and expansion and negatively modulate inflammation. In addition, extracellular HMGB1 has been shown to be able to provoke inflammation, to regulate functions of macrophage and T lymphocytes, and to mediate immune function of DCs and Tregs. *HMGB1* high-mobility group box-1 protein, *DC* dendritic cell, *Tregs* regulatory T cells, *NKT cell* natural killer T cell, *NF-κB* nuclear factor-κB, *IFN-γ* interferon-γ

apoptosis as well as significantly improved survival in cecal ligation and puncture (CLP)-induced sepsis [18]. The extrinsic pathway of apoptosis can be initiated by different FasL and Fas-associated receptors. Finally, activation of caspase-9 initiates the aspase-3 and caspase-8 that induces apoptosis, it has been shown that CLP-induced septic mice with downregulation of caspase-8 could decrease T lymphocyte apoptosis and improve survival of mice. In addition to the intrinsic and extrinsic pathways, endoplasmic reticulum (ER) stress-mediated pathway is also involved in the apoptotic process of T lymphocytes via activation of numerous overlapping cascades (caspase-12) [19], while the precise regulatory mechanism has yet to be fully elucidated. More recently, the cross talk of apoptotic pathways during sepsis has been implicated in both intrinsic pathways and the ER-mediated or extrinsic pathway [20]. T cell apoptosis can affect innate immune response, including a decrease in IFN-γ and IL-17 within the first 24 h, reduction of ability to limit macrophage phagocytosis of dead T cells, and the subsequent production of IL-10 as well as transforming growth factor (TGF)-β. Septic thymocyte apoptosis can be promoted by glucocorticoid, and it has been regarded as the most clinically relevant treatment for sepsis. The cell death in the adaptive immune system is beneficial to the host by downregulating the inflammatory response to sepsis, but the extensive loss of immune cells may compromise the ability of the host to eliminate the invading pathogens and finally lead to septic death, illustrating that an increased apoptosis in T lymphocytes plays a critical role in the adverse outcome of sepsis.

3.3 The Effect of Dendritic Cells on Immune Function in Sepsis

DCs play important roles in host resistance and immunogenicity and exhibit different expression of TLRs and cytokines by stimulation of various pathogens. Stimulation of TLRs may activate myeloid differentiation primary response protein 88 (MyD88)-dependent signaling pathways, which enhance the release of a range of proinflammatory cytokines. DCs, which are differentiated from peripheral monocytes, express TLR1, 2, 4, 5, and 8, and upon LPS stimulus, they strongly produce TNF-α and IL-6 [21]. TLR2 and TLR4 are involved in the mechanisms leading to depletion of splenic DC following polymicrobial sepsis [22]. Immature $CD11c^+$ DCs express predominantly TLR1, 2, and 3, and they secrete high levels of IL-10 following activation and induce T-regulator type 1 cells both in vivo and in vitro [23]. Following severe trauma, burns, and sepsis, in addition to TLRs, DCs have the remarkable capacity to present processed antigens via major costimulatory molecules (CD80, CD86) and major histocompatibility complex (MHC) to initiate the development of innate as well as adaptive immune responses. Recently, several reports have elucidated that endoplasmic reticulum stress (ERS) response was involved in differentiation and plasticity of T cells and also development and maturation of DC [24]. The endoplasmic reticulum, one of the most important organelles in eukaryotic cells, is extremely sensitive to alterations in homeostasis. In response to ERS, mammalian cells trigger unfolded protein response (UPR) signaling pathways to cope with stressful conditions and to monitor the protein folding capacity of the ER and to transmit that information to mechanisms that can modulate the ER environment, thus regulating various aspects of cellular metabolism and even influencing the fate of the cell [25]. Zhu et al. [25] demonstrated that HMGB1 might induce maturation and activation of DCs and regulate its function by modulating the ERS response as well as UPR pathway, thus providing intensive insights into the critical mechanism for endogenous sources of cellular stress in the central role of DCs in immunity during sepsis. More recently, the potential regulatory function of a DC subset, characterized by its particular surface marker expression of $CD11c^{low}CD45RB^{high}$, has also been investigated. Regulatory $CD11c^{low}CD45RB^{high}$ DCs may affect the immunological activity through releasing high levels of suppressor cytokines (IL-10) but low levels of proinflammatory cytokines (IL-12), thereby exerting immunoregulatory effects. Fujita and others found that these regulatory $CD11c^{low}CD45RB^{high}$ DCs generated in vitro by culturing bone marrow cells obtained from mice protected against septic response to microbial pathogens in innate immunity [26]. Increased regulatory capacity of CD11c^{low}CD45RB^{high} DC can be associated with uncontrolled inflammatory responses followed by tissue and organ destruction. Nevertheless, the potential roles of different sets of DCs and their exact molecular mechanisms in pathologic conditions such as sepsis have not yet fully been clarified.

4 Activation of Coagulation During Sepsis

The coagulation system is finely balanced in health so that there is neither excessive coagulation nor excessive hemorrhage. Sepsis is associated with multiple alterations in procoagulating and anticoagulating mechanisms [27]. These alterations may disturb the fine balance and lead to full-blown disseminated intravascular coagulation. During sepsis the activation of coagulation is primarily driven by tissue factor (TF), which is an essential mediator of coagulation and a potent stimulator of the extrinsic coagulation cascade leading to increased levels of coagulation factor Va and VIIIa. The increased Va and VIIIa ultimately trigger the conversion of fibrinogen to fibrin in the microcirculation. Activation of the coagulation system and ensuing thrombin generation are dependent on expression of TF and the simultaneous downregulation of endothelial-bound anticoagulant mechanisms and endogenous fibrinolysis during endotoxemia and sepsis. However, coagulation also considerably affects inflammatory activity; activated coagulation proteases, such as the tissue factor–factor a complex, factor Xa, and thrombin, can bind to protease-activated receptors on inflammatory cells and endothelial cells; and the ensuing intracellular signaling leads to increased production of proinflammatory cytokines and chemokines and thereby modulates the inflammatory response [28].

Procoagulant events are controlled by a family of anticoagulant proteins, including tissue factor pathway inhibitor (TFPI), antithrombin III, and protein C [28]. Protein C is activated by the binding of thrombin-α to thrombomodulin and binding of protein C to the endothelial protein C receptor. During severe sepsis, there is a rapid and profound deficiency of protein C as well as a decrease in endothelial protein C receptor, thereby protein C converts to the much more potent activated protein C(APC). APC plays pivotal roles in severe sepsis because it inactivates factors Va and VIIIa and inhibits the synthesis of plasminogen activator inhibitor-1 (PAI-1). In sepsis, in addition to APC, the activity of TFPI and antithrombin is impaired, which together with enhanced TF-dependent coagulation results in a shift toward a net procoagulant state.

5 Dysfunction of Neuroendocrine System in the Development of Sepsis

The immune system and the central nervous system are able to affect each other, and reciprocal interactions between them play an important role in the host response in septic shock. The expression of corticotropin-releasing hormone (CRH) in the hypothalamus and adrenocorticotropic hormone (ACTH) in the pituitary gland can be induced by proinflammatory cytokines [29]. Thus, enhanced adrenal release of cortisol suppresses the activation of NF-κB and activates anti-inflammatory cytokines. The inflammatory cytokines may either suppress cortisol response to adrenocorticotropin [30] resulting in insufficient adrenal output, which is coined as "relative adrenal insufficiency," or compete with intracellular glucocorticoid receptor function, leading to peripheral tissue glucocorticoid resistance. The central nervous system can also control inflammation through the cholinergic anti-inflammatory pathway and the efferent arm of the inflammatory reflex. It consists of the efferent vagus nerve, the neurotransmitter acetylcholine, and the nicotinic acetylcholine receptor alpha7 subunit. Probably, the transmission of information takes place at postsynaptic sympathetic fibers in the celiac plexus which terminate in the spleen and act on splenic immune cells [29].

The central nervous system controls a wide range of physiological functions that are crucial in regulating the immune system at all levels: innate immunity, adaptive immunity, and maintenance of immune tolerance, as well as maintain neuroendocrine and autonomic levels [31]. Disturbances in any of these adaptive functions may actively contribute to the pathogenesis of sepsis. For example, the autonomic nervous system regulates cytokine production through neural pathways. It was reported that an infusion of epinephrine could increase TNF-α, IL-6, and IL-1β contents in muscle [32]; stimulation of the efferent vagus nerve could regulate the levels of TNF-α, HMGB1, and other cytokines during endotoxemia via "the cholinergic anti-inflammatory pathway," a mechanism involving the vagus nerve and its major neurotransmitter, acetylcholine, through a process dependent on the α7 subunit of the nicotinic acetylcholine receptor [33]. Compelling studies with animal models have demonstrated that some neuropeptides may be effective in treating inflammatory disorders, such as sepsis, and T helper 1-driven autoimmune diseases, like Crohn's disease and rheumatoid arthritis [34].

The endocrine response to sepsis is complex, such as disruption of the hypothalamic-pituitary-adrenal axis and vasopressin deficiency, both of which are common features in severe sepsis. Indeed, it is now recognized that, in sepsis, adrenal insufficiency partly accounts for reduced vascular sensitivity to vasopressors and an increased risk for death, and circulating vasopressin levels also affect the course of septic shock [29]. So, in septic shock, correcting the disorder of adrenal axis and vasopressin may improve septic status and survival. Dysfunction of neuroendocrine system is a common complication of sepsis. Better understanding of several pathogenetic factors and mechanisms of neuroendocrine dysfunction may provide appealing strategies for alleviating hypermetabolism and hyperinflammatory immune response in severe sepsis.

6 Conclusions

Severe sepsis and septic shock occur with a high incidence in emergency departments and intensive care units, and they may result in both widespread activation and dysfunction of the innate as well as adaptive responses in immune system; disturb the fine balance of pro-inflammation and anti-inflammation, coagulation, and anticoagulation; and profoundly alter the neuroendocrine response. The early diagnosis of the septic condition is important for clinicians because adequate treatment antibiotics and other adjunctive and supportive therapies must be rapidly administered. Understanding the different mechanisms involved in severe sepsis and wise use of laboratory test will provide an opportunity to accurately evaluate the patient's pathophysiological status and to develop interventional strategies for septic complications as early as possible.

Acknowledgments

This work was supported, in part, by grants from the National Natural Science Foundation (81130035, 81372054, 81071545, 81272089, 81121004), the National Basic Research Program of China (2012CB518102), and the Medical Research Foundation of Chinese PLA (AWS11J008, BWS12J050).

References

1. Bone RC (1991) The pathogenesis of sepsis. Ann Intern Med 115:457–469
2. Bone RC, Sprung CL, Sibbald WJ (1992) Definitions for sepsis and organ failure. Crit Care Med 20:724–726
3. Parrillo JE, Parker MM, Natanson C et al (1990) Septic shock in humans: advances in the understanding of pathogenesis, cardiovascular dysfunction, and therapy. Ann Intern Med 113:227–242
4. Wang J, Hu Y, Deng WW et al (2009) Negative regulation of Toll-like receptor signaling pathway. Microbes Infect 11:321–327
5. Chuang TH, Ulevitch RJ (2004) Triad3A, an E3 ubiquitin-protein ligase regulating Toll-like receptors. Nat Immunol 5:495–502
6. Divanovic S, Trompette A, Atabani SF et al (2005) Negative regulation of Toll-like receptor 4 signaling by the Toll-like receptor homolog RP105. Nat Immunol 6:571–578
7. An H, Hou J, Zhou J et al (2008) Phosphatase SHP-1 promotes TLR- and RIG-I-activated production of type I interferon by inhibiting the kinase IRAK1. Nat Immunol 9: 542–550
8. Scaffidi P, Misteli T, Bianchi ME (2002) Release of chromatin protein HMGB1 by necrotic cells triggers inflammation. Nature 418:191–195
9. Rouhiainen A, Kuja-Panula J, Wilkman E et al (2004) Regulation of monocyte migration by amphoterin (HMGB1). Blood 104:1174–1182
10. Zhu XM, Yao YM, Liang HP et al (2009) The effects of high mobility group box-1 protein on splenic dendritic cell maturation in rats. J Interferon Cytokine Res 29:677–686

11. Huang LF, Yao YM, Zhang LT et al (2009) The effect of mobility group box-1 protein on activity of regulatory cells after thermal injury in rats. Shock 31:322–329
12. Souza HP, Lima-Salgado T, da Cruz Neto LM (2010) Toll-like receptors in sepsis: a tale still being told. Endocr Metab Immune Disord Drug Targets 10:285–291
13. Chang KC, Unsinger J, Davis CG et al (2007) Multiple triggers of cell death in sepsis: death receptor and mitochondrial-mediated apoptosis. FASEB J 21:708–719
14. Prakash PS, Caldwell CC, Lentsch AB et al (2012) Human microparticles generated during sepsis in patients with critical illness are neutrophil-derived and modulate the immune response. J Trauma Acute Care Surg 73:401–406
15. Andriantsitohaina R, Gaceb A, Vergori L et al (2012) Microparticles as regulators of cardiovascular inflammation. Trends Cardiovasc Med 22:88–92
16. Ma T, Han L, Gao Y et al (2008) The endoplasmic reticulum stress-mediated apoptosis signal pathway is involved in sepsis-induced abnormal lymphocyte apoptosis. Eur Surg Res 41:219–225
17. Weber SU, Schewe JC, Lehmann LE et al (2008) Induction of Bim and Bid gene expression during accelerated apoptosis in severe sepsis. Crit Care 12:R128
18. Schwulst SJ, Muenzer JT, Peck-Palmer OM et al (2008) Bim siRNA decreases lymphocyte apoptosis and improves survival in sepsis. Shock 30:127–134
19. Jimbo A, Fujita E, Kouroku Y et al (2003) ER stress induces caspase-8 activation, stimulating cytochrome c release and caspase-9 activation. Exp Cell Res 283:156–166
20. Oakes SA, Lin SS, Bassik MC (2006) The control of endoplasmic reticulum initiated apoptosis by the BCL-2 family of proteins. Curr Mol Med 6:99–109
21. Kadowaki N, Ho S, Antonenko S et al (2001) Subsets of human dendritic cell precursors express different Toll-like receptors and respond to different microbial antigens. J Exp Med 194:863–869
22. Pène F, Courtine E, Ouaaz F et al (2009) Toll-like receptors 2 and 4 contribute to sepsis-induced depletion of spleen dendritic cells. Infect Immun 77:5651–5658
23. Akdis M, Akdis CA (2007) Mechanisms of allergen-specific immunotherapy. J Allergy Clin Immunol 119:780–791
24. Kamimura D, Bevan MJ (2008) Endoplasmic reticulum stress regulator XBP-1 contributes to effector $CD8^+$ T cell differentiation during acute infection. J Immunol 181:5433–5441
25. Zhu XM, Yao FH, Yao YM et al (2012) Endoplasmic reticulum stress and its regulator XBP-1 contributes to dendritic cell maturation and activation induced by high mobility group box-1 protein. Int J Biochem Cell Biol 44:1097–1105
26. Fujita S, Seino K, Sato K et al (2006) Regulatory dendritic cells act as regulators of acute lethal systemic inflammatory response. Blood 107:3656–3664
27. Levi M, van der Poll T (2010) Inflammation and coagulation. Crit Care Med 38:S26–S34
28. Levi M (2010) The coagulant response in sepsis and inflammation. Hamostaseologie 30(10–12):14–16
29. Weismüller K, Bauer M, Hofer S et al (2010) The neuroendocrine axis and the pathophysiology of sepsis. Anasthesiol Intensivmed Notfallmed Schmerzther 45:574–578
30. Berczi I, Quintanar-Stephano A, Kovacs K (2009) Neuroimmune regulation in immunocompetence, acute illness, and healing. Ann N Y Acad Sci 1153:220–239
31. Sharshar T, Hopkinson NS, Orlikowski D et al (2005) Science review: the brain in sepsis—culprit and victim. Crit Care 9:37–44
32. Lang CH, Nystrom G, Frost RA (2008) Beta-adrenergic blockade exacerbates sepsis-induced changes in tumor necrosis factor alpha and interleukin-6 in skeletal muscle and is associated with impaired translation initiation. J Trauma 64:477–486
33. Wang H, Yu M, Ochani M et al (2003) Nicotinic acetylcholine receptor alpha7 subunit is an essential regulator of inflammation. Nature 421:384–388
34. Ameri P, Ferone D (2012) Diffuse endocrine system, neuroendocrine tumors and immunity: What's new? Neuroendocrinology 95:267–276

Chapter 3

Clinical Aspects of Sepsis: An Overview

Giacomo Monti, Giovanni Landoni, Daiana Taddeo, Francesca Isella, and Alberto Zangrillo

Abstract

Sepsis is one of the oldest and most elusive syndromes in medicine. With the confirmation of germ theory by Semmelweis, Pasteur, and others, sepsis was considered as a systemic infection by a pathogenic organism. Although the germ is probably the beginning of the syndrome and one of the major enemies to be identified and fought, sepsis is something wider and more elusive. In this chapter clinically relevant themes of sepsis will be approached to provide an insight of everyday clinical practice for healthcare workers often not directly involved in the patient's management.

Key words Sepsis, Severe sepsis, Septic shock, Sites of infection

1 Definition

Severe sepsis or septic shock is a complex syndrome defined in a consensus conference of many different and important scientific societies in 2001 [1]. This syndrome includes infection, suspected or documented, and presence of any of the diagnostic criteria shown in Table 1. This table has been adapted from the last available version of "Surviving Sepsis Campaign," published in 2013 [2]. Both former SIRS (systemic inflammatory response syndrome) criteria and organ dysfunction criteria are present.

These items are based both on clinical and laboratory parameters. Severe sepsis is defined as sepsis with an organ dysfunction.

Interestingly, over the years, tissue perfusion variables, especially lactatemia, have received great attention, and a threshold value of hyperlactatemia is nowadays included in the definition of severe sepsis, defining sepsis as "severe" also without clear organ involvement.

Septic shock is defined as persistent hypotension, with systolic blood pressure <90 mmHg or mean arterial blood pressure <70 mmHg, after adequate fluid resuscitation.

Nicasio Mancini (ed.), *Sepsis: Diagnostic Methods and Protocols*, Methods in Molecular Biology, vol. 1237,
DOI 10.1007/978-1-4939-1776-1_3, © Springer Science+Business Media New York 2015

Table 1
Criteria for severe sepsis and septic shock

Infection (proved or suspected) and any of the following
General variables
Fever (>38.3 °C) or hypothermia (core temperature <36 °C)
Heart rate > 90 bpm or >2 SD normal value for age
Tachypnea
Altered mental status
Significant edema or positive fluid balance (>20 ml/kg in 24 h)
Hyperglycemia (plasma glucose >140 mg/dl or 7.7 mmol/l) without diabetes
Inflammatory variables
Leukocytosis (WBC >12,000 cells/microL^{-1}) or leukopenia (WBC < cells/microL^{-1} 4,000)
Normal WBC with >10 % immature forms
Plasma C-reactive protein >2 SD above normal value
Plasma procalcitonin >2 SD above normal value
Hemodynamic variables
Arterial hypotension (SBP<89 mmHg, MAP<70 mmHg, or a SBP decrease >40 mmHg)
Organ dysfunction variables
Arterial hypoxemia ($PaO_2/FiO_2 < 300$)
Acute oliguria (urine output <0.5 ml/kg/h for at least 2 h despite adequate fluid resuscitation)
Creatinine increase >0.5 mg/dl or 44.2 micromol/l
Coagulation abnormalities (INR > 1.5 or aPTT > 60 s)
Ileus (absent bowel sounds)
Thrombocytopenia (PLT < 100,000 microL^{-1})
Hyperbilirubinemia (>4 mg/dl or 70 micromol/l)
Tissue perfusion variables
Hyperlactatemia (>1 mmol/l)
Decrease capillary refill or mottling

Sepsis is defined by infection (suspected or documented) and general or inflammatory variables. *Severe sepsis* requires at least one organ dysfunction

Septic shock is defined by persistent arterial hypotension despite adequate fluid resuscitation that requires inotropes or vasopressors

SD standard deviation, *WBC* white blood cells, *SBP* systolic blood pressure, *MAP* mean arterial pressure, PaO_2 arterial partial pressure of oxygen, FiO_2 fraction of inspired oxygen, *INR* international normalized ratio, *aPTT* activated partial thromboplastin time, *PLT* platelets. Adapted from Dellinger et al [2]

Despite this highly standardized definition of sepsis, there is up to 20 % variability in the incidence of severe sepsis and septic shock, due to variations in interpretation of SIRS criteria [3].

2 Epidemiology

According to Kumar [4], the rate of hospitalization for severe sepsis in the United States has increased from 143 to 343 every 100,000 people from 2000 to 2007. Mortality rates for severe sepsis exceed those of common medical conditions, such as myocardial infarction and stroke. In the last 20 years, mortality rates for severe sepsis decreased from over 50 % to almost under 30 % [5], with an odds reduction similar to that observed for other severe conditions, like congestive heart failure or surgery for intracerebral hemorrhage.

An apparently surprising observation is that, despite the reduction in mortality, nowadays hospitalized patients have higher rates of organ failure (respiratory, renal, and cardiovascular failure being the most commonly diagnosed) and also a higher probability of experiencing septic shock than only severe sepsis [4]. Nevertheless, mortality has decreased. This is probably related to advances in supportive care for the critically ill such as implementation of bundled care processes, low tidal volume ventilation for acute respiratory distress syndrome, and possibly extracorporeal membrane oxygenation.

The trend of increase in sepsis is expected to increase due to aging of the population, enhanced survival to chronic health conditions, and a wider access to advanced treatments, like high invasive surgery, transplant program, chemotherapy, and immunosuppressive therapy.

It is unclear if the trend in mortality reduction will continue also. Several important trials studying new therapeutic weapons failed to show survival benefits, either regarding new drugs [6, 7] or use of old drugs with new indications [8]. Even a promising drug as drotrecogin alfa or recombinant activated protein C, after some published efficacy data, failed to confirm its efficacy in two randomized controlled trials and has been removed from the market by the producer [9, 10]. Some reports of a similar drug (but in the zymogen form) are interesting, but high-quality evidence is still missing [11, 12]. Hospital-acquired infections have increased and account for 4.5% of admissions [13]. An alarming scarcity of new antibiotic classes in the pipelines of the pharmaceutical industry reduces availability of new molecules and has forced the healthcare community to optimize the therapeutic potential of currently available antibiotics [14], but pan-drug-resistant bacteria are reported.

3 Etiology

Etiology of sepsis is classically approached by considering the site of infection and the microbiological responsible pathogen.

1. The lungs represent the most common site of infection and pneumonia is associated with the highest mortality. According to Table 2, showing some of the most important (published after 2012 on journal with impact factor superior to 6, including at least 100 randomized patients) recent randomized studies on several sepsis treatments, the lungs were the site of

Table 2
Recently published major randomized studies on septic shock or severe sepsis

First author	Schortgen	Annane	Ranieri	Morelli	Brunkhorst	Guntupalli	Perner	Opal
Patients included	200	411	1,696	154	600	194	798	1,961
Population	SSH	SSH	SSH	SSH (a)	SeS, SSH	SeS	SeS	SeS
CA/HA	na	78/22 %	77/23 %	na	50/50 %	na	25/75 %	na
Lung	84 %	67 %	44 %	64 %	41 %	48 %	55 %	51 %
Abdomen	7 %	11 %	30 %	33 %	38 %	14 %	8 %	24 %
Genitourinary	6 %	17 %	12 %	<1 %	12 %	21 %	13 %	21 %
Bloodstream infection	na	14 %	5 %	0	3 %	0	0	5 %
Soft tissue, bones, joints	na	8 %	5 %	0	7 %	14 %	11 %	9 %
Others	14 %	8 %	8 %	0	15 %	na	10 %	8 %
Unknown site	5 %	4 %	na	0	na	na	na	na
Gram positive	26 %	40 %	na	na	54 %	51 %	na	27 %
Gram negative	41 %	44 %	na	na	49 %	27 %	na	32 %
Others (fungi, virus, anaerobia, etc.)	0 %	12 %	na	na	29 %	7 %	na	2 %
Mixed organisms	5 %	na	na	na	(b)	30 %	na	11 %
No pathogen identified	25 %	27 %	41 %	na	8 %	na	na	26 %

Type of population, major sites of infection, and causative pathogen in some of the major published randomized controlled studies on severe sepsis and septic shock
SSH septic shock, *SeS* severe sepsis, *na* not available, *CA/HA* community/hospital acquired. Causative pathogens sum may exceed 100 % because some had more than one infection site; (a) selected population; (b) already included in other groups

infection for 41 [15]–84 % [16] of enrolled patients. Secondarily, genitourinary tract, especially in young women, and intra-abdominal sepsis in surgical cohorts, account for, respectively, 1–21 % [7, 17] and 7–38 % [15, 16] of cases.

Bloodstream infections are expected to increase due to the higher number of implantable devices utilized, such as pacemakers and long-lasting central or peripheral inserted venous catheters. In some series, these could be responsible for up to 14 % of etiology of sepsis [10].

Knowing the site of infection is very important. Even though severe sepsis and septic shock are syndromes involving the whole body, the identification of the specific site of infection causes important subsequent actions.

First, it has been proved that the choice of the correct antibiotic molecule must consider its penetration and activity in the site of infection. It is well known, for example, that important molecules against severe pathogens like MRSA (*methicillin-resistant Staphylococcus aureus*), such as daptomycin, are very effective against the bloodstream infection but completely useless in lung disease, due to surfactant inactivation.

Second, the site of infection is associated with the risk of death. Urinary tract and intravascular catheter infections are less likely to be lethal than sepsis involving the lungs, abdomen, or soft tissues.

Third, some sites of infection will require adjunctive therapies other than antibiotics to obtain infection control. Drainage of abscess, revision of anastomosis, and debridement of tissue necrosis are sometimes fundamental to obtain source control in sepsis. Galeno's adage *ubi pus ibi evacua* (where there is pus, there evacuate it), from 150 b.C. is still very valid (nowadays probably not with extensive surgery, but with more accurate and conservative radiological guided procedures, as a percutaneous drainage of an abscess).

2. Many important data on the causative pathogens come from EPIC II study, an international collaborative study that enrolled 1,265 intensive care units (ICUs) all around the world in 2007 [18]. This study confirmed the respiratory tract to be the first site of infection but extensively evaluated causative microorganisms.

   Among pathogens, gram negative account for majority of isolates (62 %), with *Escherichia coli* in the prominent position (16 %). Gram positive account for 47 %, with *Staphylococcus aureus* (SA) in the first line (21 %).

   Pseudomonas species and fungi are important pathogens in sepsis nowadays (respectively, 20 % and 19 %).

   According to Table 2, it is important to notice that, even in highly controlled settings, the probability of missing the causative agent of sepsis is still too high: up to 41 % [9]. Missing pathogen identification could have important outcome on choice, tailoring, and escalation of antibiotic treatment. It is well known that missing efficacy of first-line antibiotics has a severe and important impact on patients' outcome [19].

Pathogen identification can also help to stratify patients' risk of death and advise on appropriate setting for treatment (ICU versus general ward) and intensity of clinical and laboratory monitoring.

Cohen and colleagues [20] reviewed half a thousand papers, including more than 55,000 patients with microbiologically confirmed infections. Analyzing in detail the interaction between the site of infection and the causative pathogen, they showed that SA involved in skin and soft tissue infections causes death in 0–25 % of patients, while the same pathogen in the lung causes death in 31–84 % of patients.

Therefore, the site of infection and the identification of the pathogen involved are both of paramount importance and strongly interrelated and should be considered together when approaching the evaluation or treatment of patients with severe sepsis or septic shock.

4 Overall Clinical Picture

Patients with severe sepsis suffer more than from just the consequences caused by the primary site of infection, whatever it is.

Some studies have addressed the question of how many patients with sepsis (or infection) will progress to severe sepsis or septic shock. There is great variability in this proportion, probably related to the population considered (only ICU or general ward patients): in the ICU 70 % of septic patients will develop severe sepsis and 17 % septic shock [21], and when considering general wards, 39 % of patients will develop severe sepsis [22], probably due to the less severe disease compared to the ICU.

The number of organs involved in severe sepsis is variable. An international research on severe sepsis, enrolling more than 1,900 patients [22] affected by severe sepsis within 12 h from the first organ dysfunction, showed that half of the patients had lung localization, followed by intra-abdominal and genitourinary tract infection.

In Table 3, a very common distribution of organ dysfunction is shown. About one third of patients have only one organ dysfunction. Another third of patients have two organ dysfunctions, and the last third are composed by patients with three or four organ dysfunctions. Overall, the majority of patients have a multi-organ disease.

The most common clinical picture is a patient presenting with infection and two or three organ failures (cardiovascular, renal, and respiratory dysfunction being the most frequent). Even if often not

Table 3
Distribution and frequency of organ dysfunction

Organ dysfunction	Percent	Mean SOFA score
Arterial Hypoxemia	23 %	2.7
Thrombocytopenia	16 %	0.6
Arterial Hypotension	82 %	3.3
Acute renal failure	36 %	1.9
Impaired neurological status	na	1.6
Number of organ failure	**Percent**	**Cumulative percent**
1	34 %	34 %
2	35 %	69 %
3	22 %	91 %
4	8 %	99 %
5	<1 %	100 %

Distribution of organ dysfunction in more than 1,900 patients affected by severe sepsis or septic shock and mean SOFA (sequential organ failure assessment) score for each dysfunction. For arterial hypoxemia, mean patient had a PaO2/FiO2 ration between 200 and 300, and for arterial hypoxemia, had infusion of mild to high dose of vasopressor. Mean glasgow coma scale was between 14 to 10 and platelet count more than 150,000 cell/microL1. Adapted from Opal et al. [6]

formally classified, up to 50 % of patients will also experience encephalopathy [23], representing a further failing organ.

The clinical picture often includes: impaired neurological status, varying from confusion to coma; signs of shock like hypoperfusion, oliguria or anuria, and high lactate levels; clinical signs of hypovolemia, due to temperature or to effective losses (in the third space or in the abdomen); vasodilation; and respiratory impairment even though respiratory mechanics could be normal (especially in young patients and in extrapulmonary localization) with tachypnea due to the attempts to compensate the metabolic acidosis; shock is often associated with a reduction of systemic blood pressure and a worsening of kidney function or cerebral performances; when severe cardiac impairment causes a low output syndrome, instead of the classic reddish due to vasodilation, the skin can become whitish; either fever or hypothermia can be present in sepsis.

Above this general picture, signs of the primary site of infection can be present and can guide the clinician to diagnosis. Elevated aminotransferase levels, paralytic ileus, altered glycemic control, thrombocytopenia and disseminated intravascular coagulation, euthyroid sick syndrome, and adrenal dysfunction are all common in patients with severe sepsis.

Mortality of severe sepsis and septic shock has dramatically reduced when compared to that reported 30 years ago, when they were typically lethal (often exceeding 80 %) [24]. Advances in training, better surveillance and monitoring, prompt initiation of therapy to treat the underlying infection, and support of failing organs have brought mortality down to 20–30 % in many series [25]. Numerous studies have suggested that patients who survive to hospital discharge after sepsis remain at increased risk of death in the following months and years. Those who survive often have impaired physical or neurocognitive functioning, mood disorders, and a low quality of life [26].

5 Major Sites of Infection

5.1 The Lung

5.1.1 Community-Acquired Pneumonia

Community-acquired pneumonia (CAP) [27] should be considered in any patient who has newly acquired respiratory symptoms (cough, sputum production, and/or dyspnea), especially if accompanied by fever and auscultatory findings of abnormal breath sounds and crackles. Standard posteroanterior and lateral chest radiographs are valuable in these patients and may also suggest specific etiologies or conditions such as lung abscess, tuberculosis, and acute respiratory distress syndrome. Computerized tomography of the thorax can add important sensitivity and specificity to chest X-rays and is helpful also to set mechanical ventilation (when necessary).

More recently, lung ultrasound (LUS) has gained importance in the diagnosis of CAP, being at least as accurate as chest radiography. Air bronchogram within an echo-poor area is the most important parenchymal criterion. At the same time, LUS allows the diagnosis of interstitial syndrome, showing the presence of multiple diffuse bilateral B-lines [28]. Ideally, physical examination and LUS at the bedside could allow immediate diagnosis of CAP.

In CAP every effort should be made to identify a specific etiologic pathogen in a timely manner, with focused and appropriate testing. If the etiology is identified, therapy can be focused, but this goal should account for two considerations. First of all, according to sepsis survival guidelines, therapy should be started within 1 or 3 h (if diagnostic tests lead to a delay in therapy, they are associated to adverse outcome). Even if microbiological sampling should be done before administration of the first dose of antibiotics, microbiological tests with great sensibility even after antibiotic administration exist (e.g., those based on antigen or polymerase chain reaction).

Secondly, in CAP, coinfection of a bacteria and an atypical pathogen is possible. Atypical bacteria may be harder or longer to identify; therefore consideration of a full course of effective therapy should be granted, even with negative or pending microbiology assays. When possible, a Gram stain of sputum could be useful.

Two sets of blood cultures should be drawn before initiation of antibiotics in CAP patients, as in any other patients with severe sepsis or septic shock. For patients with suspect of *Legionella* infection, measurement of urinary antigen is valuable [29].

Many invasive diagnostic techniques to obtain lower airway specimens exist (transtracheal aspiration; bronchoscopy with lavage or brush, protected or not; needle aspiration of the lung). These procedures are not indicated in most patients with CAP, but could be useful in patients whose illness is not resolving in spite of an apparently appropriate therapy.

5.1.2 Hospital-Acquired, Ventilator-Associated, and Healthcare-Associated Pneumonia

Hospital-Acquired, Ventilator-Associated, and Healthcare-Associated Pneumonia (HAP, VAP, and HCAP) are important causes of morbidity and mortality despite advances in antimicrobial therapy, better supportive care, and a wide range of preventive measures. HAP incidence varies between 5 and 15 cases per 1,000 hospital admissions. Pneumonia in ventilated patients is 6- to 20-fold higher than in non-mechanically ventilated patients.

HAP is defined as pneumonia that occurs 48 h or more after admission, which was not incubating at the time of admission. VAP refers, traditionally, to a pneumonia that arises more than 48–72 h after tracheal intubation [30]. HCAP is included in the spectrum of HAP and VAP, and patients with HCAP need therapy and, more generally, care as HAP patients. The Centers for Disease Control and Prevention (CDC) recently proposed an algorithm that uses objective, readily available data elements to identify a broad range of conditions and complications occurring in mechanically ventilated patients, including but not limited to VAP, introducing new conditions like ventilator-associated condition (VAC, an elevation in the demand of oxygen and pressure), infection-related VAC (IVAC, also an abnormal temperature or white blood cell count and the starting of a new antimicrobial agent), and, lastly, VAP that requires that patients have IVAC and laboratory and/or microbiological evidence of respiratory infection [31].

The etiology of these kinds of pneumonia is considerably different from CAP, being commonly caused by aerobic gram-negative bacilli like *Pseudomonas aeruginosa*, *Klebsiella pneumoniae*, and *Acinetobacter* species or gram-positive cocci, such as *Staphylococcus aureus*, with a great incidence of methicillin resistance.

The CDC introduced new categories also because the diagnosis of HAP is difficult, due to reduced use of cultures of protected specimen (e.g., bronchoalveolar lavage) and because chest X-ray interpretation is often challenging in patients with long hospitalization or concomitant cardiac diseases. Even adding invasive strategies to diagnostic techniques in VAP doesn't seem to affect survival [32].

Generally, two approaches can be applied for the diagnosis.

The clinical approach defines the presence of pneumonia as a new lung infiltrate plus clinical evidence that the infiltrate is of an infectious origin. This grants the starting of a new antibiotic treatment, with the execution of microbial sampling before starting the new plan of therapy.

The bacteriologic strategy uses quantitative cultures of lower respiratory secretions (endotracheal aspirate, BAL, or protected brushing) to define both etiology and presence of pneumonia. Growth below some threshold, based on the methodology of collection of sampling, is assumed to be due to colonization or contamination, and new antibiotics are administered following quantitative microbiologic results. The incompressible delay of 48–72 h for complete antimicrobial testing, including antibiotic susceptibility, has enforced the development of rapid molecular testing to optimize the choice of initial drugs and to avoid the overprescription of broad-spectrum molecules. Such tools should reliably identify both the most common pathogens and their most frequent resistance genotypes in 2–6 h. Real-time PCR, in situ DNA hybridization, and mass spectrometry are some of the leading investigation methods [33].

When therapy decisions have been based on bacteriologic strategy, fewer patients have been treated with antibiotics and a narrow spectrum of therapy used, compared to the clinical approach. Major concerns with this approach are that a false-negative culture can lead to a failure to treat the patients and that positive results, after at least 48 h of waiting, lead to a strong delay in starting new antibiotic treatment and this could worsen the outcome.

An important factor causing false-negative cultures is a recent start or change in antibiotic treatment as this can alter positivity of cultures itself or amplitude of bacterial growing. Therefore, ideally, all quantitative cultures should be obtained before any antibiotic manipulation. When this is not possible, changes in the diagnostic threshold may be helpful. Taken together these considerations imply an important alliance between the clinician and the laboratory: knowledge of the kind of sampling and history as well as timing of antibiotic treatment could be important in evaluating and interpreting microbiological results.

Guidelines are available in leading treatment of all forms of pneumonia. Therapy is complex and depends on patient's adjusted risk of atypical pathogens, multidrug-resistant pathogens, and MRSA involvement. Attention should be also paid to the pharmacodynamic and pharmacokinetic properties of every molecule and their penetration in the lung parenchyma.

Notably, pneumonia treatment, especially in patients requiring mechanical ventilation, is the way the patient receives mechanical ventilation itself [34, 35], and great attention should be done to avoid both volutrauma and barotrauma, by meticulously controlling tidal volume. In the most severe patients, with refractory hypoxemia or impossibility to use protective mechanical ventilation, extracorporeal support of oxygenation is possible today and, despite necessity of definitive evidence, results are encouraging [36, 37]. As extracorporeal support can't be delivered in every hospital due to the high complexity of this treatment, many countries have developed a system of centralization of more severe cases, in a hub-and-spoke structure.

In conclusion, an increasing burden of pneumonia, in its many different forms, can be expected in the coming years, due to many factors, like progressive aging of population, increasing of comorbidities, and intensification of cares. Both diagnosis and treatment of pneumonia remain challenging and grant an intensive work for research and development of new clinically efficient instruments [38].

5.2 Abdominal Infection

Intra-abdominal infections (IAIs) represent a wide variety of conditions that involve lesions of all intra-abdominal organs. They include also intra- and retroperitoneal abscesses and parenchymal abscesses. They are divided as uncomplicated, when localized to one organ, and complicated, when causing peritonitis. Complicated IAIs are classified according to the cause of the associated peritonitis (primary, secondary, and tertiary) and the extension of the inflammation (local or diffuse) [39]. Similarly to pneumonia, they can also be divided into community-acquired (CA-IAI) and hospital-acquired (HA-IAI) with important differences regarding antimicrobial treatment [40].

IAIs are an important cause of ICU morbidity and mortality. Mortality is approximately 30 % and up to 50 % when peritonitis arises from a complication of a previous surgical procedure or recurs during ICU admission [41].

Gastrointestinal perforation with leakage of alimentary or fecal contents in the peritoneal cavity is the main cause of IAIs. Perforation can be caused by appendicitis, diverticulitis, ulcer, cancer, trauma, and medical procedures (like colonoscopy, gastroscopy, or biliary tract procedures). A second group of IAIs is related to biliary tract diseases (e.g., acute cholecystitis, cholangitis). The third group includes postoperative intra-abdominal infections (anastomosis leakage is an important cause of HA-IAIs and correlates with a very severe prognosis) [42].

The typical clinical presentation of IAI includes abdominal pain and tenderness with signs of peritoneal irritation on physical examination. Diffuse pain suggests generalized peritonitis, while localized pain suggests a walled-off process arising from an organ in the anatomic vicinity.

The epidemiology of IAIs is largely dominated by aerobic gram-negative bacteria (AGNB). In a study of 239 patients [43], abdominal drainage cultures revealed 53 % of AGNB, with

Escherichia coli being the most frequent. Interestingly, more than 30 % of patients had isolations of more than one pathogen. The incidence of gram negative is higher in distal (like colorectal and appendix) than in proximal perforation. Fungi are often involved in IAIs, being isolated in 20 % of patients, especially in proximal (gastroduodenal) perforations. Gram positives are also represented, up to 40 %.

Treatment of IAIs is challenging due to the high demand of a multi-faced therapy: surgery (as definitive or source control), antimicrobials, and an aggressive support of organ dysfunctions that often requires ICU management and full life support.

Antimicrobial therapy itself could be challenging due to: frequent polymicrobial infections, multidrug resistance for both in-hospital and out-of-hospital patients (especially due to community-acquired extended-spectrum beta-lactamase-producing bacteria), and fungi involvement.

Appropriate microbiological sample should be taken, possibly before antimicrobial starts but avoiding any possible delay in the first dose. Dosage consideration should include using high loading doses (patients with IAIs often have higher volume of distribution) and reduced further doses, because of the frequent association of IAIs with renal dysfunction.

Apart from blood sampling that should be done as in any other septic or septic shock patients, cultures should be taken from intra-abdominal samples during surgical or interventional drainage procedures, ensuring sufficient volume (at minimum 1 ml of fluid or tissue) and using transport systems that properly handle the samples so as not to damage them or compromise their integrity.

Concluding, IAIs are an important cause of preventable morbidity and mortality. The responsible disorders are numerous. Etiology often includes gram-negative pathogens, but also gram positive and fungi can be isolated and should be considered in treatment. Close collaboration between the surgeon, the radiologist, the microbiologist, and the intensive care specialist is imperative to ameliorate outcome.

5.3 Urinary Tract Infection

Urinary tract infection includes urinary infection, acute nonobstructive pyelonephritis, and, in men, bacterial prostatitis. The urinary tract is the source of infection of up to 30 % of severe sepsis or septic shock patients in some series [44]. In Table 2, this proportion is slightly lower, varying from 6 to 21 % [7, 16].

At the same time, few patients with urinary tract infection develop severe sepsis or septic shock. In an Israeli study including women with complicated pyelonephritis, only 13 % developed severe sepsis [45]. Ideally, progression of an uncomplicated urinary infection to severe sepsis should suggest an underlying complicating factor or the presence of a severe comorbidity (e.g., poorly controlled diabetes, liver cirrhosis), immune modulation, or suppression.

An indwelling catheter is of paramount importance to differentiate between urosepsis [46].

For non-catheterized patients, an evidence of infection by culture of pathogen directly from the infected tissue (not urine) is required or from fever, urgency, localized pain, tenderness at involved site, a compatible analysis of urine (pyuria, more than 10^5 cfu/ml, positive Gram stain), or a compatible imaging study.

For the catheterized patients, criteria for diagnosis of infection are more stringent due to the possibility of contamination or colonization. A direct evidence of infection or a positive culture above certain threshold associated with clinical compatible signs is required.

Urine culture should be collected in any patient with a suspected infection and could lead to definitive diagnosis, etiology determination, and therapy guidance. Since systemic antimicrobial therapy will usually sterilize the urine within minutes, it is very important that specimen for culture should be collected before initiation of therapy. Special attention should be given when urine sampling is done through an indwelling catheter, especially if it has been in situ for more than 2 weeks [47].

Blood cultures should also be collected, are frequently positive (up to 30 %) even in patients that will not progress to severe sepsis or shock, and might identify the most important strain in patients with multiple organism isolated from urine culture with implications in the tailoring of antibiotic therapy [48].

Treatment should be done according to available guidelines, usually including an extended-spectrum cephalosporin, a fluoroquinolone with mainly renal excretion, and, sometimes, a molecule with antipseudomonal activity [49].

Severe sepsis and septic shock have a relatively low mortality (10–20 %) in urosepsis [48], probably because of a relative straightforward approach to source control and a lower impairment of vital function (e.g., ARDS) [50].

6 Sepsis Management

Surviving Sepsis Campaign is an international consortium of professional societies involved in critical care and in infectious diseases. It recently issued the third iteration of clinical guidelines for the management of severe sepsis and septic shock [2] that provides extensive information on how to treat a sick septic patient.

Since guidelines have little immediate impact on bedside behavior, tools to increase guideline adherence and to speed their application have been developed.

6.1 Clinical Management

Clinical management of sepsis is grouped into interventions (or bundles) to be completed within 6 h and management bundles to be accomplished in the ICU.

The 6-h bundle includes initial volemic resuscitation with goal-directed fluid challenge, diagnosis of infection with microbiological sampling coupled with imaging studies, treatment of infection with antibiotics (also with surgery or radiological procedures when appropriate), and hemodynamic support with vasopressors or inotropes if volemic resuscitation fails to reverse hypoperfusion defects.

The management bundle after 6 h includes optimization of organ support and monitoring, avoidance of further complications, and de-escalation of care when possible. Routine critical care support therapy should be started: management of anemia and coagulation abnormalities, ventilation according to ARDSNet protective strategy, glycemic control, renal support, deep vein thrombosis, stress ulcer prophylaxis, and feeding. The only immune-modulating therapy is, in selected circumstances, a short course of hydrocortisone.

In patients with severe sepsis and septic shock, it is important to discuss goals of care and prognosis with patients and families. The goals of care, including any end-of-life care planning or the use of palliative care principles should be accomplished as appropriate [51].

6.2 Sepsis Performance Improvement Programs [51, 52]

Guidelines in sepsis should serve as a resource document for the creation of treatment protocols that, when coupled with audit and feedback as a part of a formal hospital-based performance improvement initiative, can change bedside practice and grant a real change in patient's outcome. Therefore sepsis treatment, as described in the guidelines, is only a part of a more complex group of actions that should be taken at a higher level, usually involving the full hospital and, in some instances, also the health service.

Programs to improve the performance start with hospital-wide education initiatives, centered around early identification and familiarity with the treatment protocols that will be applied once the patient is identified. Protocols can be successful in changing bedside behaviors only with the application of education and commitment of physician, nurse, and other healthcare professionals from key areas of the hospital (ICU, emergency department, and hospital floors).

Success of severe sepsis performance improvement programs requires multidisciplinary commitment from physicians, nurses, pharmacists, and administration. Programs must be multispecialized as well and include medicine, surgery, emergency medicine, microbiology, and others. Establishing support from key ICU, emergency dept., and floor leaders is crucial. Interdepartmental communication and collaboration facilitate seamless steps in the continuum of care and give the best chance of success.

7 Conclusions

Severe sepsis and septic shock are a frequent cause of mortality and morbidity. This syndrome is increasingly diagnosed over time, caused by many pathogens with an everyday harder profile of sensibility to antibiotics, one of the main cornerstones in the treatment of sepsis. Besides that, bundle approach and organization efforts are very important issues. The lung, abdomen, and urinary tract are still the major sites of sepsis, but other sites of infection, as the skin and blood, are increasing.

Early diagnosis and expedited treatment based on evidence-based medicine can decrease sepsis morbidity and mortality. Extensive collaboration between many figures (intensivists, surgeons, infectivologists, microbiologists, pharmacists, nurses, and many others) is required to get this goal. Over that, institutions and healthcare systems are also very important players in sepsis fight.

References

1. Levy MM, Fink MP, Marshall JC, Abraham E, Angus D, Cook D et al (2003) 2001 SCCM/ESICM/ACCP/ATS/SIS International Sepsis Definitions Conference. Crit Care Med 31:1250–1256
2. Dellinger RP, Mitchell ML, Andrew R, Djillali A, HerwigGerlach SM, Opal JE et al (2013) Surviving sepsis campaign: international guidelines for management of severe sepsis and septic shock. Crit Care Med 41:580–637
3. Klein K, Peter MC, David SYO, Marc JMB, Olaf LC (2012) Classification of sepsis, severe sepsis and septic shock: the impact of minor variations in data capture and definition of SIRS criteria. Intensive Care Med 38:811–819
4. Kumar G, Nilay K, Amit T, Thomas K, Sergey T, Emily M et al (2011) Nationwide trends of severe sepsis in the 21st century (2000–2007). Chest 140:1223–1231
5. Zimmerman JE, Andrew AK, William AK (2013) Changes in hospital mortality for United States intensive care unit admissions from 1988 to 2012. Crit Care 17:R81
6. Opal S, Laterre PF, Francois B, LaRosa SP, Angus DC, Mira JP et al (2013) Effect of eritoran, an antagonist of MD2-TLR4, on mortality in patients with severe sepsis: the ACCESS randomized trial. JAMA 309:1154–1162
7. Guntupalli K, Nathan D, Peter EM, Venkata B, Benjamin M, Emanuel R et al (2013) A phase 2 randomized, double-blind, placebo-controlled study of the safety and efficacy of talactoferrin in patients with severe sepsis. Crit Care Med 41:706–716
8. Pasin L, Landoni G, Castro ML, Cabrini L, Belletti A, Feltracco P et al (2013) The effect of statins on mortality in septic patients: a meta-analysis of randomized controlled trials. PLoS One 12:e82775
9. Ranieri VM, Taylor T, Philip SB, Jean-François D, Ivor SD, Simon F et al (2012) Drotrecogin alfa (activated) in adults with septic shock. N Engl J Med 366:2055–2064
10. Annane D, Timsit JF, Megarbane B, Martin C, Misset B, Mourvillier B et al (2013) Recombinant human activated protein C for adults with septic shock: a randomized controlled trial. Am J Respir Crit Care Med 187:1091–1097
11. Crivellari M, Silvetti S, Gerli C, Landoni G, Franco A, Bove T et al (2013) Protein C zymogen in adults with severe sepsis or septic shock. Med Intensiva 38:278–282. doi:10.1016/j.medin.2013.04.005
12. Crivellari M, Della Valle P, Landoni G, Pappalardo F, Gerli C, Bignami E et al (2009) Human protein C zymogen concentrate in patients with severe sepsis and multiple organ failure after adult cardiac surgery. Intensive Care Med 35:1959–1963
13. Klevens RM, Edwards JR, Richards LR Jr, Horan TC, Gaynes RP, Pollock DA et al (2007) Estimating health care-associated infections and deaths in U.S. hospitals, 2002. Public Health Rep 122:160–166

14. Rice LB (2009) The clinical consequences of antimicrobial resistance. Curr Opin Microbiol 12:476–481
15. Brunkhorst FM, Oppert M, Marx G, Bloos F, Ludewig K, Putensen C et al (2012) Effect of empirical treatment with moxifloxacin and meropenem vs. meropenem on sepsis-related organ dysfunction in patients with severe sepsis: a randomized trial. JAMA 307:2390–2399
16. Schortgen F, Clabault K, Katsahian S, Devaquet J, Mercat A, Deye N et al (2012) Fever control using external cooling in septic shock: a randomized controlled trial. Am J Respir Crit Care Med 185:1088–1095
17. Morelli A, Ertmer C, Westphal M, Rehberg S, Kampmeier T, Ligges S et al (2013) Effect of heart rate control with esmolol on hemodynamic and clinical outcomes in patients with septic shock: a randomized clinical trial. JAMA 310:1683–1691
18. Vincent J, Rello J, Marshall J (2009) International study of the prevalence and outcomes of infection in intensive care units. JAMA 302:2323–2329
19. Ramphal R (2005) Importance of adequate initial antimicrobial therapy. Chemotherapy 51:171–176
20. Cohen J, Cristofaro P, Carlet J, Opal S (2004) New method of classifying infections in critically Ill patients. Crit Care Med 32:1510–1526
21. Rangel-Frausto MS, Pittet D, Costigan M, Hwang T, Davis CS, Wenzel RP (1995) The Natural History of the Systemic Inflammatory Response Syndrome (SIRS). A prospective study. JAMA 273:117–123
22. Sundararajan V, Macisaac C, Presneill JJ, Cade JF, Visvanathan K (2005) Epidemiology of sepsis in Victoria, Australia. Crit Care Med 33:71–80
23. Eidelman LA, Putterman D, Putterman C, Sprung CL (1966) The spectrum of septic encephalopathy. definitions, etiologies, and mortalities. JAMA 275:470–473
24. Friedman G, Silva E, Vincent JL (1998) Has the mortality of septic shock changed with time. Crit Care Med 26:2078–2086
25. Angus DC, van der Poll T (2013) Severe sepsis and septic shock. NEJM 369:840–851
26. Angus DC, Carlet J, 2002 Brussels Roundtable Participants (2003) Surviving intensive care: a report from the 2002 Brussels Roundtable. Intensive Care Med 29:368–377
27. American Thoracic Society (2001) Guidelines for the management of adults with community-acquired pneumonia. http://www.thoracic.org/statements/resources/archive/commacq1-25.pdf. Accessed 19 Feb 2014
28. Reissig A, Copetti R (2014) Lung ultrasound in community-acquired pneumonia and in interstitial lung diseases. Respiration 87:179–189. doi:10.1159/000357449
29. Reynaldos CM, Györik T, Lora S, Gaia V, Pagnamenta A (2014) ARDS with septic shock due to Legionella longbeachae pneumonia in a patient with polymyalgia rheumatic. Heart Lung Vessel 6:114–118
30. American Thoracic Society, Infectious Diseases Society of America (2005) Guidelines for the management of adults with hospital-acquired, ventilator-associated, and healthcare-associated pneumonia. Am J Respir Crit Care Med 171:388–416
31. Magill SS, Klompas M, Balk R, Burns SM, Deutschman CS, Diekema D et al (2013) Developing a new, national approach to surveillance for ventilator-associated events. Crit Care Med 41:2467–2475
32. Cruciani M (2011) Meta-analyses of diagnostic tests in infectious diseases: how helpful are they in the intensive care setting? HSR Proc Intensive Care Cardiovasc Anesth 3:103–108
33. Lung M, Codina G (2012) Molecular diagnosis in HAP/VAP. Curr Opin Crit Care 18:487–494
34. Gattinoni L, Protti A, Caironi P, Carlesso E (2010) Ventilator-induced lung injury: the anatomical and physiological framework. Crit Care Med 38:S539–S548
35. Ricard JD, Dreyfuss D, Saumon G (2003) Ventilator-induced lung injury. Eur Respir J 42:2s–9s
36. Zangrillo A, Biondi-Zoccai G, Landoni G, Frati G, Patroniti N, Pesenti A et al (2013) Extracorporeal membrane oxygenation (ECMO) in patients with H1N1 influenza infection: a systematicreview and meta-analysis including 8 studies and 266 patients receiving ECMO. Crit Care 17:R30
37. Zangrillo A, Landoni G, Biondi-Zoccai G, Greco M, Greco T, Frati G et al (2013) A meta-analysis of complications and mortality of extracorporeal membrane oxygenation. Crit Care Resusc 15:172–178
38. Barbier F, Andremont F, Wolff M, Bouadma L (2013) Hospital-acquired pneumonia and ventilator-associated pneumonia: recent advances in epidemiology and management. Curr Opin Pulm Med 19:216–228
39. Menichetti F, Sganga G (2009) Definition and classification of intra-abdominal infections. J Chemother 21:3–4
40. Sartelli M, Viale P, Catena F, Ansaloni L, Moore E, Malangoni M et al (2013) 2013

WSES guidelines for management of intra-abdominal infections. World J Emerg Surg 8:3
41. Marshall JC, Innes M (2008) Intensive care unit management of intra-abdominal infection. Crit Care Med 31:2228–2237
42. Moran BJ, Heald RJ (2001) Risk factors for, and management of anastomotic leakage in rectal surgery. Colorectal Dis 3:135–137
43. De Ruiter J, Weel J, Manusama E, Kingma WP, van der Voort PHJ (2009) The epidemiology of intra-abdominal flora in critically ill patients with secondary and tertiary abdominal sepsis. Infection 37:522–527
44. Peake SL, Bailey M, Bellomo R, Cameron PA, Cross A, Delaney A et al (2009) Australasian resuscitation of sepsis evaluation (ARISE): a multi-centre, prospective, inception cohort study. Resuscitation 80:811–818
45. Chen Y, Nitzan O, Saliba W, Chazan B, Colodner R, Raz R (2006) Are blood cultures necessary in the management of women with complicated pyelonephritis? J Infect 53:235–240
46. Calandra T, Cohen J, International Sepsis Forum Definition of Infection in the ICU Consensus Conference (2005) The international sepsis forum consensus conference on definitions of infection in the intensive care unit. Crit Care Med 33:1538–1548
47. Raz R, Schiller D, Nicolle LE (2000) Chronic indwelling catheter replacement before antimicrobial therapy for symptomatic urinary tract infection. J Urol 164:1254–1258
48. Nicolle LE (2013) Urinary tract infection. Crit Care Clin 29:699–715
49. Grabe M, Bjerklund-Johansen TE, Botto H (2011) Guidelines on urological infections. Eur Assoc of Urology. Available at: http://www.uroweb.org/gls/pdf/18_Urological%20infections_LR.pdf. Accessed 13 Feb 2014
50. Sheu CC, Gong MN, Zhai R, Bajwa EK, Chen F, Thompson TB et al (2010) The influence of infection sites on development and mortality of ARDS. Intensive Care Med 36:963–970
51. Schorr CA, Zanotti S, Dellinger RP (2014) Severe sepsis and septic shock: management and performance improvement. Virulence 5:190–199
52. Townsend SR, Schorr C, Levy MM, Dellinger RP (2008) Reducing mortality in severe sepsis: the surviving sepsis campaign. Clin Chest Med 29:721–733

Chapter 4

Technical Improvements in Culturing Blood

Giacomo Pardini

Abstract

Blood culture is a laboratory test where a blood specimen, taken from a patient, is inoculated into bottles containing culture media to determine if infection-causing microorganisms (bacteria or fungi) have invaded the patient's bloodstream. This test is an important investigation with major implications for the diagnosis and treatment of patients with bloodstream infections and possible sepsis. Moreover, blood culture will also provide the etiologic agent for antimicrobial susceptibility testing, enabling optimization of antibiotic therapy with significant impact on the outcome of the disease. Even if the potential beneficès of blood culture are well known, critical factors mainly in pre- and post-analytical phases can reduce the clinical value of this test.

Key words Bloodstream infection, Blood culture, Sepsis, Culture media, Sample collection, Diagnostic tests, Gram staining, Preliminary tests

1 Introduction

Blood culture is an essential tool and a validated procedure to detect the presence of microorganisms (bacteria and fungi) in the bloodstream, and it leads clinicians to appropriate antibiotic therapy [1, 2]. This test is crucial to help the microbiologist in the management of patients with sepsis, endocarditis, infections related to intravascular catheters, fever of unknown origin, or localized infections such as pneumonia and septic arthritis [3]. The presence of microorganisms in the blood can be transient (presence of microorganisms for a short time), intermittent (or recurrent transient, associated with localized or systemic infections), or continuous (typical of intravascular infection, [4]). The isolation of bacteria or fungi from the blood either establishes or confirms that there is an infectious etiology for the patient's illness [5, 6]. Moreover, blood cultures have a high prognostic value and provide the etiologic agent for antimicrobial susceptibility testing, enabling optimization of antibiotic therapy with significant impact on the outcome of the disease. Even if the literature is unanimous in

Nicasio Mancini (ed.), *Sepsis: Diagnostic Methods and Protocols*, Methods in Molecular Biology, vol. 1237, DOI 10.1007/978-1-4939-1776-1_4, © Springer Science+Business Media New York 2015

attributing the high diagnostic value to the blood culture, the timing of this test is still too long compared to the clinical needs [1, 7]. Many factors, largely depending on a policy of appropriateness at different stages, can decrease or increase the effectiveness of blood culture. The complete blood culture process consists of proper collection of the sample, detection, isolation, and identification of microorganisms causing bloodstream infections to provide an antibiotic susceptibility test result for the clinicians. A workflow organized or reorganized in association with new available technology can reduce the execution time and the transmission of results to the clinician. Automatic systems have quickly replaced manual systems for the benefits they offer in terms of standardization of outcome and of reduction of time to results. For this reason, only automated methods will be described in this chapter.

Blood culture can contribute to the management of patients with bloodstream infection only if clinicians and microbiologists are both involved. A good internal communication between these two experts is a prerequisite for the improvement of this clinical process and for the patient safety. Clinicians and microbiologists, together, can define and implement a proper diagnostic guideline including the formulation of a clinical suspicion and the decision to carry out the test, the well-defined laboratory approach, and the treatment decisions based on the results of blood cultures.

Despite the appearances, blood culture is a laboratory test with critical steps during preparation, execution, and interpretation. Standard and alternative methods contribute to improve the diagnostic capability of the laboratory, but the role of the microbiologist exercised also in the pre- and post-analytical phases is still fundamental. Good knowledge and continuous internal communication are key elements to prevent the waste of resources available and to improve the patient care.

2 Materials

Microorganisms causing bloodstream infections are highly varied (aerobes, anaerobes, fungi, fastidious microorganisms) and, in addition to nutrient elements, may require specific growth factors and/or a special atmosphere. Mainly, blood culture media used in clinical routine contain peptones, yeast extract, sugars and/or casein, and all elements to allow a good growth in liquid medium. Currently all media available on the market for automatic systems are dispensed with added CO_2, and anaerobic media are prereduced and dispensed with CO_2 and N_2. The blood drawn should be divided equally between the aerobic and anaerobic bottles. A blood culture medium must be sensitive enough to detect a broad range of clinically relevant microorganisms, even the most fastidious (e.g., *Neisseria*, *Haemophilus*) or microorganisms releasing small amounts of CO_2 (e.g., *Brucella*, *Acinetobacter*) and versatile to provide a result for all types of sample collection: adults, infants, and patients receiving antibiotic therapy mainly. It is therefore important to use a blood culture medium able to sustain microbial growth in the presence of antibiotics.

Several elements play a critical role in the reduction of antibiotic activity:

1. Dilution factor: Each blood specimen (and antibiotic contained inside) is diluted in medium in ratio of 1:5–1:10. This dilution reduced the antibiotic activity on the growth of bacteria.
2. Molecules such as charcoal and/or resins (nonionic adsorbing resin and cationic exchange resin) interfere with antibiotics and enhance the growth of bacteria in the broth. The use of resins for antibiotic neutralization appears to be more versatile in association with modern tools for the rapid identification of pathogens directly from blood culture like PCR methods or matrix-assisted laser desorption/ionization time-of-flight mass spectrometry (MALDI-TOF MS).

Since bacteria and fungi are not constantly present in the bloodstream, to increase the sensitivity of blood culture, it is recommended to collect two/three sets (two bottles/set) per patient. A single blood culture set should never be drawn from adult patients, since this practice will result in an inadequate volume of blood cultured and a substantial number of bacteremia may be missed. In a recent study it has been observed that the cumulative yield of pathogens from three blood culture sets with a blood volume of 20 mL in each set (10 mL/bottle) was 65–80 % with the first set, 80–88 % with the first two sets, and 96–99 % with the first three sets [8].

A contaminant will usually be present in only one bottle of a set of blood culture bottles, in contrast to a true bloodstream infection, in which multiple blood culture sets from separate anatomical sites will be positive [9]. This further underlines the importance of collecting more than one blood culture set and taking each set from a separate anatomical site [10]. It is therefore generally recommended to collect two, or preferably three, blood culture sample sets from separate anatomical sites for each septic episode [4].

To detect all microorganisms involved in bloodstream infections with different growth rates, the current recommendation for routine blood cultures performed by continuous-monitoring blood culture systems is 5 days. However, published data suggest that 3 days may be adequate to recover up to 95–97 % of clinically significant microorganisms [11]. Riedel et al. showed the number of significant microorganisms isolated per day for 35,500 consecutive blood cultures collected over 30 months, of which 2,609 were clinically significant isolates and 1,097 were contaminants [12].

Another recent study carried out by Cockerill et al. demonstrated that when using a continuous-monitoring blood culture system, 99.5 % of non-endocarditis bloodstream infections and 100 % of endocarditis episodes were detected within 5 days of incubation [10]. This data suggests that extended incubation periods previously recommended for detection of the fastidious microorganisms that sometimes cause endocarditis are usually no longer necessary when using modern continuous-monitoring blood culture systems.

3 Methods, Sepsis, and Host Immune Response

3.1 Sample Collection Using Winged Blood Collection Set

1. Before touching the patient, wash hands with soap and water and then dry, or apply an alcohol hand rub. Confirm patient identification (for the clinical symptom that can lead to blood culture prescription; *see* **Note 1**).
2. Gather blood collection kit in a cleaned trolley. Remove the plastic cap from the blood culture bottles and disinfect the septum using an appropriate disinfectant. Use a fresh swab/applicator for each bottle. Allow bottle tops to dry in order to fully disinfect. Check expiry date for each bottle and mark 10 mL above the broth for fill level. Do not use bottles which show any signs of damage, deterioration, or contamination.
3. Apply a disposable tourniquet to palpate and to identify the adequate vein.
4. Wash hands and wear gloves to protect the operator. Sterile gloves are not necessary, unless it is necessary to re-touch the skin already disinfected for the detection of the vein. If necessary, disinfect gloves with chlorhexidine.
5. Using solution adopted in the hospital, disinfect the venipuncture site using a scrubbing motion (one fresh swab for each scrub). Use 2–3 scrubs. Do this for a total of 1–2 min, allowing the site to dry (approximately 30 s; *see* **Note 2**).
6. Prepare winged infusion set and vacutainer. To prevent contaminating the puncture site, do not re-palpate the prepared vein before inserting the needle. Insert the needle into the prepared site. Release tourniquet during procedure where appropriate.
7. Place 10 mL blood per bottle (20 mL/set), keeping blood culture bottle upright, and use the graduation lines to accurately gauge sample volume. Each adult routine blood culture set shall mandatory include paired aerobic and anaerobic blood culture bottles ([13]; *see* **Notes 3** and **4**). The drawn blood should be split equally in aerobic and anaerobic bottles. The aerobic bottle should be filled first to prevent air transfer from the device into the anaerobic bottle (*see* **Note 5**).

8. Apply cotton ball and pressure to site (where possible obtain patient assistance to hold and apply pressure); repeat procedure for the second set of blood culture at a different peripheral site, maintaining aseptic technique; invert bottles gently several times to prevent clotting.
9. Discard the winged collection set into a sharps container and cover the puncture site with an appropriate dressing. Remove gloves and wash hands before recording the procedure, including indication for culture, time, site of venipuncture, and any complications. Ensure additional labels do not cover the bottle barcodes and that the tear-off barcode labels are not removed.
10. Take two/three sets of blood culture in rapid succession with intervals of 5–10 min. In case of suspected endocarditis in which there is a continuous bacteremia, it is recommended to take specimen with intervals of 30–60 min (*see* **Note 6**). If the first two sets are negative, repeat sampling after 24 h (*see* **Note** 7). In suspicion of CVC-related infection, take blood samples from a peripheral vein and from the catheter(s) at the same time and with the same amount of blood. Disinfect the connection with alcohol solution, if compatible with the material of the CVC, without discarding the first amount of blood. Use only one aerobic bottle for CVC (*see* **Note 8**). For peripheral vein, follow the protocol described above and use two bottles (aerobic and anaerobic). Insert the same amount of blood in each vial (from CVC and vein).

3.2 Sample Collection Using Needle and Syringe

1. Before touching the patient, wash hands with soap and water and then dry, or apply an alcohol hand rub.
2. Prepare blood collection kit: Gather all materials before beginning the procedure. Ensure the blood culture bottles are within date. Do not use bottles which show any signs of damage, deterioration, or contamination.
3. Prepare bottles for inoculation: Wash hands with soap and water and then dry, or apply an alcohol hand rub. Remove the plastic "flip-cap" from the blood culture bottles and disinfect the septum using an appropriate disinfectant, such as 2 % chlorhexidine in 70 % isopropyl alcohol, 70 % isopropyl alcohol, or iodine in swab or applicator form. Use a fresh swab/applicator for each bottle. Allow bottle tops to dry in order to fully disinfect.
4. Confirm patient identification. If skin is visibly soiled, clean with soap and water. Apply a disposable tourniquet. Palpate to identify the vein and cleanse using an appropriate disinfectant, such as 2 % chlorhexidine in 70 % isopropyl alcohol, 70 % isopropyl alcohol, or iodine in swab or applicator form. The venipuncture site is not fully clean until the disinfectant has fully evaporated.

5. Wash hands again or use an alcohol hand rub and apply clean examination gloves. Sterile gloves are not necessary.
6. Attach a winged blood collection set to a collection adapter cap. To prevent contaminating the puncture site, do not re-palpate the prepared vein before inserting the needle. Insert the needle into the prepared site.
7. Collect the sample. Transfer the blood into the culture bottles, starting with the anaerobic bottle. Hold the bottle upright and use the graduation lines to accurately gauge sample volume. Add up to 10 mL of blood per adult bottle and up to 4 mL per pediatric bottle.
8. Discard the needle and syringe into a sharps container and cover the puncture site with an appropriate dressing. Remove gloves and wash hands before recording the procedure, including indication for culture, time, site of venipuncture, and any complications. Ensure additional labels do not cover the bottle barcodes and that the tear-off barcode labels are not removed.

3.3 Blood Culture Processing and Result Interpretation

3.3.1 Gram Stain

When a blood culture is flagged as positive by the instrument, additional tests have to be performed to produce the final report. Here, only Gram stain will be described, as the following chapters are focused to direct identification from positive blood culture using MALDI-TOF MS technology. The first notification of a positive blood culture is typically based on the Gram stain result [14]. At this time, 12–20 % of the patients may not have started antibiotic treatment, and in another 30–45 % of patients, the Gram stain result is followed by a change in the empirical treatment [14]. Even if a range of other promising direct tests for rapid identification has been described in recent years, Gram staining remains an inexpensive, fast, and highly accurate technology:

1. Fix the slide by passing it over a heat source or using alcohol solution (ethanol or methanol).
2. Flood the fixed smear with crystal violet solution and allow to remain for 1 min.
3. Rinse off the crystal violet solution with distilled water and flood the slide with iodine solution. Allow to remain for 1 min.
4. Rinse off the iodine solution with distilled water and flood the slide with decolorizer for 10 s.
5. Rinse off the decolorizer with distilled water.
6. Flood the slide with safranin and allow to remain for 1 min.
7. Rinse off the safranin with distilled water, dry the slide on bibulous paper or absorbent paper, and place in an upright position.

8. If the Gram stain confirms the blood culture to be positive, the morphology of the result should be reported immediately and subcultures performed for further organism identification and antibiotic susceptibility testing. If a sample is Gram stain negative, no report is made to the clinician unless there is growth on subculture. Clinically relevant results must be reported as soon as available, due to the immediate impact on patient care decisions.

3.3.2 Subculture of Positive Blood Cultures

All positive blood cultures must be streaked on an appropriate set of culture media, like Trypticase Soy Agar + 5 % sheep blood agar or chocolate agar in aerobic and anaerobic conditions. Chocolate agar has to be incubated in CO_2. New chromogenic media can be used to improve identification after overnight incubation or on the basis of the result of microscopy.

3.3.3 Direct Susceptibility Testing

The purpose of a blood culture is to provide the etiologic agent for antimicrobial susceptibility testing, enabling optimization of antibiotic therapy with significant impact on the outcome of the disease. As traditional workflow from sample collection to result needs several days, preliminary antibiotic susceptibility test can be performed to allow the early administration of adequate antibiotic therapy and to reduce mortality:

1. Prepare the inoculums using ten drops of positive blood cultures. Ideally, the initial concentration should have a turbidity of 0.5 McFarland. Both agar diffusion and gradient strip methods have to be performed using a homogeneous inoculum as well as a standardized concentration, so this point is a very crucial point in the protocol [15, 16].
2. Soak a sterile, nontoxic swab in the inoculum suspension and remove excess fluid by pressing it against the inside wall of the test tube.
3. Remove more fluid when streaking a 90-mm plate and less for a 150-mm plate.
4. Carefully streak the entire agar surface three times, rotating the plate 60° each time to evenly distribute the inoculum.
5. Allow excess moisture to be absorbed for approximately 15–20 min so that the surface is completely dry before applying the disks or the gradient strips.
6. In function of plate diameter and method use, the number of antibiotics to test can change.
7. Read the plates after 24 h of incubation at 37 °C.

3.3.4 Result Interpretation

The results of blood cultures, positive or negative, have a crucial impact in the outcome of the patient. For this, they should be promptly reported to the clinician, as they become available, keeping track of report production and delivery. Even if the communication between clinicians and microbiologists depends, obviously, on the hospital organization, few guidelines may be provided:

1. Status of the sample: The clinician should be able to know at any time if:
   - Blood cultures have been prescribed.
   - Specimen has been collected.
   - Microbiology lab received the blood cultures, at what time and what the delta between sample collection and check-in of the lab is.
   - Other investigations have been requested.
   - There has been growth of bacteria/fungi.
   - What results are available (preliminary or final reports).
   - Some microbiologists provide on daily basis information like "negative after 24 h of incubation," "negative after 48 h of incubation," "investigation ongoing," or "no results at this time."
2. Preliminary reports: Positive blood culture and Gram results should be promptly communicated to the clinic. The way to communicate these results, orally and/or in writing, has to be compliant with the UNI EN ISO 9001 to ensure a complete traceability and a rapid/accurate reception by the clinicians.

   Preliminary reports (antibiotic susceptibility tests and/or direct identification) must be sent by paper or electronic format. In both cases, it is mandatory that the preliminary report must be identified as "preliminary" and not confused with final report. The final report must include the definitive identification and susceptibility testing of the microorganism. In this report, any conflicting data with the preliminary results have to be indicated.
3. Contaminants: Microorganisms isolated from blood cultures may not have an etiologic role, but they can be contaminants coming from several sources like the patient's skin, the equipments used to collect the sample, the hands of phlebotomists, or the environment. The contamination, even in the best case studies, is more than 2 % [17].

   Informational trainings to promote, explain and verify the correct procedures to collect and process the specimen, can help to reduce the rate of contaminated blood cultures [18].

   Roth et al. showed that these trainings, performed by microbiologists, can have significant effects on the level of contaminated blood cultures, when nurses and auxiliary nurses conduct phlebotomies [18].

Contaminants such as *Bacillus* spp., *Corynebacterium* spp., *Propionibacterium* spp., and coagulase-negative *Staphylococci* may play the role of true pathogens in certain situations (e.g., catheter-associated infections):

- Microbiologist should adopt an interpretive algorithm to detect contaminants and periodically assess the rates of contamination.
- Microbiologist should minimize the identification of contaminants taking into account the immune status of the patient.
- Microbiologist does not perform sensitivity testing on contaminants or perform the tests without notification of results in the final report.
- Microbiologist should store contaminants for a few days to have the possibility to perform other investigations in case of subsequent isolation of the germ from the patient.
- Microbiologist must always perform identification of species and antibiotic susceptibility test in case of multiple isolations from the same patient.
- Microbiologist has to comment on the possible role of contaminant.
- Microbiologist should report, each year, data on pathogens, rates of contamination, and other quality indicators, according to the Department of Provenance (*see* **Note 8**).

4 Notes

1. Blood culture should always be required when a bloodstream infection or sepsis is suspected. Clinical symptoms which may lead to a suspicion of a bloodstream infection are fever (>36 °C) or hypothermia (<36 °C), severe local infections, shock, chills, rigors, abnormally raised heart rate, low or raised blood pressure, and raised respiratory rate. It is important to note that fever alone is not a useful indicator of bloodstream infection and one or more symptoms have to be analyzed. Blood cultures should be collected as soon as possible after the onset of clinical symptoms. Ideally, they should be obtained prior to the administration of antimicrobial therapy. If the patient is already on antimicrobial therapy, blood cultures should be collected immediately before administering the next dose.
2. Clean the skin covering an area of 7–8 cm in diameter using a gauze or a swab containing 70 % isopropyl alcohol. During the procedures to disinfect the skin, using a spiral motion, clean from the proposed puncture site to the peripheral area, and allow to dry. It is mandatory to leave the disinfectant for the

time necessary: chlorhexidine in alcohol solution (2 %) need at least 30 s in time, while the iodine compounds require more than 90 s. In all cases, allow to dry antiseptic, without removing the excess [19, 20].

3. It is generally recommended that 2–3 sets (two bottles/set) of blood culture should be obtained over a brief time period (e.g., within 1 h). Drawing blood at spaced intervals, such as 1–2 h apart, is only recommended to monitor continuous bacteremia/fungemia in patients with suspected infective endocarditis or other endovascular (i.e., catheter-related) infections. For the other aims, it has been shown no significant difference in yield between multiple blood cultures obtained simultaneously or those obtained at intervals. Within a 24-h period, increased yields appear to be a sole function of the overall volume of blood cultured [21].

4. For pediatric blood culture on infant/small child, use one pediatric aerobic bottle and fill adequate blood culture volume (0.5 mL for patients 1 month of age, 1.0 mL for patients between 1 month and 36 months of age, 4.0 mL for patients 36 months of age, [22]). If the child is less than 2 months of age, use only 70 % alcohol swabs to disinfect the skin. Several studies suggest to use alcohol solution in association with 0.5 % chlorhexidine instead of 2 % concentration. Using a spiral motion, clean from the proposed puncture site outward and use a fresh swab for each spiral. Iodine compounds are prohibited because the absorption from topical iodine-containing antiseptics causes disturbances in thyroid function in premature infants. Do this for 1–2 min and allow to dry. As for adults, 2–3 blood cultures should be collected within a 24-h period.

5. Some studies showed that the use of factors that inactivate antimicrobial agents has improved recovery and time to detection of yeasts. In general, special media formulated for the recovery of yeasts are unnecessary [4].

6. For endocarditis disease more sets have to be collected, depending on the degree of illness:

   Acute infective endocarditis: When suspected, the severity of this disease requires blood cultures to be drawn immediately to avoid unnecessary delays in treatment. Multiple blood culture sets should be drawn during a 30-min period prior to administration of empiric antimicrobial therapy.

   Subacute infective endocarditis: If subacute infection is suspected, there is usually not an urgent need to initiate empiric therapy. It is more important to attempt to establish the microbiological diagnosis. Multiple blood culture sets should be obtained prior to initiation of antimicrobial therapy, with sets spaced 30 min to 1 h apart. This may help document a continuous bacteremia and could be of additional clinical value [4].

Fungal infective endocarditis: A rare occurrence in the past, nowadays the incidence of fungal endocarditis is increasing considerably [23]. *Candida* is the most common fungal pathogen involved in infective endocarditis [24]. If optimum collection conditions are observed, the yield for positive blood cultures in fungal endocarditis for *Candida* spp. is 83–95 % [25].

7. In case of persistent negative result, consider the following points:
   (a) False-negative result due to antibiotic therapy.
   (b) Presence of microorganisms which do not grow (or have grown slowly) in blood culture media (*Tropheryma whippelii*, *Rickettsia* spp., *Bartonella* spp.).
   (c) Use of other diagnostic methods (molecular biology, serology, etc.) to research the microorganism.
8. A quality indicator is a tool that enables the user to quantify the quality of a selected aspect of care by comparing it with a criterion. It may be defined as an objective measure that evaluates critical healthcare domains as defined by the Institute for Quality in Laboratory Medicine (patient safety, effectiveness, equity, patient-centeredness, timeliness, and efficiency), based on evidence associated with those domains, and can be implemented in a consistent and comparable manner across settings and over time. Each microbiology lab can choose their own quality indicators for blood culture. Potential indicators can be the number of bottles/set per episode, the blood volume per episode, the mean time between blood collection and incubation or between positivity and result report, the positivity rate (per patient), the percentage of blood culture bottles inoculated with a volume minor of 5 mL or major than 10 mL, the percentage of cases where specimen is sent with a correct set of information, the contamination rate, the agreement between Gram staining and final identification, and the mean time for Gram staining notification after positivity. All these potential indicators should be notified by microbiologists to the departments with monthly cadence.

References

1. Paolucci M, Landini MP, Sambri V (2010) Conventional and molecular techniques for the early diagnosis of bacteraemia. Int J Antimicrob Agents 36(Suppl 2):S6–S16
2. Towns ML, Jarvis WR, Hsueh PR (2010) Guidelines on blood cultures. J Microbiol Immunol Infect 43(4):347–349
3. Munson E, Diekema DJ, Beekmann SE, Kimberle C, Chapin KC, Doern GV (2003) Detection and treatment of bloodstream infection: laboratory reporting and antimicrobial management. J Clin Microbiol 41: 495–497
4. Wayne PA (2007) Principles and procedures for blood cultures; approved guideline, CLSI document M47-A. Clinical and Laboratory Standards Institute (CLSI)
5. Garey KW, Rege M, Pai MP, Mingo DE, Suda KJ, Turpin RS, Bearden DT (2006) Time to initiation of fluconazole therapy impacts

mortality in patients with candidemia: a multi-institutional study. Clin Infect Dis 43(1):25–31

6. Khatib R, Saeed S, Sharma M, Riederer K, Fakih MG, Johnson LB (2006) Impact of initial antibiotic choice and delayed appropriate treatment on the outcome of Staphylococcus aureus bacteremia. Eur J Clin Microbial Infect Dis 25(3):181–185
7. Saito T, Iinuma Y, Takakura S, Nagao M, Matsushima A, Shirano M, Ichiyama S (2009) Delayed insertion of blood culture bottles into automated continuously monitoring blood culture systems increases the time from blood sample collection to the detection of microorganisms in bacteremic patients. J Infect Chemother 15:49–53
8. Weinstein MP (2003) Blood culture contamination: persisting problems and partial progress. J Clin Microbiol 41:2275–2278
9. Baron EJ, Weinstein MP, Dunne WM, Yagupsky P, Welch DF, Wilson DM (2005) Cumitech 1C, Blood Cultures IV. In: Baron EJ (ed). ASM, Washington, DC
10. Cockerill FR 3rd, Wilson JW, Vetter EA, Goodman KM, Torgerson CA, Harmsen WS, Schleck CD, Ilstrup DM, Washington JA 2nd, Wilson WR (2004) Optimal testing parameters for blood cultures. Clin Infect Dis 38: 1724–1730
11. Bourbeau PP, Foltzer M (2005) Routine incubation of BacT/ALERT FA and FN blood culture bottles for more than 3 days may not be necessary. J Clin Microbiol 43:2506–2509
12. Riedel S, Bourbeau P, Swartz B, Brecher S, Carroll KC, Stamper PD, Dunne WM, McCardle T, Walk N, Fiebelkorn K, Sewell D, Richter SS, Beekmann S, Doern GV (2008) Timing of specimen collection for blood cultures from febrile patients with bacteremia. J Clin Microbiol 46(4):1381–1385
13. Riley JA, Heiter BJ, Bourbeau PP (2003) Comparison of recovery of blood culture isolates from two BacT/ALERT FAN aerobic blood culture bottles with recovery from one FAN aerobic bottle and one FAN anaerobic bottle. J Clin Microbiol 41:213–217
14. Søgaard M, Nørgaard M, Schønheyder HC (2007) First notification of positive blood cultures and the high accuracy of the gram stain report. J Clin Microbiol 45(4):1113–1117
15. Fraser A, Paul M, Almanasreh N, Tacconelli E, Frank U, Cauda R, Borok S, Cohen M, Andreassen S, Nielsen AD, TREAT Study Group (2006) Benefit of appropriate empirical antibiotic treatment: thirty-day mortality and duration of hospital stay. Am J Med 119: 970–976
16. Kumar A, Roberts D, Wood KE, Light B, Parrillo JE, Sharma S, Suppes R, Feinstein D, Zanotti S, Taiberg L, Gurka D, Kumar A, Cheang M (2006) Duration of hypotension before initiation of effective antimicrobial therapy is the critical determinant of survival in human septic shock. Crit Care Med 34(6): 1589–1596
17. Dunne WM, Nolte FS, Wilson ML (1997) Cumitech 1B, Blood Cultures III. In: Hindler JA (ed) ASM, Washington, DC
18. Roth A, Wiklund AE, Pålsson AS, Melander EZ, Wullt M, Cronqvist J, Walder M, Sturegård E (2010) Reducing blood culture contamination by a simple informational intervention. J Clin Microbiol 48(12):4552–4558
19. Washer LL, Chenoweth C, Kim HW, Rogers MA, Malani AN, Riddell J 4th, Kuhn L, Noeyack B Jr, Neusius H, Newton DW, Saint S, Flanders SA (2013) Blood culture contamination: a randomized trial evaluating the comparative effectiveness of 3 skin antiseptic interventions. Infect Control Hosp Epidemiol 34(1):15–21
20. Suwanpimolkul G, Pongkumpai M, Suankratay C (2008) A randomized trial of 2 % chlorhexidine tincture compared with 10 % aqueous povidone-iodine for venipuncture site disinfection: effects on blood culture contamination rates. J Infect 56(5):354–359
21. Li J, Plorde J, Carlson L (1994) Effects of volume and periodicity on blood cultures. J Clin Microbiol 32:2829–2831
22. Connell TG, Rele M, Cowley D, Buttery JP, Curtis N (2007) How reliable is a negative blood culture result? Volume of blood submitted for culture in routine practice in a children's hospital. Pediatrics 119(5):891–896
23. Rubenstein E, Lang R (1995) Fungal endocarditis. Eur Heart J 16(Suppl B):84–89
24. Ellis ME, Al-Abdely H, Standridge A, Greer W, Venturea W (2001) Fungal endocarditis: evidence in the world literature, 1965–1995. Clin Infect Dis 32:50–62
25. McLeod R, Remington JS (1978) Fungal endocarditis. In: Rahimtoola SH et al (eds) Infective endocarditis. Gune & Stratton, New York, NY, pp 211–290

Chapter 5

Matrix-Assisted Laser Desorption/Ionization Time of Flight Mass Spectrometry (MALDI-TOF/MS)-Based Identification of Pathogens from Positive Blood Culture Bottles

Philippe Lagacé-Wiens

Abstract

Since the expansion of commercial use of MALDI-TOF/MS instruments for the identification of bacteria from culture which has occurred over the past 5–8 years, techniques for the identification of bacteria directly from positive blood cultures have been developed (Lagace-Wiens et al., J Clin Microbiol 50:3324–3328, 2012; Martiny et al., Eur J Clin Microbiol Infect Dis 31:2269–2281, 2012; Moussaoui et al., Clin Microbiol Infect 16:1631–1638, 2010). These techniques have the potential to provide definitive identification of pathogens causing sepsis 18–48 h earlier than conventional methodologies, and implementation of these methods has been shown to impact morbidity and hospital costs in a positive way (Martiny et al., Clin Microbiol Infect 19:E568–E581, 2013; Loonen et al., Eur J Clin Microbiol Infect Dis 31:1575–1583, 2012). Although many methods for purification of bacterial cells have been developed, including differential centrifugation, centrifuge lysis, and preincubation on sold media (March-Rossello et al., Eur J Clin Microbiol Infect Dis 32:699–704, 2013; Saffert et al., Diagn Microbiol Infect Dis 73:21–26, 2012; Schubert et al., J Mol Diagn 13:701–706, 2011), we will describe the method by which intact bacterial cells are extracted from positive blood culture bottles using a commercially available kit (SepsiTyper™) which is based on the centrifuge lysis methodology (Lagace-Wiens et al., J Clin Microbiol 50:3324–3328, 2012; Buchan et al., J Clin Microbiol 50:346–352, 2012).

Key words Sepsis, Rapid diagnosis, Hemoculture, Mass spectrometry, Identification, Diagnostics, Blood culture

1 Introduction

Although the principle of using mass spectrometry for the identification of bacteria dates to the early 1970s [10], recent refinements, including matrices that allow analysis of large intact proteins (e.g., 2-Cyano-3-(4-hydroxyphenyl) acrylic acid (HCCA)), development of rapid and reliable information technology systems and computers, and progressive miniaturization of mass spectrometers have led to a revolution in clinical microbiology laboratories [11–13]. Clinical microbiology labs can now use mass spectrometers to

Nicasio Mancini (ed.), *Sepsis: Diagnostic Methods and Protocols*, Methods in Molecular Biology, vol. 1237,
DOI 10.1007/978-1-4939-1776-1_5,

identify a wide range of organisms, including those that were previously very difficult to identify, within minutes of observing growth on a wide variety of media. Furthermore, the applicability of MALDI-TOF MS extends to the identification of yeasts, mycelial fungi (mould), and mycobacteria, and even more novel applications include the identification of protozoa, insects, and plants [14]. The principle of MALDI-TOF MS identification of bacteria is relatively simple. An unknown organism is placed onto a suitable surface, overlaid with a matrix that allows the preservation of the large protein structures during ionization (typically HCCA) and subject to laser desorption and ionization. The proteome of the organism is released and ionized by the laser, and the charged particles are subjected to acceleration. The mass-to-charge ratios (m/z) are determined using the time of flight method. The m/z spectrum is then compared to a library of spectra known as bacterial organisms, and the closest match is assumed to be the unknown pathogen. If no close matches exist in the library, the organism identification cannot occur. This may be the result of a poor sample, a mixed sample, insufficient material, or that the organism is novel to the library. A recent comprehensive review summarizes many of the principles and applications of MALDI-TOF to microbiology [12].

Among the applications considered most valuable for the management of patients with septic shock has been the analysis of positive blood cultures by MALDI-TOF MS [1, 3, 15]. Since blood cultures are typically monomicrobial, with most studies reporting rates of 85 % or greater being monomicrobial, MALDI-TOF/MS can readily be applied to the identification of the bacterial species present in the broth. Unfortunately, the presence of human blood, cells, serum, and broth, which all contain protein, will interfere with the analysis by introducing unexpected peaks into the mass spectrum [8]. Thus, the key step in the analysis of positive blood cultures by MALDI-TOF is the creation of a cell pellet that is relatively free of extraneous (human and broth) material that would otherwise hamper the accurate identification of the organism.

2 Materials

1. Chemical hood (*see* **Note 1**).
2. 100% ethanol (HPLC/MS grade) (*see* **Note 2**).
3. 100% formic acid (HPLC/MS grade) (*see* **Note 2**).
4. 100% acetonitrile (HPLC/MS grade) (*see* **Notes 2** and **3**).
5. 100% trifluoroacetic acid (HPLC/MS grade) (*see* **Notes 2** and **3**).
6. Water (HPLC/MS grade) (*see* **Note 2**).
7. Chemical safe gloves (*see* **Note 1**).

8. Volumetric glass pipettes (if available), or pipettes of different sizes (for pipetting 1–1,000 μL), and appropriate pipette tips.
9. 10–50 mL amber glass bottles with chemical safe lids.
10. 50 mL glass bottles.

2.1 Matrix (2-Cyano-3-(4-Hydroxyphenyl) Acrylic Acid: HCCA) Preparation

1. HCCA solvent: 5 % v/v trifluoroacetic acid, 50 % v/v acetonitrile in HPLC grade water (*see* **Note 4**).
2. 2-cyano-3-(4-hydroxyphenyl) acrylic acid (available in proportioned 2.5 mg amounts from Bruker Daltonics™ (part # 8255344)) (*see* **Note 5).**
3. Add 250 μL of HCCA solvent (*see* Subheading 2.2) to 2.5 mg preportioned 2-cyano-3-(4-hydroxyphenyl) acrylic acid (HCCA). Vortex for 1 min.

2.2 Formic Acid Extraction Solvents

Formic acid extraction is used to extract proteins from the bacteria in the cell pellet extracted from positive blood cultures. Four separate solvents are used: (1) 70 % formic acid, (2) 100 % acetonitrile, (3) 100 % ethanol, and (4) HPLC/MS grade water.

1. 70 % formic acid:
   Measure 14 mL 100 % formic acid in a 50 mL amber bottle, and add 6 mL HPLC/MS grade water. Store at room temperature and stable for at least six months.
2. 100 % acetonitrile (*see* **Note 6**):
   Measure 5 mL of 100 % acetonitrile in a 10 mL amber glass bottle.
3. 100 % ethanol (*see* **Note** 7):
   Measure 20 mL of 100 % ethanol in a glass bottle.
4. HPLC/MS grade water:
   Measure 50 mL of water into a glass bottle (*see* **Note 8**).

2.3 Bacterial Separation and Formic Acid Extraction

1. Bruker SepsiTyper™ kit (Bruker Daltonics™ part # 8270170) containing:
   - Lysis buffer (LB).
   - Washing buffer (WB).
   - MALDI-quality reaction tubes.
2. Class II biosafety cabinet (*see* **Note 9**).
3. Pipettes of different sizes (for pipetting 1–1,000 μL) and appropriate pipette tips.
4. Benchtop centrifuge.
5. 70 % v/v formic acid.
6. 100 % acetonitrile.
7. 100 % ethanol.

8. HPLC grade water.
9. Vortex.
10. Microfuge tube rack.
11. Positive blood culture bottle (e.g., BACTEC collection tube (10 mL) BD # 44226 0 or BacT/Alert® blood collection tube (10 mL), bioMerieux # 259789). Do not use bottles containing activated charcoal (*see* **Note 10**).
12. 3 mL syringe.

2.4 Sample Analysis

1. Bruker MALDI Biotyper™ instrument and database.
2. 96 spot polished steel target plate (Bruker Daltonics # 280800) (*see* **Note 11**).
3. Formic acid extract of positive blood culture (*see* Subheading 3.1 for preparation).
4. HCCA matrix (*see* Subheading 2.2 for preparation) (*see* **Note 12).**
5. 1 μL pipette and appropriate tips.

3 Methods

3.1 Blood Culture Extraction and Formic Acid Extraction (All Sample Manipulation Steps Performed in Biosafety Cabinet)

1. Disinfect the septum of the blood culture bottle with 70 % ethanol.
2. Collect 2–3 mL blood culture fluid using the syringe.
3. Transfer 1 mL blood culture fluid to a MALDI-quality microcentrifuge tube.
4. Add 200 μL lysis buffer and mix by vortexing for 10 s (*see* **Note 13**).
5. Centrifuge for 2 min at 16,000 × *g*.
6. Remove the supernatant by pipetting and discard.
7. Suspend pellet in 1 mL wash buffer by pipetting up and down.
8. Centrifuge for 1 min at 16,000 × *g*.
9. Remove the supernatant by pipetting and discard.
10. Resuspend the pellet in 300 μL HPLC grade water and add 900 μL 100 % ethanol.
11. Centrifuge the suspension at 16,000 × *g* and decant and discard the supernatant.
12. Centrifuge the pellet for 2 min at 16,000 × *g*, and remove residual ethanol using a pipette (*see* **Note 14**).
13. Allow the pellet to dry at room temperature for few minutes (*see* **Note 14**).

14. Add 2–50 μL1 70 % formic acid to the pellet, and mix thoroughly by pipetting up and down (*see* **Note 15**).
15. Add an equal volume of 100 % acetonitrile to the tube and mix carefully (*see* **Note 15**).
16. Centrifuge the tube at maximum speed for 2 min. The supernatant is the extract required for analysis.

3.2 MALDI-TOF MS Identification

1. Place 1 μL formic acid extract (supernatant from **step 16**, above) onto the target plate spot.
2. Once the spot is completely dry, overlay the spot with 1 μL HCCA matrix (*see* **Note 16**).
3. Perform MALDI Biotyper™ analysis following the manufacturer's instructions (*see* **Note 17**).

4 Notes

1. Always prepare solvents and solvent-containing solution in a certified chemical hood and always wear chemical safe gloves.
2. Ensure that all solvents and water used during each step are of HPLC or MS grade. Reagents should be stored at room temperature unless otherwise stated. Some reagents are photosensitive—always follow manufacturer's recommendations for storage. Some reagents are flammable or highly corrosive. Always follow local MSDS information sheets. Disposal should be done in accordance with local regulations.
3. Note that acetonitrile is highly flammable and trifluoroacetic acid is highly corrosive. Manipulations should always be performed in the fume hood.
4. Store at room temperature and stable for at least 6 months. It must be tightly closed and should not be left open for long periods as acetonitrile is highly volatile and evaporation will alter solvent ratios. Always aliquot working solutions of acetonitrile and never use supplier's bottle to store working solutions as they may become contaminated with bacterial proteins and adversely affect results.
5. HCCA powder is available from a variety of sources, but we recommend the Bruker product as it is pre-aliquoted and simple to prepare, and the working solution can be stored in its original bottle. Each bottle is sufficient for ~250 determinations and is the same matrix used for routine identification of organisms using the Bruker MALDI Biotyper™.
6. Store at room temperature and stable for at least 6 months. It must be tightly closed and should not be left open for long periods as acetonitrile is highly volatile and evaporation will

alter solvent ratios. Always aliquot working solutions of acetonitrile and never use supplier's bottle to store working solutions as they may become contaminated with bacterial proteins and adversely affect results.

7. Store at room temperature and stable for at least 6 months; always aliquot ethanol and never use supplier's bottle to store working solutions as they may become contaminated with bacterial proteins and adversely affect results.
8. Store at room temperature and stable for one month; replace immediately if water appears cloudy. Never use supplier's bottle to store working solutions as they may become contaminated with bacterial proteins and adversely affect results.
9. A certified biosafety cabinet is recommended for all manipulations of biological materials/samples. Follow local biosafety policies.
10. Aerobic, anaerobic, and pediatric bottles can be used. Laboratories should always verify the performance of their systems when implementing new identification methodologies. Charcoal will interfere with mass spectra, and bottles containing activated charcoal should not be used with this procedure as results will be significantly affected [16]. Procedures are available for removing charcoal residues from sample [16]. Contact the MALDI-TOF instrument manufacturer for details. Positive blood cultures should always have a Gram stain performed and reported prior to performing MALDI-TOF. Only blood cultures that appear monomicrobial by Gram stain should routinely be analyzed by MALDI-TOF. Accurate analysis of polymicrobial cultures is not possible currently [1, 3, 17]. If polymicrobial cultures are analyzed, one of the following outcomes may occur: one of the pathogens may be accurately identified, multiple pathogens present in the blood culture will be listed in the possible results, or no identifiable spectrum will be detected. Some laboratories have reported that customized databases (e.g., create separate databases for Gram-positive and Gram-negative organisms) may allow accurate identification of polymicrobial cultures, but insufficient evidence exists at this time to support this approach.
11. 24, 48, and 96 spot targets are available in polished steel or ground steel. Both plate types can be used with this procedure.
12. Once reconstituted, store in the dark at room temperature for up to 5 days. Matrix is frequently the first reagent to become unstable or to result in poor performance. If test standards or quality control is not working, always consider preparing fresh matrix and trying test standards again.

13. At low temperatures, lysis buffer may contain precipitates. These will dissolve when the buffer is at room temperature.
14. It is critical that all the ethanol be removed from the sample as it will interfere with analysis. Ensure as much as possible is pipetted off and allow to dry completely. Extend drying time as required.
15. The volume of formic acid and acetonitrile added to the pellet should be proportional to the size of the pellet. For very small pellets, the volume can be reduced down to 2 μL.
16. The spot should be completely dry before overlaying matrix. Results will be affected if the spot is not allowed to dry completely.
17. Follow the procedures outlined by the manufacturer. User should be cognizant of the limitation of their instrument. For the Bruker Biotyper™, a confidence score of >1.7 from a blood culture is considered accurate to species, while a score 1.5–1.699 is considered accurate to genus (information provided by manufacturer). Using these breakpoints, approximately 80–90 % of blood cultures can be accurately identified using this procedure. Specific limitations need to be considered and are detailed in manufacturer's instructions. These include but are not limited to:
    (a) Members of the *Streptococcus mitis* group and *S. pneumoniae* may not be accurately differentiated from each other. Additional testing (e.g., bile solubility) is required for accurate identification. Some laboratories have reported algorithms for the accurate identification of *S. pneumoniae* by MALDI-TOF [18, 19].
    (b) *Shigella* spp. are not in the database as it cannot be differentiated from *E. coli*. This is unlikely to be relevant to blood culture analysis. Some laboratories have reported algorithms for the accurate identification of *Shigella* spp. by MALDI-TOF [20].
    (c) Members of closely related complexers (e.g., *Enterobacter cloacae* complex) are difficult to differentiate from each other. This is unlikely to be relevant to blood culture analysis.
    (d) *Salmonella enterica* subspecies *enterica* serotypes (including serotypes Typhi and Paratyphi) cannot be differentiated from each other, and determination of typhoidal and non-typhoidal strains requires additional testing (serotyping or biochemical analysis). However, some laboratories have reported methods by which these can be accurately identified by MALDI-TOF [21, 22].

(e) Although the routine database of the Biotyper™ contains a comprehensive list of yeasts and bacteria, additional databases may be required for the accurate identification of mycobacteria and organisms of public health importance (e.g., *Bacillus anthracis, Francisella tularensis, Brucella* spp., *Yersinia pestis*).

References

1. Lagace-Wiens PR, Adam HJ, Karlowsky JA, Nichol KA, Pang PF, Guenther J, Webb AA, Miller C, Alfa MJ (2012) Identification of blood culture isolates directly from positive blood cultures by use of matrix-assisted laser desorption ionization-time of flight mass spectrometry and a commercial extraction system: analysis of performance, cost, and turnaround time. J Clin Microbiol 50(10):3324–3328. doi:10.1128/JCM.01479-12, JCM.01479-12 [pii]
2. Martiny D, Dediste A, Vandenberg O (2012) Comparison of an in-house method and the commercial Sepsityper kit for bacterial identification directly from positive blood culture broths by matrix-assisted laser desorption-ionisation time-of-flight mass spectrometry. Eur J Clin Microbiol Infect Dis 31:2269–2281
3. Moussaoui W, Jaulhac B, Hoffmann AM, Ludes B, Kostrzewa M, Riegel P, Prevost G (2010) Matrix-assisted laser desorption ionization time-of-flight mass spectrometry identifies 90 % of bacteria directly from blood culture vials. Clin Microbiol Infect 16(11):1631–1638
4. Martiny D, Debaugnies F, Gateff D, Gerard M, Aoun M, Martin C, Konopnicki D, Loizidou A, Georgala A, Hainaut M, Chantrenne M, Dediste A, Vandenberg O, van Praet S (2013) Impact of rapid microbial identification directly from positive blood cultures using matrix-assisted laser desorption/ionization time-of-flight mass spectrometry on patient management. Clin Microbiol Infect 19:E568–E581. doi:10.1111/1469-0691.12282
5. Loonen AJ, Jansz AR, Stalpers J, Wolffs PF, van den Brule AJ (2012) An evaluation of three processing methods and the effect of reduced culture times for faster direct identification of pathogens from BacT/ALERT blood cultures by MALDI-TOF MS. Eur J Clin Microbiol Infect Dis 31(7):1575–1583
6. March-Rossello GA, Munoz-Moreno MF, Garcia-Loygorri-Jordan de Urries MC, Bratos-Perez MA (2013) A differential centrifugation protocol and validation criterion for enhancing mass spectrometry (MALDI-TOF) results in microbial identification using blood culture growth bottles. Eur J Clin Microbiol Infect Dis 32(5):699–704. doi:10.1007/s10096-012-1797-1
7. Saffert RT, Cunningham SA, Mandrekar J, Patel R (2012) Comparison of three preparatory methods for detection of bacteremia by MALDI-TOF mass spectrometry. Diagn Microbiol Infect Dis 73(1):21–26
8. Schubert S, Weinert K, Wagner C, Gunzl B, Wieser A, Maier T, Kostrzewa M (2011) Novel, improved sample preparation for rapid, direct identification from positive blood cultures using matrix-assisted laser desorption/ionization time-of-flight (MALDI-TOF) mass spectrometry. J Mol Diagn 13(6):701–706
9. Buchan BW, Riebe KM, Ledeboer NA (2012) Comparison of the MALDI Biotyper system using Sepsityper specimen processing to routine microbiological methods for identification of bacteria from positive blood culture bottles. J Clin Microbiol 50(2):346–352
10. Anhalt JP, Fenselau C (1975) Identification of bacteria using mass spectrometry. Anal Chem 47(2):219–225
11. Vargha M, Takats Z, Konopka A, Nakatsu CH (2006) Optimization of MALDI-TOF MS for strain level differentiation of Arthrobacter isolates. J Microbiol Methods 66(3):399–409
12. Clark AE, Kaleta EJ, Arora A, Wolk DM (2013) Matrix-assisted laser desorption ionization-time of flight mass spectrometry: a fundamental shift in the routine practice of clinical microbiology. Clin Microbiol Rev 26(3):547–603. doi:10.1128/CMR.00072-12, 26/3/547 [pii]
13. Bizzini A, Greub G (2010) Matrix-assisted laser desorption ionization time-of-flight mass spectrometry, a revolution in clinical microbial identification. Clin Microbiol Infect 16(11):1614–1619
14. Croxatto A, Prod'hom G, Greub G (2012) Applications of MALDI-TOF mass spectrometry in clinical diagnostic microbiology. FEMS Microbiol Rev 36(2):380–407
15. Foster AG (2013) Rapid identification of microbes in positive blood cultures by use of the vitek MS matrix-assisted laser desorption

ionization-time of flight mass spectrometry system. J Clin Microbiol 51(11):3717–3719. doi:10.1128/JCM.01679-13, JCM.01679-13 [pii]

16. Wuppenhorst N, Consoir C, Lorch D, Schneider C (2012) Direct identification of bacteria from charcoal-containing blood culture bottles using matrix-assisted laser desorption/ionisation time-of-flight mass spectrometry. Eur J Clin Microbiol Infect Dis 31(10):2843–2850. doi:10.1007/s10096-012-1638-2
17. Kok J, Thomas LC, Olma T, Chen SC, Iredell JR (2011) Identification of bacteria in blood culture broths using matrix-assisted laser desorption-ionization Sepsityper and time of flight mass spectrometry. PLoS One 6(8):e23285
18. Dubois D, Segonds C, Prere MF, Marty N, Oswald E (2013) Identification of clinical Streptococcus pneumoniae isolates among other alpha and nonhemolytic streptococci by use of the Vitek MS matrix-assisted laser desorption ionization-time of flight mass spectrometry system. J Clin Microbiol 51(6):1861–1867. doi:10.1128/JCM.03069-12, JCM.03069-12 [pii]
19. Werno AM, Christner M, Anderson TP, Murdoch DR (2012) Differentiation of Streptococcus pneumoniae from nonpneumococcal streptococci of the Streptococcus mitis group by matrix-assisted laser desorption ionization-time of flight mass spectrometry. J Clin Microbiol 50(9):2863–2867. doi:10.1128/JCM.00508-12, JCM.00508-12 [pii]
20. Khot PD, Fisher MA (2013) Novel approach for differentiating shigella species and Escherichia coli by matrix-assisted laser desorption ionization-time of flight mass spectrometry. J Clin Microbiol 51(11):3711–3716. doi:10.1128/JCM.01526-13, JCM.01526-13 [pii]
21. Kuhns M, Zautner AE, Rabsch W, Zimmermann O, Weig M, Bader O, Gross U (2012) Rapid discrimination of Salmonella enterica serovar Typhi from other serovars by MALDI-TOF mass spectrometry. PLoS One 7(6):e40004. doi:10.1371/journal.pone.0040004, PONE-D-12-12439 [pii]
22. Dieckmann R, Malorny B (2011) Rapid screening of epidemiologically important Salmonella enterica subsp. enterica serovars by whole-cell matrix-assisted laser desorption ionization-time of flight mass spectrometry. Appl Environ Microbiol 77(12):4136–4146

Chapter 6

Bacterial and Fungal DNA Extraction from Positive Blood Culture Bottles: A Manual and an Automated Protocol

Minna Mäki

Abstract

When adapting a gene amplification-based method in a routine sepsis diagnostics using a blood culture sample as a specimen type, a prerequisite for a successful and sensitive downstream analysis is the efficient DNA extraction step. In recent years, a number of in-house and commercial DNA extraction solutions have become available. Careful evaluation in respect to cell wall disruption of various microbes and subsequent recovery of microbial DNA without putative gene amplification inhibitors should be conducted prior selecting the most feasible DNA extraction solution for the downstream analysis used. Since gene amplification technologies have been developed to be highly sensitive for a broad range of microbial species, it is also important to confirm that the used sample preparation reagents and materials are bioburden-free to avoid any risks for false-positive result reporting or interference of the diagnostic process. Here, one manual and one automated DNA extraction system feasible for blood culture samples are described.

Key words Sample preparation, Blood culture, DNA extraction, Fungi, Bacteria

1 Introduction

BacT/ALERT® (bioMérieux Inc, Durham, USA) and BACTEC™ (BD Diagnostics, Sparks, USA) are widely used automated continuously monitoring instruments for culturing blood samples in standard aerobic and anaerobic media. Formulations of these soybean-casein digest-based media differ in supplements and the anticoagulant sodium polyanetholesulfonate (SPS) concentrations [1, 2]. SPS is a potent inhibitor of gene amplification technologies and resistant to removal by some DNA extraction methods [3]. Hence, it needs to be removed efficiently prior using a blood culture sample as a specimen in gene amplification-based sepsis diagnostics. Incomplete removal of SPS or other inhibitors can be, however, accomplished by adding the V fraction of 96 % BSA to the gene amplification reaction [4, 5]. Feasible sample preparation method involves not only the removal of potent inhibitors but also efficient cell wall disruption of a microbe and subsequent recovery

Nicasio Mancini (ed.), *Sepsis: Diagnostic Methods and Protocols*, Methods in Molecular Biology, vol. 1237,
DOI 10.1007/978-1-4939-1776-1_6, © Springer Science+Business Media New York 2015

of microbial DNA. The sample preparation should be capable of breaking equally well sepsis-causing bacterial and more challenging fungal cell walls and to extract high-quality bacterial and fungal DNA [6–8].

In recent years, a number of in-house and commercial DNA extraction solutions for both bacterial and fungal targets have become available. A comprehensive review on the principles of the most common DNA extraction methods and commercial kits has been recently written by Anandika Dhaliwal [9]. The DNA extraction step has profound influence on the sensitivity and overall performance of the downstream analysis used, and therefore, careful evaluation should be conducted prior selecting the most feasible DNA extraction solution for the used diagnostic process. In addition to the performance characteristics (i.e., removal of inhibitors, efficient cell wall disruption of a microbe, and recovery of microbial DNA), the DNA extraction solutions are typically compared in terms of reproducibility, turnaround time, hands-on time, costs, and how many samples can be processed simultaneously. Also, the applicability of the DNA extraction solution to other routine diagnostic processes, the need of additional instruments, instrument footprints, and environmental issues such as waste management are also emphasized. Several comparative studies on the performance of various in-house and/or commercial DNA extraction methods using blood culture samples have been conducted and published. It has been argued that commercial, automated DNA extraction systems often provide a more standardized solution, with better traceability [5, 10–14].

Gene amplification technologies can be highly sensitive, detecting also inherent contamination, i.e., microbial bioburden originating from the used reagents and materials. Therefore, when adapting a gene amplification-based method in clinical diagnostics, reagents and materials used in the process should also be studied and confirmed to be bioburden-free. Some reports have demonstrated that bacterial or fungal bioburden can be detected from the DNA extraction reagents [4, 15–18]. Bioburden may interfere with the diagnostic process by decreasing the sensitivity, or, in the worst case scenario, may cause a false-positive test result if the bioburden load is high enough and microbial species causing the bioburden is included in the target panel of the used identification test. Reagents and materials can be validated using negative test controls, which are recommended to include in the test series every time when any of the reagents or the production lot is changed. The negative test controls also monitor the potential risk of carry-over contaminations. It should be noted here that the blood culture bottle can also be a source of bioburden [19].

No traces of bacterial or fungal bioburden have been observed from the current production version of the commercially available automated DNA extraction device NucliSENS®

easyMAG® (bioMérieux, Marcy-l'Etoile, France). The performance of NucliSENS® easyMAG® extraction device in routine clinical settings for sepsis diagnostics has been studied in a thorough manner together with the commercially available Prove-it™ Sepsis assay [20]. Prove-it™ sepsis assay analyzes blood culture, having undergone DNA extraction, through a PCR and microarray platform. In the multicenter study, the first version of Prove-it™ sepsis assay, consisting of the detection over 50 bacterial species, achieved a sensitivity and specificity of 95 % and 99 %, respectively, on 3,318 blood culture samples. Both the BacT/ALERT 3D and BACTEC 9240 blood culture instruments and corresponding blood culture bottles were used in the study. Lately, the current generation of the Prove-it™ Sepsis v2.0 assay, consisting of a pathogen panel that covers over 60 Gram-negative and Gram-positive bacterial species and 13 fungal species, was also validated using the NucliSENS® easyMAG® extraction device. The fungal speciation in turn was 99 % sensitive and 97 % specific, with no deterioration in bacterial target performance [21]. Other studies have also demonstrated the suitability of NucliSENS® easyMAG® extraction device for blood culture samples and gene amplification-based downstream analysis [14, 22, 23].

The automated version of NucliSENS® easyMAG® extraction device and its manual version NucliSENS® miniMAG® (bioMérieux, Marcy-l'Etoile, France) are introduced here. NucliSENS® easyMAG® and NucliSENS® miniMAG® are generic extraction systems for DNA and RNA from a variety of sample types and volumes. Of note is that NucliSENS® easyMAG® has been labelled for in vitro diagnostics. Nucleic acid extraction in both systems is based on bioMérieux's proprietary Boom® technology, with magnetic silica particles. Shortly, the sample is first lysed with a chaotropic lysis buffer, after which magnetic silica particles are added to the sample/lysis solution. The magnetic unit of miniMAG® or easyMAG® is then introduced to the silica particles, enabling the system to separate the silica particles from cellular components and to purify nucleic acids trough washing steps. After washing, the elution buffer releases DNA from the silica particles, after which it is ready to be used in gene amplification-based applications. In easyMAG®, 1–24 samples can be run simultaneously, and the turnaround time for 24 samples is 1 h. In miniMAG®, 1–12 samples can be run simultaneously and the turnaround time for 12 samples is 1 h and for 24 samples 90 min [1]. Several studies have emphasized that these systems perform well with various specimen types and are easy to use, and moreover, easyMAG® requires little hands-on time [24–27]. It has also been demonstrated that miniMAG® can yield high quantity and quality of nucleic acids and its performance is comparable to, or even better than, some commercially available automated DNA extraction devices, especially in terms of reproducibility [28].

2 Materials

1. Disposable gloves and laboratory coats.
2. Adjustable micropipettes.
3. Nucleic acid and nuclease-free, aerosol-resistant pipette tips.
4. Sterile, nucleic acid-free 1.5 ml microcentrifuge tubes.
5. Racks for tubes.
6. A vortex mixer.
7. A thermal shaker with a microcentrifuge tube adapter.
8. Distilled water.
9. NucliSENS® miniMAG® workstation.
10. NucliSENS® nucleic acid extraction reagents for miniMAG®:
    (a) NucliSENS® magnetic extraction reagents.
    (b) NucliSENS® lysis buffer (2.0 ml).
    (c) NucliSENS® 1.5 ml micro tubes with caps.

Or

11. NucliSENS® easyMAG® platform.
12. NucliSENS® nucleic acid extraction reagents for easyMAG®:
    (a) EasyMAG® magnetic silica.
    (b) EasyMAG® disposables.
    (c) EasyMAG® lysis buffer.
    (d) EasyMAG® wash buffers 1, 2, and 3.

3 Methods

3.1 Manual Protocol: Extraction of DNA from a Blood Culture Sample with NucliSENS® miniMAG®

NucliSENS® miniMAG® workstation is recommended to be used according to the manufacturer's instructions and recommendations [1].

3.1.1 Lysis of the Cells

1. Centrifuge NucliSENS® lysis buffer tube (2 ml) for 10 s at 1,500 × *g*.
2. Add 100 μl of a blood culture to NucliSENS® lysis buffer.
3. Mix well the sample/lysis buffer mixture.
4. Incubate for 10 min at RT.

3.1.2 Binding of DNA to Magnetic Silica Particles

1. Add 50 μl of the magnetic silica particles to the sample/lysis buffer mixture.
2. Mix well and incubate for 10 min at RT.

3.1.3 Washing the Magnetic Silica Particles

1. Centrifuge a sample/lysis buffer/silica particles tube for 2 min at 1,500 × *g*.
2. Remove the supernatant.
3. Add 400 μl of wash buffer 1 and transfer the mixture to microcentrifuge tube.
4. Place the microcentrifuge tube into the NucliSENS® miniMAG® workstation.
5. Wash for 30 s at **step 1** in the NucliSENS® miniMAG® workstation, with the magnet on.
6. Remove the supernatant (the workstation magnet on).
7. Turn the workstation magnet off.
8. Add 400 μl of wash buffer 1 and repeat **steps 11** and **12**.
9. Turn the workstation magnet off.
10. Add 500 μl of wash buffer 2 and repeat **steps 11** and **12**.
11. Repeat **step 16**.
12. Turn the workstation magnet off.
13. Add 500 μl of wash buffer 3 and wash for 15 s at **step 1** in the NucliSENS® miniMAG® workstation, with the magnet on.
14. Remove the supernatant (the workstation magnet on).

3.1.4 Elution

1. Add 50 μl of elution buffer.
2. Incubate for 5 min at 60 °C in a thermal shaker with 700 rpm agitation.
3. Transfer the supernatant to a clean storage tube for the use in gene amplification applications.

3.2 Automated Protocol

NucliSENS® easyMAG® instrument is recommended to be used according to the manufacturer's instructions and recommendations [1]:

3.2.1 Extraction of DNA from a Blood Culture Sample with NucliSENS® easyMAG®

1. Switch the instrument on.
2. Select the protocol Generic 2.0.1 and start the off-board lysis extraction protocol by adjusting the elution volume to 55 μl.
3. Add 100 μl of a blood culture to EasyMAG® lysis buffer.
4. Mix well the sample/lysis buffer mixture.
5. Incubate for 10 min at RT.
6. Insert aspiration tips into the instrument.
7. Pipette the sample/lysis buffer mixture into the one well of the 8-well vessel.
8. Mix 50:50 magnetic silica particles and distilled water, e.g., 550 μl:550 μl.

9. Add 100 μl of the EasyMAG® magnetic silica mixture to the well of the 8-well vessel containing the sample/lysis buffer mixture and mix well.
10. Place the 8-well vessel into the instrument.
11. Start the run.
12. Instrument checks first if there are sufficient amount of reagents placed in the instrument and continues if passed the check.
13. The run time is 40 min after which the eluted DNA can be moved from the 8-well vessel to a clean storage tube for the use in gene amplification applications.

4 Notes

1. Always wear protective gloves and laboratory coats during the procedure.
2. Blood culture samples should be considered as potentially infectious and handled with safe laboratory procedures.
3. Handling of blood culture samples that give rise to infectious aerosols must be conducted in a microbiological safety cabinet.
4. DNA extraction should be performed in a separated area than gene amplification steps to avoid any risk of contamination with microbial organisms or nucleic acids or previous gene amplification products.
5. DNA extraction area should have its own dedicated laboratory equipment.
6. Always follow the workflow from the DNA extraction area to the pre-/post-gene amplification areas.
7. Avoid contacting any material from the post-amplification area with that of the DNA extraction area.
8. Always use appropriate controls in each DNA extraction run:
   (a) Negative control, e.g., molecular grade water.
   (b) Positive control(s), e.g., a fungal and/or bacterial isolate.
9. Always use a negative control if the lot of any reagent in the procedure is changed.
10. Proceed to the gene amplification step immediately after the DNA extraction step.

References

1. http://www.biomerieux.com
2. http://www.bd.com/ds/
3. Fredricks DN, Relman DA (1998) Improved amplification of microbial DNA from blood cultures by removal of the PCR inhibitor sodium polyanetholesulfonate. J Clin Microbiol 36:2810–2816
4. Maaroufi Y, De Bruyne JM, Duchateau V et al (2004) Early detection and identification of commonly encountered *Candida* species from simulated blood cultures by using a real-time PCR-based assay. J Mol Diagn 6(2):108–114
5. Hindiyeh M, Smollan G, Grossman Z (2011) Rapid detection of blaKPC carbapenemase genes by internally controlled real-time PCR assay using bactec blood culture bottles. J Clin Microbiol 49(7):2480–2484
6. Fredricks DN, Smith C, Meier A (2005) Comparison of six DNA extraction methods for recovery of fungal DNA as assessed by quantitative PCR. J Clin Microbiol 43(10): 5122–5128
7. Maaroufi Y, Ahariz N, Husson M et al (2004) Comparison of different methods of isolation of DNA of commonly encountered *Candida* species and its quantitation by using a real-time PCR-based assay. J Clin Microbiol 42(7): 3159–3163
8. Dhaliwal A (2013) DNA extraction and purification. Mater Methods 3:191
9. Hogg GM, McKenna JP, Ong G (2008) Rapid detection of methicillin-susceptible and methicillin-resistant *Staphylococcus aureus* directly from positive BacT/Alert blood culture bottles using real-time polymerase chain reaction: evaluation and comparison of 4 DNA extraction methods. Diagn Microbiol Infect Dis 61(4):446–452
10. Metwally L, Fairley DJ, Coyle PV et al (2008) Improving molecular detection of *Candida* DNA in whole blood: comparison of seven fungal DNA extraction protocols using real-time PCR. J Med Microbiol 57:296–303
11. Villumsen S, Pedersen R, Krogfelt KA et al (2010) Expanding the diagnostic use of PCR in leptospirosis: improved method for DNA extraction from blood cultures. PLoS One 5(8): e12095. doi:10.1371/journal.pone.0012095
12. Regan JF, Furtado MR, Brevnov MG et al (2012) A sample extraction method for faster, more sensitive PCR-based detection of pathogens in blood culture. J Mol Diagn 14(2): 120–129
13. Pillet S, Bourlet T, Pozzetto B (2012) Comparative evaluation of the QIAsymphony RGQ system with the easyMAG/R-gene combination for the quantitation of cytomegalovirus DNA load in whole blood. Virol J 9:231
14. Laakso S, Kirveskari J, Tissari P (2011) Evaluation of high-throughput PCR and microarray-based assay in conjunction with automated DNA extraction instruments for diagnosis of sepsis. PLoS One 6(11):e26655. doi:10.1371/journal.pone.0026655
15. Loeffler J, Hebart H, Bialek R et al (1999) Contaminations occurring in fungal PCR assays. J Clin Microbiol 37:1200–1202
16. Mohammadi T, Reesink HW, Vandenbroucke-Grauls CM et al (2005) Removal of contaminating DNA from commercial nucleic acid extraction kit reagents. J Microbiol Methods 61(2):285–288
17. Evans GE, Murdoch DR, Anderson TP et al (2003) Contamination of Qiagen DNA extraction kits with *Legionella* DNA. J Clin Microbiol 41:3452–3453
18. Van der Zee A, Crielaard JW (2002) Qiagen DNA extraction kits for sample preparation for *Legionella* PCR are not suitable for diagnostic purposes. J Clin Microbiol 40:1126
19. Millar BC, Xu J, Moore JE (2002) Risk assessment models and contamination management: implications for broad-range ribosomal DNA PCR as a diagnostic tool in medical bacteriology. J Clin Microbiol 40(5):1575–1580
20. Tissari P, Zumla A, Tarkka E et al (2010) Accurate and rapid identification of bacterial species from positive blood cultures with a DNA-based microarray platform: an observational study. Lancet 375:224–230
21. Aittakorpi A, Kuusela P, Koukila-Kähkölä P (2012) Accurate and rapid speciation of *Candida* fungemia by PCR and microarray-based Prove-it™ Sepsis assay. J Clin Microbiol 50(11):3635–3640
22. Loonen AJ, Jansz AR, Kreeftenberg H et al (2011) Acceleration of the direct identification of *Staphylococcus aureus* versus coagulase-negative staphylococci from blood culture material: a comparison of six bacterial DNA extraction methods. Eur J Clin Microbiol Infect Dis 30(3):337–342
23. Järvinen AK, Laakso S, Piiparinen P et al (2009) Rapid identification of bacterial pathogens using a PCR- and microarray-based assay. BMC Microbiol 9:161–177

24. Loens K, Bergs K, Ursi D et al (2007) Evaluation of NucliSens easyMAG for automated nucleic acid extraction from various clinical specimens. J Clin Microbiol 45(2): 421–425
25. Dundas N, Leos NK, Mitui M et al (2008) Comparison of automated nucleic acid extraction methods with manual extraction. J Mol Diagn 10(4):311–316
26. Perandin F, Pollara PC, Gargiulo F et al (2009) Performance evaluation of the automated NucliSens easyMAG nucleic acid extraction platform in comparison with QIAamp Mini kit from clinical specimens. Diagn Microbiol Infect Dis 64(2):158–165
27. Wiesinger-Mayr H, Jordana-Lluch E, Martró E et al (2011) Establishment of a semi-automated pathogen DNA isolation from whole blood and comparison with commercially available kits. J Microbiol Methods 85(3):206–213
28. Tang YW, Sefers SE, Li H (2005) Comparative evaluation of three commercial systems for nucleic acid extraction from urine specimens. J Clin Microbiol 43(9):4830–4833

Chapter 7

Broad-Range PCR in the Identification of Bacterial and Fungal Pathogens from Positive Blood Culture Bottles: A Sequencing Approach

Yoshitomo Morinaga and Katsunori Yanagihara

Abstract

Rapid identification of causative bacteria in patients with sepsis can contribute to appropriate selection of antibiotics and improvement of patients' prognosis. Genotypic identification is an emerging technology that may provide an alternative method to, or complement, established phenotypic identification procedures.

Sequence analysis of the 16S rRNA gene is a widely accepted tool for molecular identification of bacteria. Pyrosequencing is a DNA sequencing technique that is based on the detection of pyrophosphate that is released during DNA synthesis. Pyrosequencing can provide sequence information rapidly by reading short sequences; therefore, it may contribute to a rapid identification and lead to a great help in improving the outcome of sepsis. The DNA pyrosequencing-based identification from positive blood culture samples basically consisted of the following four steps: (1) DNA extraction, (2) amplification of target genes, (3) DNA pyrosequencing, and (4) homology searching.

Key words Pyrosequence, Genetic identification, 16S rRNA, V1, V3, Sepsis

1 Introduction

Rapid identification of causative bacteria in patients with sepsis can lead to the appropriate selection of antibiotics and the improvement of prognosis. Bacterial identification based on genetic methods can provide information that is useful for the selection of targeted antibiotics. The new methods such as polymerase chain reaction, mass spectrometry, and microarrays are highly expected to improve diagnosis processes and treatment outcomes. In the Surviving Sepsis Campaign Guideline 2012, these methods are also introduced as useful tools for a quicker identification of pathogens [1].

For patients with sepsis, the rapid identification of causative bacteria is important; however, the conventional phenotyping-based identification requires an extra day after blood culture becomes positive. Thus, the direct identification of bacteria from

Nicasio Mancini (ed.), *Sepsis: Diagnostic Methods and Protocols*, Methods in Molecular Biology, vol. 1237,
DOI 10.1007/978-1-4939-1776-1_7, © Springer Science+Business Media New York 2015

blood culture-positive bottles can give a lot of benefits for the management of sepsis. In this chapter, pyrosequencing-based identification is introduced as a rapid procedure for the detection of blood culture-positive pathogens [2]. The process of pyrosequencing identification can be completed within approximately 4 h.

Pyrosequencing is a DNA sequencing technique that is based on the detection of pyrophosphate that is released during DNA synthesis. The length of the sequence that can be obtained by pyrosequencing is fairly short and limited to about 30–60 bases. Although the read length of new-generation readers is much longer than 30–60 bp, carefully designed applications by pyrosequencing can provide information that is sufficient for the differentiation of gene sequences in a short time. The sequence analysis of the 16S rRNA gene is a widely accepted tool for molecular identification of bacteria [3, 4]. Because the 16S rRNA includes variable regions such as V1 and V3, pyrosequencing these target genes can provide rapid identification [5].

2 Materials

1. 1–10 mL syringe.
2. 22–26G needle.
3. BiOstic bacteremia DNA isolation kit (MO BIO Laboratories, Inc., Carlsbad, CA).
4. Ampdirect (Shimadzu Co., Kyoto, Japan).
5. AmpliTaq Gold DNA polymerase LD (Life Technologies, Carlsbad, CA).
6. PCR primers for V1 and V3 (Table 1).
7. Sequencing primers.
8. Vacuum prep tool (Qiagen, Valencia, CA).
9. PyroMark ID instrument (Qiagen, Valencia, CA).

3 Methods

3.1 Sample Collection

Sample collection should be performed with extreme attention, since contamination in this step can lead to incorrect interpretation.

1. Remove blood culture-positive bottles (e.g., BacT/ALERT FA, BacT/ALERT FN (bioMérieux, Hazelwood, MO)) from an automated microbial detection system (BacT/ALERT 3D (bioMérieux, Hazelwood, MO)).
2. Extract > 1 mL samples from the bottle by using 1–10 mL syringe and 22–26G needle and collect 1 mL sample into collection tube. As necessary, put a drop of extracted samples

Table 1
Primer sequences

	Sequences
V1	
Forward	Bio-pBR5 (5′-biotin-GAAGAGTTTGATCATGGCTCAG-3′)
Reverse	pBR-V1 (5′-TTACTCACCCGTCCGCCACT-3′)
V3	
Forward	Bio-B-V3 (5′-biotin-ACGACAGCCATGCAGCACCT-3′)
Reverse	pJBS.V3 (5′-GCAACGCGAAGAACCTTACC-3′)

Bio biotin labeled

on a glass slide for microscopic observation and culture the suspected bacteria on the appropriate agar-based culture plates for subsequent identification.

3.2 DNA Extraction

The extraction of DNA direct from the blood culture-positive bottles is the first important step in the pyrosequencing-based identification. The blood culture bottle contains a lot of materials such as resins and charcoal to neutralize antibiotics or capture antibodies. In addition, cell debris of human origin is also present in the blood culture-positive bottle. Because these contents can inhibit the amplification of target genes, it is required to exclude PCR inhibitors as much as possible in this step.

The BiOstic bacteremia DNA isolation kit is a product designed for the extraction of bacterial DNA from the blood culture bottles. This kit helps us prepare template DNA samples with an adequate quality to be used in the subsequent PCR:

1. Centrifuge the collection tube at 13,000 × *g* for 2 min to pellet the bacteria and remove the supernatant.
2. Perform the DNA extraction using the BiOstic bacteremia DNA isolation kit according to the manufacturer's instructions.

3.2.1 Cell Lysis

1. Add 450 μL of solution CB1 and resuspend the pellet.
2. Transfer the lysate into the provided MicroBead Tube.
3. Vortex for 10 s.
4. Place in a heat block at 70 °C for 15 min.
5. Vortex at a maximum speed for 10 min.
6. Centrifuge the tube at 10,000 × *g* for 1 min.
7. Transfer the supernatant to a fresh 2 mL tube.

3.2.2 Inhibitor Removal

1. Add 100 μL of solution CB2 and vortex.
2. Incubate for 5 min at room temperature.
3. Centrifuge at 10,000 × *g* for 1 min and transfer the supernatant to a provided 2 mL collection tube.

3.2.3 Bind DNA

1. Add 1 mL of solution CB3. Vortex shortly and spin down.
2. Apply 600 μL of lysate onto the spin filter.
3. Centrifuge at 10,000 × *g* for 1 min
4. Discard the flow through and place the spin filter back into the 2-mL collection tube.
5. Repeat **steps 2–4** twice.
6. Transfer the spin filter to a new 2 mL collection tube.

3.2.4 Wash

1. Add 500 μL of solution CB4.
2. Centrifuge at 10,000 × *g* for 1 min.
3. Discard the flow through and place the spin filter back into the 2 mL collection tube.
4. Repeat **steps 1–3**.
5. Centrifuge at 13,000 × *g* for 2 min and transfer the spin filter to a new 2 mL collection tube.

3.3 DNA Elution

Elute chromosomal DNA in a final volume of 50 μL of elution buffer. After incubation at room temperature for 5 min, centrifuge at 10,000 × *g* for 1 min. As needed, check the concentration and quality of the extracted samples using an appropriate spectrophotometer (NanoDrop 2000, Thermo Scientific).

3.4 Polymerase Chain Reaction (PCR)

The extracted DNA samples can still contain some PCR inhibitors. To amplify the target genes successfully, the samples should be treated as crude samples for PCR. As well as the DNA extraction, PCR buffer and polymerase for crude samples are commercially available. For the PCR buffer, Ampdirect is recommended. For the polymerase, AmpliTaq Gold DNA polymerase LD is suitable.

In the case of amplification error, a lot of PCR inhibitors can remain in the template samples. In such cases, samples should be diluted and then used as the template. In our experience, 10- to 100-fold dilutions lead to successful amplification, while the original concentration was not amplified.

The targets for sequencing are designed in the variant regions, V1 and V3, of the 16S rRNA genes (Table 1) [5] (*see* **Note 1**). The amplicon sizes of V1 and V3 are 115 bp and 81 bp, respectively. All reagents in this step should be mixed with ice (Table 2).

Because efficient amplification is generally varied according to the thermal cycler, the appropriate conditions should be verified in each laboratory. Representative conditions are shown in Table 3.

Table 2
PCR mixture

	Final conc.	Volume(/tube)
2 ×Ampdirect Plus	1×	25 μL
AmpliTaq Gold DNA polymerase LD (5U/μL)	1.25U/tube	0.25 μL
Forward primer(10μM)	0.2–0.5 μM	1–2.5 μL
Reverse primer(10μM)	0.2–0.5 μM	1–2.5 μL
Distilled water		14.75–17.75 μL
Template		5 μL
Total		50 μL

Table 3
PCR condition

Step	Temperature and period	Cycle number
Initial denaturation	95 °C, 10 min	
Amplification	94–95 °C, 30–40 s 55 °C, 40–60 s 72 °C, 60 s	35–40 cycles
Final extension	72 °C, 60 s	

Our protocol was performed using the GeneAmp® PCR system 9700 (Life Technologies, Carlsbad, CA) or the Veriti® Thermal Cycler (Life Technologies, Carlsbad, CA).

Appropriate positive control and negative control should be amplified in the run. In our original study, DNAs extracted from the clinical isolates including *Staphylococcus aureus*, *Bacillus cereus*, and *Escherichia coli* were used for positive controls. Each PCR product can be used for the subsequent pyrosequencing. As needed, secure amplification should be verified by agarose gel electrophoresis. The samples without single band can be amplified after 10- or 100-fold dilution.

3.5 DNA Pyrosequencing

DNA pyrosequencing is separated into two processes: preparation and analyzing. The amplified V1 and V3 products were prepared for pyrosequencing by using the recommended protocol for the vacuum prep tool (VPT):

1. For the preparation of each reaction, 40 μL of the biotinylated PCR product was used.

2. Resuspend PCR product in 43 μL of binding buffer and 3 μL of streptavidin beads.
3. Shake the PCR tube for 10 min at room temperature to disperse beads.
4. Perform the following processes while waiting:
   (a) Prepare following buffers in each tray for the VPT
      - (Tray 1) 180 mL of 70 % ethanol
      - (Tray 2) 120 mL of denaturation solution (0.2 M NaOH)
      - (Tray 3) 180 mL of washing buffer
      - (Tray 4) 180 mL of H_2O

   (b) To prepare the pyroplate, apply 40 μL of annealing buffer and 1.0–1.6 μL sequencing primer (10pmol/μL) in each well. The sequencing primers for V1 and V3 regions are pBR-V1 and pJBS.V3, respectively.
   (c) Turn on the vacuum system and wash the VPT in the Tray 4 with gentle rocking for 20 s.
5. Suck the samples by use of the VPT immediately after immobilization. Additional vortex (2–3 s) is acceptable if the beads are seen at the bottom.
6. Translocate the VPT through Tray 1 for 10 s, Tray 2 for 10 s, and Tray 3 for 15–20 s.
7. Tip the VPT beyond the vertical to drain liquid completely.
8. Turn the vacuum off.
9. Move the VPT into the pyroplate prepared in **step 4**(b) and gently tap within the wells to release the beads.
10. Heat the pyroplate at 80 °C for 2 min and then cool at room temperature to anneal the template to the sequencing primer.

DNA pyrosequencing is performed on the PyroMark ID instrument. It is important to set the dNTP dispensation to develop the reaction rapidly:

1. Select SQA on the main menu and click the SQA Entries button.
2. Enter necessary information and type "8(ACTG)" in the field of entered dispensation order to apply dNTPs with 8 cycles of a repetitive ACTG dispensation (Eight or more cycles are acceptable).
3. Click SQA run button and enter the necessary information.
4. Prepare the reaction reagents (enzyme mixture, substrate mixture, and each dNTP) and apply appropriate wells of reagent cartridge.
5. Set the PSQ plate and the regent cartridge.

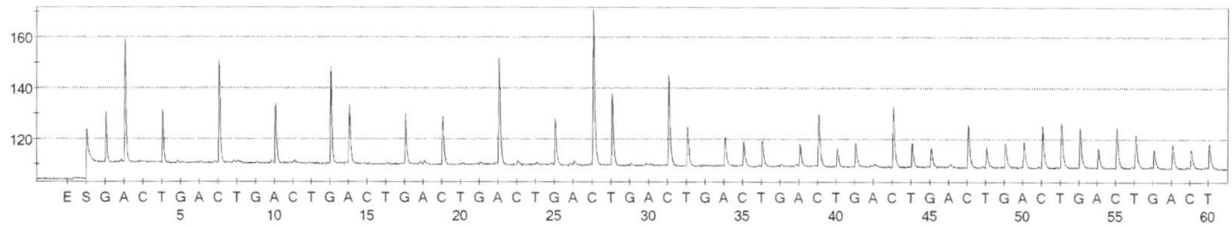

Fig. 1 Pyrogram after pyrosequencing. dNTPs are applied in the following order: G, A, C, and T. The software of pyrosequence automatically read the sequence as GAATCCAGGA GCAAGCCCCT TCCTACTGCC TCGACTGCTG ACT

6. Run.
7. Confirm the sequence results and pyrograms (Fig. 1). When reading errors are found, careful correction of the sequence may be required.

3.6 Homology Searching

Sequence homology of PCR products is compared using the homology searching programs published on the Internet. Several programs are available (some programs are free of charge):

1. DDBJ search program (http://www.ddbj.nig.ac.jp/)
2. BLAST (http://blast.ncbi.nlm.nih.gov/Blast.cgi)
3. EzTaxon (http://eztaxon-e.ezbiocloud.net/)
4. Bioinformatic bacterial identification (BIBI) (http://pbil.univ-lyon1.fr/bibi/)
5. Nucleotide sequence database at the European Molecular Biology Laboratory (EMBL-Bank) (http://www.ebi.ac.uk/ena/)
6. Ribosomal Database Project II (RDP II) (http://rdp.cme.msu.edu/seqmatch/seqmatch_intro.jsp)

These databases are useful for identification of clinical blood culture isolates [6]. A strain with >99 % sequence homology is considered as an isolated strain (*see* **Notes 2–4**).

4 Notes

1. In many samples, the sequences of V1 and V3 represent similar results but also sometimes show different characteristics in some specific bacteria. V1 can effectively classify genus *Enterococcus* into *E. faecalis* or *E. faecium*, and V3 can have the advantage of detecting *S. epidermidis* and *E. coli.* These suggested that the sequencing of V1 and V3 improved the accuracy of diagnosis. However, the best combination of variable regions of 16S rRNA for diagnosis has been a controversial issue [7, 8].
2. Pyrosequencing can fail to separate distinct bacteria which have similar sequences because it only reads short sequence

lengths. The genera *Aeromonas*, *Bacillus*, and *Staphylococcus* are typical genera which have similar sequences in the target gene in each genus. Therefore, organisms which belong to these genera are not effectively identified at the species level but show good agreement with culture results at the genus level.

3. In polymicrobial infections, pyrosequencing may not identify all of the bacteria. Thus, when a sample for pyrosequencing contains polymicrobial genes, the result obtained from sequencing can consist of a mix of sequences from those organisms. Therefore, pyrosequencing may not effectively detect organisms in patients with polymicrobial infection. In intra-abdominal and urinary tract infections, polymicrobial infections are often observed. Therefore, the samples that include *E. cloacae*, *E. faecalis*, *B. fragilis*, and *B. thuringiensis* may be polymicrobial.
4. The 23S rRNA gene is another useful target for pyrosequencing-based identification [9]. Targeting 23S rRNA gene may improve the identification efficiency of some specific bacteria such as *Enterobacteriaceae* and *Streptococcus* species.

Acknowledgments

We thank Maiko Motoshima and Sayaka Mori for technical assistance.

References

1. Dellinger RP, Levy MM, Rhodes A et al (2013) Surviving Sepsis Campaign: international guidelines for management of severe sepsis and septic shock, 2012. Crit Care Med 41:580–637
2. Motoshima M, Yanagihara K, Morinaga Y et al (2012) Identification of bacteria directly from positive blood culture samples by DNA pyrosequencing of the 16S rRNAgene. J Med Microbiol 61:1556–1562
3. Patel JB (2001) 16S rRNA gene sequencing for bacterial pathogen identification in the clinical laboratory. Mol Diagn 6:313–321
4. Woese CR (1987) Bacterial evolution. Microbiol Rev 51:221–271
5. Luna RA, Fasciano LR, Jones SC et al (2007) DNA pyrosequencing-based bacterial pathogen identification in a pediatric hospital setting. J Clin Microbiol 45:2985–2992
6. Park KS, Ki CS, Kang CI et al (2012) Evaluation of the GenBank, EzTaxon, and BIBI services for molecular identification of clinical blood culture isolates that were unidentifiable or misidentified by conventional methods. J Clin Microbiol 50: 1792–1795
7. Sundquist A, Bigdeli S, Jalili R et al (2007) Bacterial flora-typing with targeted, chip-based Pyrosequencing. BMC Microbiol 7:108
8. Wang Q, Garrity GM, Tiedje JM et al (2007) Naive Bayesian classifier for rapid assignment of rRNA sequences into the new bacterial taxonomy. Appl Environ Microbiol 73: 5261–5267
9. Jordan JA, Jones-Laughner J, Durso MB (2009) Utility of pyrosequencing in identifying bacteria directly from positive blood culture bottles. J Clin Microbiol 47:368–372

Chapter 8

Identification of Bacterial and Fungal Pathogens from Positive Blood Culture Bottles: A Microarray-Based Approach

Teresa Raich and Scott Powell

Abstract

Rapid identification and characterization of bacterial and fungal pathogens present in the bloodstream are essential for optimal patient management and are associated with improved patient outcomes, improved antimicrobial stewardship, improved infection control, and reduced healthcare costs. Microarrays serve as reliable platforms for the identification of these bloodstream pathogens and their associated antimicrobial resistance genes, if present. Nanosphere's (Nanosphere, Inc., Northbrook, IL, USA) Verigene Gram-Positive Blood Culture Nucleic-Acid Test (BC-GP) is one such microarray-based approach for the detection of bacteria that cause bloodstream infection. Here, we describe the design of the microarray-based Verigene BC-GP Test, the steps necessary for performing the test, and the different components of the test including nucleic acid extraction and hybridization of target nucleic acid to a microarray.

Key words Bloodstream infection, Microarray, Verigene BC-GP, Oligonucleotide, Extraction, Hybridization, Sample-to-result

1 Introduction

Until the past few years, there have been few viable molecular multiplex diagnostic assays that could provide the accurate and reliable results necessary to shift microbiology laboratories away from culture-based phenotypic identification. By targeting the genetic nucleic acids unique to each target organism (molecular identification), these molecular tests can deliver test results much quicker and more accurately than culture-based techniques that rely upon the growth of the organism (phenotypic identification). These multiplex tests can target anywhere from 5 to 50 different clinically relevant pathogens at one time from one specimen, allowing for streamlined diagnostic testing and clinical laboratory workflow.

The field of microarray technology has exploded over the last 20 years with many publications describing both routine and novel applications in the life sciences, clinical research, and clinical

Nicasio Mancini (ed.), *Sepsis: Diagnostic Methods and Protocols*, Methods in Molecular Biology, vol. 1237, DOI 10.1007/978-1-4939-1776-1_8, © Springer Science+Business Media New York 2015

diagnostics arenas. Microarrays have been developed to detect DNA, RNA, and proteins and can distinguish single-nucleotide polymorphisms, small sequences, large sequences, cDNA, chromosomes, amino acids, proteins, ligands, antibodies, antigens, and tissues, just to name a few of the many generic applications. These applications include qualitative detection of nucleic acid targets for discovery, mapping, screening, and diagnostic purposes, semiquantitative/ quantitative detection in gene expression and analysis and profiling, and cell comparisons utilizing comparative genomic hybridization. Molecular microarrays can be manufactured through a variety of methods; two common DNA microarray methods include the "spotting" of oligonucleotides (synthesized off-line) onto the microarray surface and then "binding" covalently the oligonucleotide to the surface through a chemical process, while photolithography methods involve the synthesizing of oligonucleotides directly onto the microarray surface. There are also many different methods of "spotting" oligonucleotides onto the surface of microarrays, and these methods include piezoelectric dispensing, inkjet spotting, and simple pin spotting. Overall, microarrays remain a popular technological platform to immobilize captures to interrogate targets in a multiplex assay.

Detection of bloodstream pathogens is a new application of multiplexed molecular diagnostics. Bloodstream infections are initiated when a viable bacterium or fungi reach the bloodstream. Time to appropriate therapy has been proven to be a critical determinant of patient outcomes for patients with bloodstream infection, as survival rates decrease by 7.6 % for each hour that optimal therapy is delayed following the onset of sepsis-related hypotension [1]. Conventional culture-based diagnostics, which remain the gold standard for identification of the bloodstream pathogen(s), are not ideal as they are associated with very slow turnaround times, sometimes taking over 3 days. With the long time to identification associated with conventional culture-based diagnostics, a patient might remain on the inappropriate empiric therapy, significantly increasing the patient's risk of mortality. In fact, patients in the ICU receiving inadequate antimicrobial treatment for bloodstream infection have an associated mortality rate of 61.9 %, while those receiving appropriate therapy have an associated mortality rate of 28.4 % [2]. Rapid multiplexed diagnostic tests for the detection of bloodstream infection like the Verigene BC-GP Test can provide identification of the causative pathogen of a bloodstream infection and associated antimicrobial resistance 1–2 days faster than conventional culture-based diagnostics. These rapid results allow for the patient to be placed on the optimal therapy much earlier, resulting in improved patient outcome, improved antimicrobial stewardship, improved infection control, and reduced healthcare costs [3–5]. Similar tests for gram-negative bacteria, including Nanosphere's Verigene Gram-Negative Blood Culture

Nucleic-Acid Test (BC-GN), have been shown to have a similar clinical impact as the rapid blood culture tests for gram-positive bacteria (*see* **Note 1**).

2 Materials

The Verigene Gram-Positive Blood Culture Nucleic-Acid Test (BC-GP) (FDA cleared, CE-IVD) is performed using the Verigene System, which is comprised of single-use test consumables and shared instrumentation. The Verigene instrumentation is a bench-top sample-to-result molecular diagnostics workstation consisting of two modules: the Verigene Processor *SP* and the Verigene Reader (Fig. 1).

2.1 Material Provided (Fig. 2)

1. Verigene BC-GP Test Cartridge.
   Each Test Cartridge is composed of two components: the Reagent Pack and the Test Substrate (Fig. 3). The Reagent Pack comes preloaded with all required reaction solutions, including wash solutions, oligonucleotide probe solution, and signal amplification solutions, to generate a test result. The Test Substrate is located beneath the Reagent Pack and is composed of a microarray that has 368 wells spotted across the surface that contain oligonucleotides designed to specifically bind complementary to a conserved genetic region of a target bacterium or resistance gene (Fig. 4).

Fig. 1 The Verigene System is composed of the Verigene Reader (*left*) and Verigene Processor *SP* (*right*). Up to 32 separate Verigene Processor *SP*s can be networked with one Verigene Reader

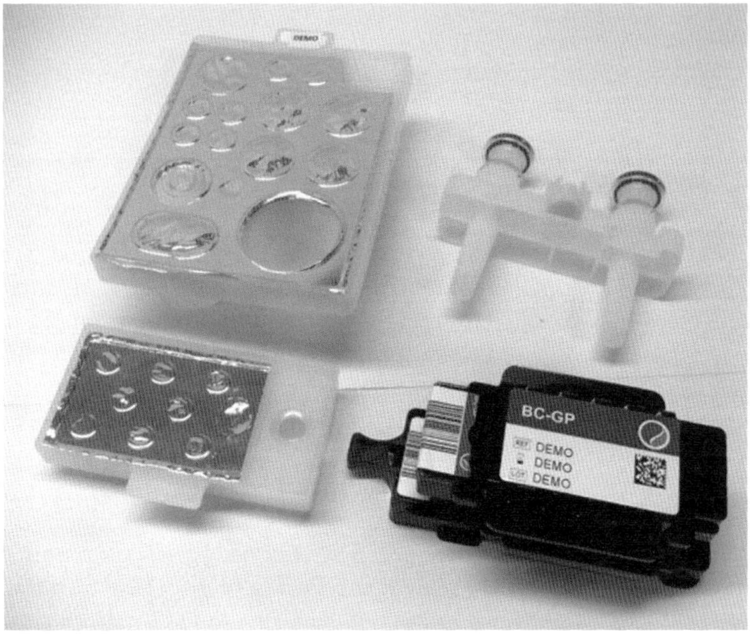

Fig. 2 The four consumables used for the Verigene BC-GP Test: Extraction Tray (*top left*), Tip Holder Assembly (*top right*), Utility Tray (*bottom left*), and BC-GP Test Cartridge (*lower right*)

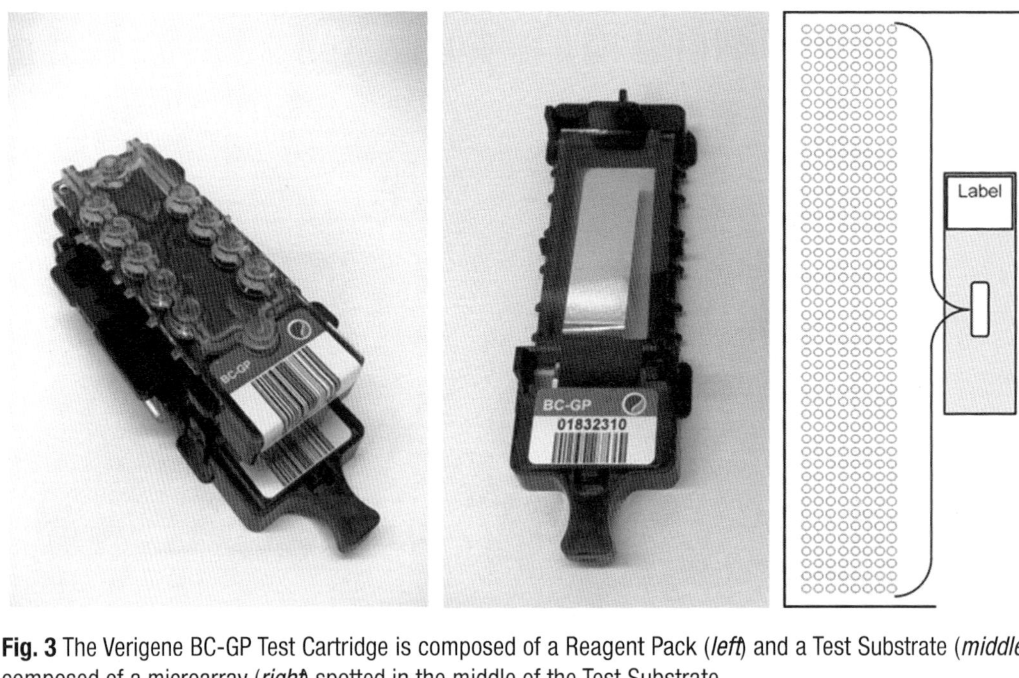

Fig. 3 The Verigene BC-GP Test Cartridge is composed of a Reagent Pack (*left*) and a Test Substrate (*middle*), composed of a microarray (*right*) spotted in the middle of the Test Substrate

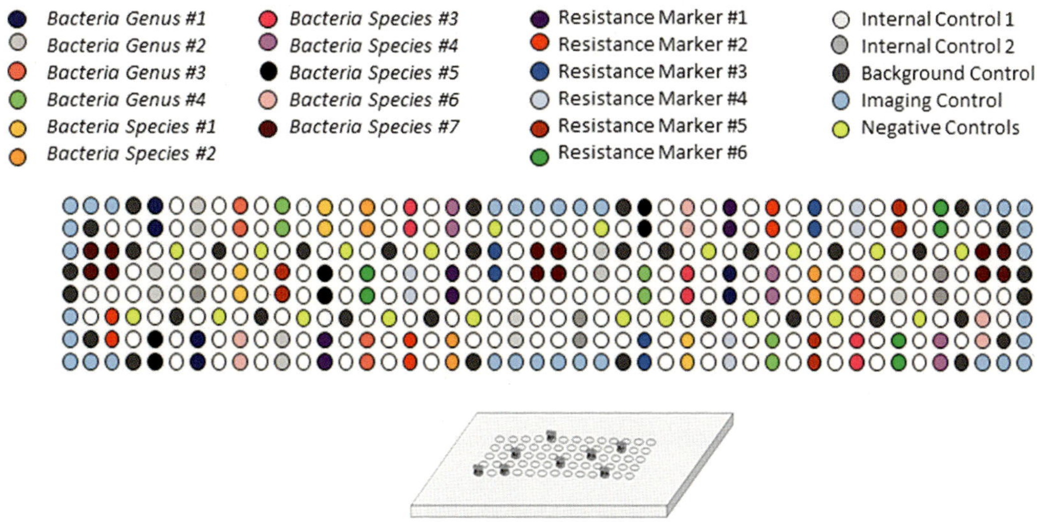

Fig. 4 Example of microarray layout used in tests like the Verigene BC-GP Test. This microarray is composed of 368 different wells that contain a variety of oligonucleotide designed to complementary bind a specific genetic sequence unique to a given target bacterium or resistance gene

2. Verigene BC-GP Extraction Trays (with Tip Holder Assembly). Each Extraction Tray comes preloaded with all required solutions, including lysis/binding buffer, digestion enzymes, wash solutions, and buffer solutions, necessary to extract nucleic acids and generate a test result.
3. Verigene BC-GP Utility Tray.
   Each Utility Tray comes preloaded with all required solutions, including digestion enzymes and the *Bacillus subtilis* Internal Processing Control, necessary to extract nucleic acids and generate a test result.

2.2 Materials Needed but Not Provided

Instruments and Equipment

1. 2–8 °C refrigerator.
2. Automated blood culture monitoring system.
3. Micro-pipettors and tips.
4. Vortex mixer and microcentrifuge.

Consumables and Reagents

1. Blood culture bottles.
2. Gram staining reagents.

3 Methods

The Verigene® BC-GP is performed using the sample-to-result Verigene System (*see* **Note 2**) and is a qualitative, multiplexed in vitro diagnostic test for the simultaneous detection and identification of potentially pathogenic gram-positive bacteria which may cause bloodstream infection (BSI). BC-GP is performed directly on blood culture bottles identified as positive by a continuous monitoring blood culture system and which contain gram-positive bacteria. BC-GP detects and identifies the following bacterial genera and species in less than 2½ h:

Staphylococcus spp.	*Streptococcus* spp.	*Enterococcus faecalis*
Staphylococcus aureus	*Streptococcus pneumoniae*	*Enterococcus faecium*
Staphylococcus epidermidis	*Streptococcus pyogenes*	*Listeria* spp.
Staphylococcus lugdunensis	*Streptococcus agalactiae*	
	Streptococcus anginosus group	

In addition, BC-GP detects the *mecA* resistance marker, inferring *mecA*-mediated methicillin resistance in staphylococcal species, and the *vanA* and *vanB* resistance markers, inferring *vanA/vanB-mediated* vancomycin resistance in enterococcal species. The analytical sensitivity, or limit of detection, for each BC-GP analyte is as follows:

Staphylococcus spp.	2.9×10^6 to 4.0×10^6 CFU/mL
Staphylococcus aureus	1.9×10^5 to 5.7×10^5 CFU/mL
Staphylococcus epidermidis	2.0×10^6 to 7.5×10^6 CFU/mL
Staphylococcus lugdunensis	3.4×10^6 to 4.0×10^6 CFU/mL
Streptococcus spp.	1.8×10^6 to 1.2×10^8 CFU/mL
Streptococcus pneumoniae	1.8×10^6 to 9.9×10^6 CFU/mL
Streptococcus pyogenes	9.5×10^6 to 6.3×10^7 CFU/mL
Streptococcus agalactiae	1.2×10^7 to 2.2×10^7 CFU/mL
Streptococcus anginosus group	1.4×10^7 to 1.2×10^8 CFU/mL
Enterococcus faecium	2.4×10^6 to 3.7×10^7 CFU/mL
Enterococcus faecalis	1.1×10^7 to 5.7×10^7 CFU/mL
Listeria spp.	7.5×10^6 to 1.2×10^7 CFU/mL

In mixed growth, BC-GP does not specifically attribute *van-mediated* vancomycin resistance to either *E. faecalis* or *E. faecium* or *mecA*-mediated methicillin resistance to either *S. aureus* or *S. epidermidis*. BC-GP is indicated for use in conjunction with other clinical and laboratory findings to aid in the diagnosis of bacterial bloodstream infections; however, it is not to be used to monitor these infections. Subculturing of positive blood cultures is necessary to recover organisms for susceptibility testing, identification of organisms not detected by BC-GP, differentiation of mixed growth, association of antimicrobial resistance marker genes to a specific organism, or epidemiological typing (*see* **Note 3**).

3.1 Specimen Collection and Storage

1. Draw blood using aseptic techniques into the blood culture bottle following manufacturer's instructions.
2. Incubate bottle in automated blood culture monitoring system until the bottle is flagged positive for microbial growth following manufacturer's instructions (*see* **Note 4**).
3. When the bottle is positive for microbial growth, perform a Gram stain.
4. For gram-positive bacteria, test 350 μL of the blood culture media using BC-GP. Ensure the blood culture bottle is thoroughly mixed by inverting several times (>4) before retrieving test sample volume.
5. Subculturing of positive blood cultures is necessary to recover organisms for susceptibility testing, identification of organisms not detected by BC-GP, differentiation of mixed growth, association of the *mecA* gene to an organism, and/or association of the *vanA/vanB* gene to an organism.
6. Positive blood culture media may be stored at room temperature (18–24 °C) for up to 12 h or remain in the automated blood culture monitoring system at 35 °C for up to 8 h prior to testing.
7. Inadequate or inappropriate specimen collection, storage, or transport may yield false-negative results (*see* **Note 5**).

3.2 Test Procedure

3.2.1 Preparing the Work Area for Testing

Sanitize vortex mixers, centrifuges, pipettes, countertops, and any other equipment used for sample processing with a lint-free decontaminating cloth before and after sample preparation.

3.2.2 Test Setup

1. Remove the Extraction Tray, Utility Tray, Tip Holder Assembly, and Test Cartridge from the refrigerator. If the Utility Tray was stored in the freezer, thaw at room temperature for 10 min. Begin test run within 30 min or store the Utility Tray at <8 °C until ready to initiate testing.
2. Open the Drawer Assembly by pressing the black open/close button located on the front of the Verigene Processor *SP*. Open the Drawer Clamp by pressing in the silver latch and lifting the clamp prior to loading the consumables. Figure 5 shows the empty Verigene Processor *SP* tray.

3.2.3 Loading the Extraction Tray

1. Prior to loading the Extraction Tray, thoroughly shake the tray to resuspend the magnetic beads which have settled during storage. Check for complete resuspension by visually inspecting the

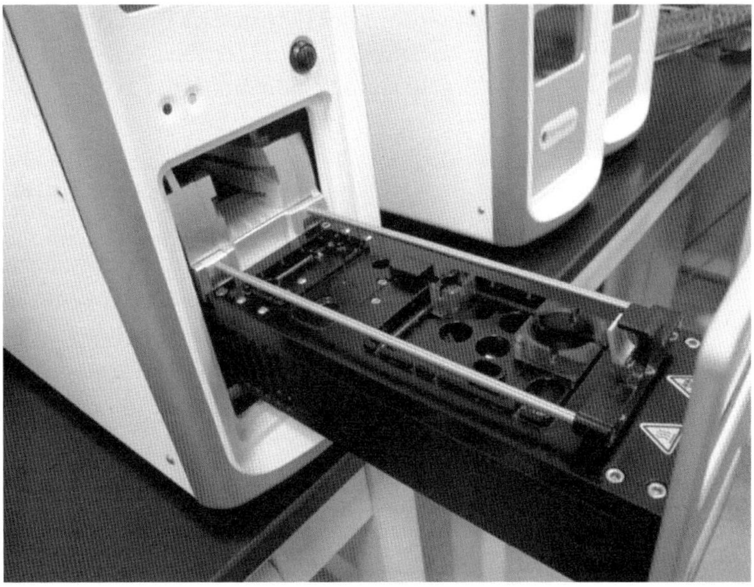

Fig. 5 Empty tray of the Verigene Processor *SP*

well containing the beads. The well containing the magnetic beads is easily distinguished as the beads are black in color. Following adequate resuspension, gently tap the tray on the bench to ensure that the reagents settle to the bottom of each well.

2. The Extraction Tray can only be loaded in one direction in the Drawer Assembly. When loaded correctly, the Sample Well is located in the front right-hand corner of the Drawer Assembly. Place the Extraction Tray in the Drawer Assembly and press down on the corners of the tray to ensure it is level.

3.2.4 Loading the Tip Holder Assembly

1. The Tip Holder Assembly is a plastic holder that contains two pipette tips and a rubber Tip Seal. Each pipette tip contains an O-ring on top.
2. Before using the Tip Holder Assembly, check the top of each pipette tip for the O-ring and check for the rubber Tip Seal sitting straight and flush between the tips. If either is missing, replace with a new Tip Holder Assembly.
3. Insert the Tip Holder Assembly into the Drawer Assembly. The tip assembly can only be loaded in one direction in the Drawer Assembly. For orientation, there are two holes on the deck of the Drawer Assembly that fit each pipette tip, and the opening to the Tip Seal should face away from Processor *SP*.

3.2.5 Loading the Utility Tray

1. Gently vortex the Utility Tray and gently tap the tray on the bench to settle the reagents. Remove and save the cap from the *B. subtilis* Process Control (PC) Tube and fully insert the

PC Tube into the Utility Tray. Visually inspect the tube to ensure the *B. subtilis* pellet is seated in the lower half of the PC Tube as shown in the picture below.

2. Insert the Utility Tray into the Drawer Assembly. The Utility Tray can only be loaded in one direction in the Drawer Assembly. When loaded properly, the tray sits flat.
3. Lower and latch the Drawer Clamp over the trays while supporting the drawer with the opposite hand. The Drawer Clamp will latch onto the Drawer Assembly when closed properly, and the user will be unable to lift the Drawer Clamp without pressing in the silver latch.

3.2.6 Ordering a Test

1. All tests must be ordered through the Verigene Reader. No tests can be processed on the Verigene Processor *SP* without the user entering the Test Cartridge ID and sample ID to the Verigene Reader.
   i. Login to the Verigene Reader as a "user."
   ii. If the user would like to start a new session, proceed to the next **step** (**iii**). If the user would like to order a test in a previously created session, they can select the desired session from the drop-down "SESSION" menu and then proceed to **step** (**v**). Up to 60 cartridges can be entered into a single session.
   iii. From the Menu Bar, SESSION tab, select Start New Session where the Session Setup window will appear.
   iv. Touch Session ID button and enter information by using the data entry keyboard. This can be any unique identifier in a format defined by the laboratory. The operator ID is automatically entered as the currently logged in "user."
   v. Touch the Processing option on the Navigation Bar at the bottom of the screen.
2. Enter the Test Cartridge ID by scanning the barcode using the barcode scanner attached to the Reader. The user may manually enter in the Test Cartridge ID by selecting MENU and "Enter Barcode" and then keying in the Test Cartridge ID number with the Reader's keyboard.

3.2.7 Loading a Test Cartridge

1. Hold the Test Cartridge by the handle with one hand; using the other hand, apply pressure with the palm of the hand and remove the cartridge cover by bending the cover away and over the Reagent Pack edge. Ensure that the valve plate is not moved during cover removal (*see* illustration below). Do not remove the Test Cartridge cover until immediately prior to inserting the Test Cartridge into the Processor *SP*.
2. The user must settle the reagents in the cartridge before loading into the Verigene Processor *SP*. The optimal method for

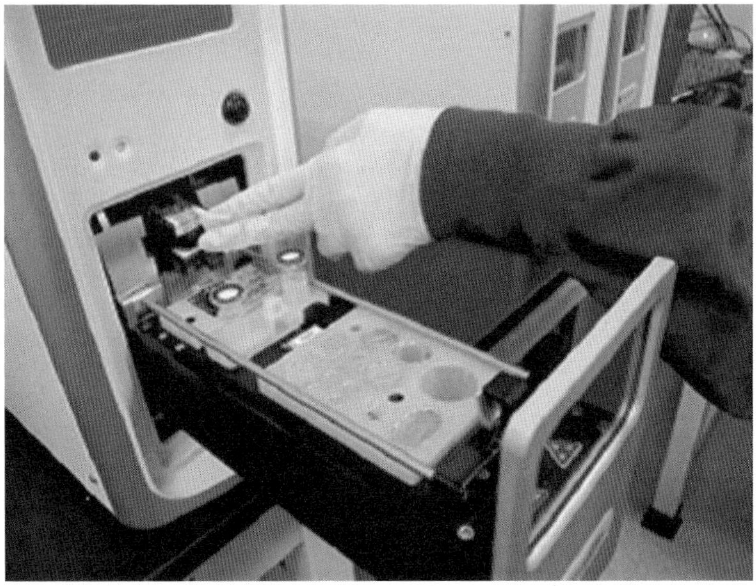

Fig. 6 User loading the consumable and BC-GP Test Cartridge into the Processor *SP*

settling the reagents is to hold the Test Cartridge's reagent container on the side opposite the handle and tap the reagent container's barcode with your index finger. When tapping the cartridge, allow the force of the tapping to move the cartridge and your right hand. The tapping is more effective when the cartridge is held in the air so that it moves slightly.

3. Insert the Test Cartridge into the Hybridization Module of the Verigene Processor *SP* until it reaches a stopping point. Figure 6 shows the user loading a Test Cartridge into the Verigene Processor *SP*.

3.2.8 Loading the Sample

1. At the Reader enter the sample number/ID by scanning or using the reader's touch-screen keyboard. Press Yes to confirm the sample ID (*see* image below). Ensure Hybridization and Extraction options are selected.
2. In the subsequent dialogue box, select or deselect bacteria species or resistance markers from the list to activate or deactivate results reporting for those targets. Press "Yes" to confirm. The Verigene Reader will automatically default to the previously selected targets.
3. Gently vortex the gram-positive blood culture sample and pipette 350 μL of the gram-positive blood culture sample into the bottom of the Sample Well in the bottom right of the Extraction Tray.
4. Close the Drawer Assembly by pressing the open/close button on the Processor *SP*. The processor will automatically verify

that each consumable is properly loaded and begin sample processing.

5. Confirm countdown has started on the Processor *SP* display screen before leaving the area.
6. In order to set up additional tests on other Processor *SP* instruments, follow the same procedure. To avoid contamination and sample mix-ups, only set up one test at a time, change gloves after handling a sample, and decontaminate pipettes and sample tubes between tests.

3.2.9 Upon Completion of a Test Run (See **Note 6**)

1. The Verigene Reader will ring to notify the user when the test is completed and the Processor *SP* will display a message indicating the test is finished. The Test Cartridge should be removed from the Processor *SP* upon completion of the test.
2. Open the Drawer Assembly by pressing the OPEN/CLOSE button. Cap the PC Tube for disposal.
3. Remove the Test Cartridge and immediately orient to the side.
4. While keeping the Test Cartridge on its side, separate the Reagent Pack and keep the substrate on its side for 30–60 s after removal as illustrated below to allow the final rinse to dry away from the analysis area.

3.2.10 Analyzing Results

1. Remove the protective tape from the back of the Substrate Holder.
2. Use the Reader's barcode scanner to read the barcode on the substrate and immediately insert the Substrate Holder into the Reader.
3. When the barcode is accepted, a prompt to load the Substrate Holder will display.
4. Scanning the barcode ensures that the test result is associated with the correct sample. When the load substrate prompt occurs, it will only display for 20 s. The analysis will only start if the substrate is loaded during the animated prompt.
5. To properly insert the substrate into the reader, hold the substrate by the handle with the barcode facing away from you. Next, insert the Substrate Holder into the substrate compartment. The compartment is designed to place the holder in the correct position. Do not force the holder in, but do insert it into the compartment as far as it will go comfortably. Close the door of the substrate compartment (Fig. 7).
6. The analysis will automatically begin. A small camera icon will appear on the reader letting the user know analysis has begun (*see* **Note** 7).
7. The analysis is completed by the reader when the camera icon is replaced with an upward-facing arrow and the reader rings.

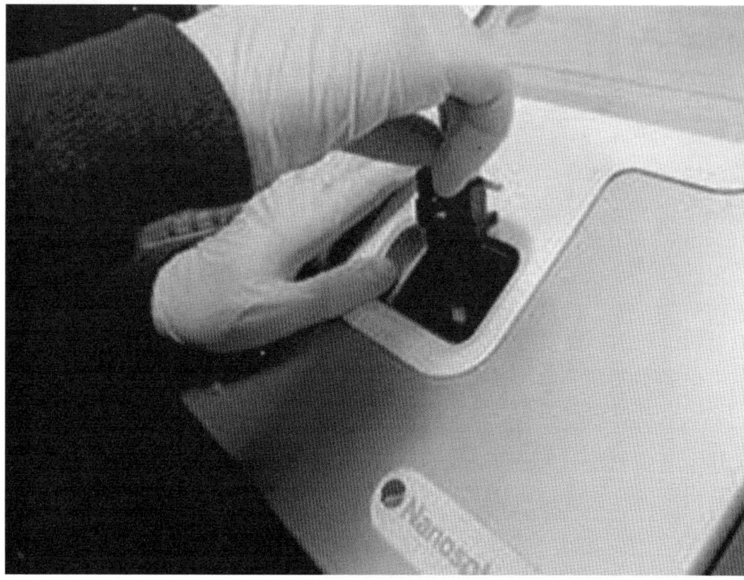

Fig. 7 Inserting the substrate from the BC-GP Test Cartridge into the Verigene Reader

8. Confirm that a result other than "No Call—No GRID" has been generated by touching the substrate icon for the test. A substrate producing a "No Call—No GRID" result should be rescanned and reanalyzed. Use Subheading 3.3 to analyze results.
9. Once the scan is complete, dispose of used Test Substrate.

3.3 Interpretation of Test Results

BC-GP provides a qualitative result for the presence ("Detected") or absence ("Not Detected") for all bacterial analytes in the test panel. The panel analytes are represented by target-specific spots on the Test Substrate (microarray). The image analysis of the Test Substrate provides image intensities for each panel analyte as well as imaging controls (IC) and negative controls (NC). Intensities at the panel analytes are required to be above an empirically determined "Noise Threshold" after which they are normalized to generate "Ratio-to-IC" and "Ratio-to-NC" values. Cutoffs for both the normalized ratios were determined by ROC curve analysis (*see* **Note 8**).

Two Internal Controls, INT CTL 1 (extraction control) and INT CTL 2 (hybridization control), guide decisions regarding the validity of the test process. Both INT CTL 1 and INT CTL 2 are treated as unique targets (or panel members), and their presence is verified in order for a valid result to be generated. If the Internal Controls fail, a No Call—INT CTL 1 (for INT CTL 1 failure), a No Call—INT CTL 2 (for INT CTL 2 failure), or a No Call—INT CTL (for failure of both INT CTL 1 and INT CTL 2) is provided. If the Internal Controls are verified, the presence or absence of

Table 1
Call algorithm for valid results on the Verigene BC-GP Test

Organism/gene	Test result reported as "detected"		
	Genus	Species	Resistance marker
Staphylococcus spp.	*Staphylococcus*	–	–
Staphylococcus aureus	*Staphylococcus*	*S. aureus*	–
Staphylococcus epidermidis	*Staphylococcus*	*S. epidermidis*	–
*Staphylococcus aureus, mec*A	*Staphylococcus*	*S. aureus*	*mec*A
*Staphylococcus epidermidis, mec*A	*Staphylococcus*	*S. epidermidis*	*mec*A
Staphylococcus lugdunensis	*Staphylococcus*	*S. lugdunensis*	–
Enterococcus faecalis	–	*E. faecalis*	–
*Enterococcus faecalis, van*A	–	*E. faecalis*	*van*A
*Enterococcus faecalis, van*B	–	*E. faecalis*	*van*B
Enterococcus faecium	–	*E. faecium*	–
*Enterococcus faecium, van*A	–	*E. faecium*	*van*A
*Enterococcus faecium, van*B	–	*E. faecium*	*van*B
Streptococcus spp.	*Streptococcus*	–	–
Streptococcus agalactiae	*Streptococcus*	*S. agalactiae*	–
Streptococcus anginosus group	*Streptococcus*	*S. anginosus* group	–
Streptococcus pneumoniae	*Streptococcus*	*S. pneumoniae*	–
Streptococcus pyogenes	*Streptococcus*	*S. pyogenes*	–
Listeria spp.	*Listeria*	–	–
All analytes "not detected"	–	–	–

individual bacteria is reported based on the cutoff criteria. Both INT CTL 1 and INT CTL 2 signal intensities must meet the detection criteria for a "valid call" to be generated, and this call is made only after both INT CTL 1 and INT CTL 2 are verified during analysis of each test. This signifies that the extraction and hybridization processes were performed correctly.

1. Call for valid results
   Table 1 lists the possible test results generated by BC-GP, representing identification of bacterial nucleic acid sequences/ targets, when the Internal Controls INT CTL 1 and INT CTL 2 are verified as "Detected." An initial "Not Detected" test result may be repeated once, at the discretion of the user,

Table 2
Invalid BC-GP results and the associated recourse

Error call	Reason	Recourse
No Call—INT CTL 1	INT CTL 1 not detected. Processing and/or lysis/extraction issues	Repeat *BC-GP* from original blood culture specimen
No Call—INT CTL 2	INT CTL 2 not detected. Inhibition during the target hybridization procedure	Repeat *BC-GP* from original blood culture specimen
No Call—INT CTL, INT CTL 1, and INT CTL 2 Not	Detected. Processing and/or lysis/extraction issues and inhibition during target hybridization	Repeat *BC-GP* from original blood culture specimen
No Call—NO GRID	Reader unable to image Test Substrate	Ensure protective silver tape has been removed from back of Test Substrate. Ensure Test Substrate is seated properly in the Substrate Holder. Repeat image analysis by selecting "Menu" and "Enter Barcode" and then scanning the substrate barcode. If the No Call persists, repeat *BC-GP* from original blood culture specimen
No Call—VARIATION No Call—BKGD No Call—NEG CTL	Inability to obtain the test result because of high variability in the target-specific signals	Repeat *BC-GP* from original blood culture specimen
Processing error	Pre-analytical error Internal checks within the Processor *SP* detected an unexpected event	Power cycle Processor *SP* and repeat *BC-GP* from original blood culture specimen

in order to confirm the initial result. Should a "Detected" result be obtained upon repeat testing, it is appropriate to consider this latter test result reportable.

2. Calls for invalid results: error calls and recourse
   Error calls related to an invalid test are listed in Table 2, together with the appropriate recourse which should be taken by the user.

3.4 Quality Control

Quality control, as a component of an overall quality assurance program, consists of tests and procedures for monitoring and evaluating the analytical performance of a measurement system to ensure the reliability of patient test results.

3.4.1 Verigene System

The Verigene System uses a series of automated online quality measurements to monitor instrument functionality, software performance, fluidics, test conditions, reagent integrity, and procedural steps each time a test is performed. A series of automated online procedural checks guide the user through the testing process each time a test is performed. BC-GP Test barcode and sample information are linked upon entry into the Verigene Reader to help prevent misreporting of results.

3.4.2 Assay Controls

1. Several levels of controls are built into BC-GP to ensure that ailures at any procedural step of BC-GP are identified during the procedure.
2. Internal Controls
   An Internal Processing Control, designated "INT CTL 1," comprises a nontarget organism *Bacillus subtilis*, a gram-positive bacterium with an intact genome. It is automatically added to each sample in the processor immediately prior to sample extraction. The INT CTL 1 functions as a complete assay control, the primary purpose of which is to monitor failures likely to be attributable to the sample preparation step (i.e., lysis and nucleic acid extraction); it also functions as nontarget hybridization/detection control.
   A second Internal Processing Control, designated "INT CTL 2," comprises an assay-specific single-stranded DNA target present in the Sample Hybridization Mix reagent and is added by the system to each sample as a means to monitor hybridization inhibition (due to sample- or process-related inhibitors or reagent failures).
   For each test performed, both controls (INT CTL 1 and INT CTL 2) must yield correct results to enable reporting of a valid test result.
3. External controls
   It is highly recommended that known culture-confirmed blood culture specimens positive for each of the BC-GP panel organisms be tested routinely as defined by the user's laboratory's standard operating procedures on a rotating basis using 3–4 smaller groups of organisms and/or under the following circumstances (*see* **Note 9**):

   Instrument installation, test validation, and when troubleshooting is necessary.

   During performance verification for receipt of a new set/lot of consumables.

   When the integrity of consumables or the device is in question.

4 Notes

1. The Verigene Gram-Negative Blood Culture Nucleic-Acid Test (BC-GN) is a rapid multiplexed test that identifies eight gram-negative bacteria (*Acinetobacter* spp., *Citrobacter* spp.,

Enterobacter spp., *Proteus* spp., *Escherichia coli*, *Klebsiella oxytoca*, *Klebsiella pneumoniae*, *Pseudomonas aeruginosa*) and six beta-lactamases, including one extended-spectrum beta-lactamase (ESBL) (CTX-M) and five carbapenemases (KPC, NDM, VIM, IMP, OXA). Like BC-GP, BC-GN is run on the sample-to-result Verigene System and provides results within 2 h of a positive blood culture. Mancini et al. recently published a study in which BC-GN showed a sensitivity of 97.9 % (94/96) for detectable organisms and a specificity of 100 % (104/104) [6]. In this study, BC-GN provided a 16 h time savings for bacterial identification and resistance detection over the comparator method, matrix-assisted laser desorption/ionization time of flight (MALDI-ToF) combined with antibiotic susceptibility testing [6].

2. The Verigene System is a benchtop sample-to-result molecular diagnostics workstation consisting of two modules: the Verigene Processor *SP* and the Verigene Reader. The Verigene Reader serves as the central control unit for the Verigene System as well as the user interface, storing and tracking information throughout the assay process and interpreting and generating test results once the assay is complete. The Verigene Processor *SP* automates (a) sample preparation, cell lysis and magnetic bead-based bacterial DNA isolation from positive blood culture specimens obtained from patients, and (b) hybridization of bacteria-specific target DNA that employs a gold nanoparticle probe-based technology in a microarray format.
3. In mixed cultures containing gram-positive bacteria and other organisms, BC-GP may not identify all the detectable organisms in the specimen, depending upon the concentration of each target present. Isolation in solid media is needed to differentiate mixed growth with other organisms and to identify positive blood cultures yielding a negative result.
4. BC-GP is FDA cleared for use on all FDA-cleared blood culture bottle types.
5. The detection of bacterial nucleic acid is dependent on proper specimen collection, handling, transport, storage, and preparation, including extraction. Failure to observe proper procedures in any of these steps could lead to incorrect results. False-negative results may occur from improper specimen collection, handling, or storage; technical error; sample mix-up; or target concentration below the analytical sensitivity of the test or below the concentration at bottle positive, which might be caused by the growth of other organism(s).
6. Bacterial DNA is extracted from the organisms present in a positive blood culture media specimen, fragmented and denatured (Fig. 8). This fragmented, single-stranded bacterial

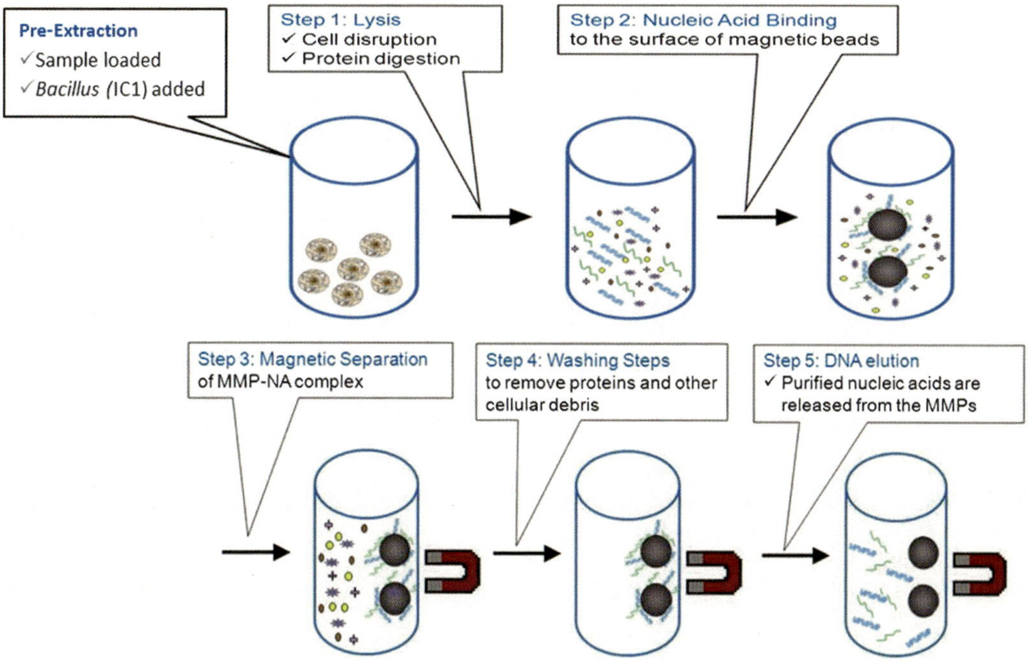

Fig. 8 Schematic of the automated sample preparation and magnetic bead-based extraction on the Verigene Processor *SP*

DNA hybridizes to complementary sequence-specific DNA oligonucleotides, known as capture oligonucleotides, arrayed on the surface of a substrate (glass slide). A second DNA oligonucleotide is then hybridized to the bacterial DNA that was captured initially. This oligonucleotide is known as a mediator oligonucleotide containing two sequence domains: one domain is complementary to the bacterial DNA target, and a second domain is complementary to a common oligonucleotide attached to a signal-generating gold nanoparticle. After washing away any DNA not affixed to the captures, the gold nanoparticle is exposed to the captured mediator/target complex where it hybridizes to any captured mediator oligonucleotides.

7. The presence of the silver-enhanced gold nanoparticle probes at a particular location on the substrate is assessed optically (Fig. 9).
8. For a "Detected" and "Not Detected" result to be generated by BC-GP, three conditions (or "filters") must be met. These conditions serve as a single set of clinical "cutoff" or detection criteria:
   (a) *Filter 1*: signal $\geq$ noise threshold
   (b) *Filter 2*: normalized "Ratio to Negative Control (NC)" >0.85
   (c) *Filter 3*: normalized "Ratio to Imaging Control (IC)" ≥ -0.4.

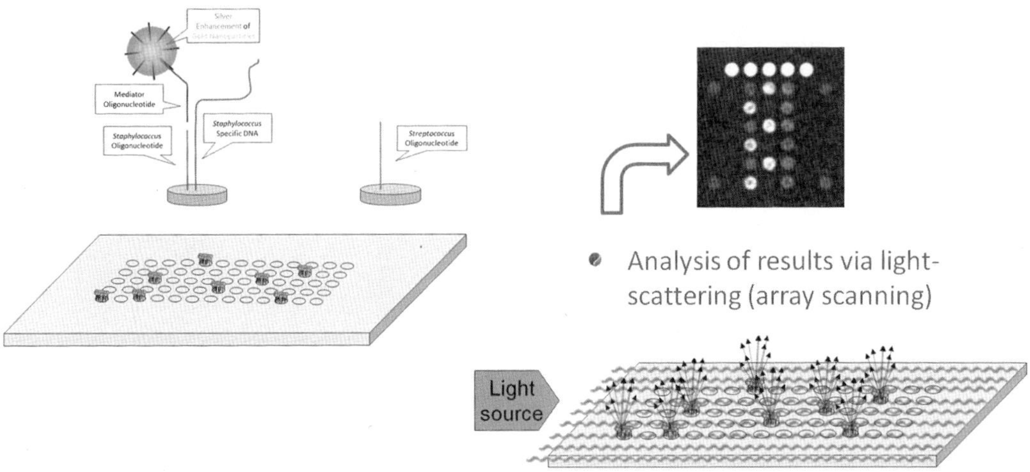

Fig. 9 A diagram depicting a *Staphylococcus*-specific DNA present in a sample being bound to the microarray by the capture and mediator oligonucleotides along with the silver-encoded gold nanoparticle and finally detected via a simple light scatter across the microarray

9. Frozen aliquots of blood cultures containing these organisms may be used for this purpose. When preparing QC material from a positive blood culture bottle, sterilize the bottle top by wiping with an alcohol wipe, invert the bottle 4–5 times to homogenize the specimen, draw fluid by using a 10 mL syringe (equipped preferably with a 16 gauge needle), and transfer to a secondary vessel. Vortex secondary vessel to homogenize specimen, dispense 500 μL aliquots into cryovials, and store the aliquots at −80 °C.

References

1. Kumar A, Roberts D, Wood KE, Light B, Parrillo JE et al (2006) Duration of hypotension before initiation of effective antimicrobial therapy is the critical determinant of survival in human septic shock. Crit Care Med 34: 1589–1596
2. Ibrahim EH, Sherman G, Ward S, Fraser VJ, Kollef MH (2000) The influence of inadequate antimicrobial treatment of bloodstream infections on patient outcomes in the ICU setting. Chest 118:146–155
3. Bauer KA, West JE, Balada-Llasat JM, Pancholi P, Stevenson KB, Goff DA (2010) An antimicrobial stewardship program's impact. Clin Infect Dis 51:1074–1080
4. Sango A, McCarter YS, Johnson D, Ferreira J, Guzman N, Jankowski CA (2013) Stewardship approach for optimizing antimicrobial therapy through use of a rapid microarray assay on blood cultures positive for *Enterococcus* species. J Clin Microbiol 51:4008–4011
5. Buchan BW, Ginocchio CC, Manii R, Cavagnolo R, Pancholi P, Swyers L, Thomson RB Jr, Anderson C, Kaul K, Ledeboer NA (2013) Multiplex identification of gram-positive bacteria and resistance determinants directly from positive blood culture broths: evaluation of an automated microarray-based nucleic-acid test. PLoS Med 10(7):e1001478
6. Mancini N, Infurnari L, Ghidoli N, Valzano G, Clementi N, Burioni R, Celmenti M (2014) Potential impact of a microarray-based nucleic-acid assay for the rapid detection of Gram-negative bacteria and resistance markers in positive blood cultures. J Clin Microbiol 52(4):1242–1245. doi:10.1128/JCM.00142-14

Chapter 9

Detection of Carbapenemases Using Matrix-Assisted Laser Desorption/Ionization Time-of-Flight Mass Spectrometry (MALDI-TOF MS) Meropenem Hydrolysis Assay

Jaroslav Hrabák

Abstract

Matrix-assisted laser desorption/ionization time-of-flight mass spectrometry (MALDI-TOF MS) has been recently introduced to many diagnostic microbiological laboratories. Besides the identification of bacteria and fungi, that technique provides a potentially useful tool for the detection of antimicrobial resistance, especially of that conferred by β-lactamases. Here, we describe an assay allowing a detection of meropenem hydrolysis in clinical isolates of Enterobacteriaceae, *Pseudomonas* spp., and *Acinetobacter baumannii* using MALDI-TOF MS. This method is able to confirm carbapenemases within 3 h. The results are important for proper and fast intervention to limit the spread of carbapenemase-producing bacteria and provide information for appropriate initial therapy of the infections caused by these microbes.

1 Introduction

Carbapenemase-producing bacteria represent a serious threat that can complicate further the development of current medicine (e.g., surgery and intensive care). For epidemiological purposes as well as for proper initial therapy of infections caused by Gram-negative bacteria (especially of Enterobacteriaceae family), there is an urgent need to detect the production of carbapenemases.

In 2011, a fast, sensitive, and specific method for the detection of carbapenemase-producing bacteria was developed [1]. This assay is able to detect carbapenemase activity by using matrix-assisted laser desorption/ionization time-of-flight mass spectrometry (MALDI-TOF MS). Several formats of this test have been proposed and evaluated [2, 3]. In all of them, a fresh bacterial culture is mixed with a carbapenem solution (meropenem or ertapenem). After incubation at 35–37 °C for 2–4 h, the reaction mixture is centrifuged, and the supernatant is measured by MALDI-TOF MS. The visualization of the carbapenem molecule and its salts

Nicasio Mancini (ed.), *Sepsis: Diagnostic Methods and Protocols*, Methods in Molecular Biology, vol. 1237,
DOI 10.1007/978-1-4939-1776-1_9, © Springer Science+Business Media New York 2015

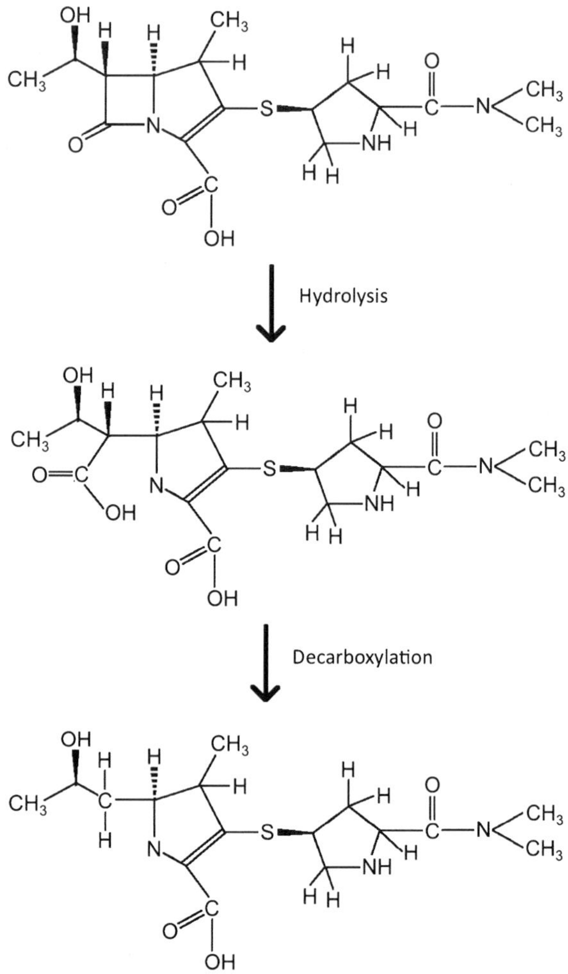

Fig. 1 Meropenem and its degradation by carbapenemases

represents negative results, while in carbapenemase-producing bacteria, degradation products of meropenem can be detected (Fig. 1). A study published recently reported on some modifications, which can minimize false-positive and false-negative results [4].

2 Materials

All solutions should be prepared using deionized water applicable for mass spectrometry. All chemicals should be of the highest purity (e.g., Sigma-Aldrich, St. Louis, MO, USA). Meropenem can be obtained from Sigma-Aldrich (St. Louis, MO, USA). Other medicinal products provided by pharmaceutical companies can be also used (e.g., Merrem (meropenem), AstraZeneca, Macclesfield, UK).

2.1 Buffers and Solutions Used for the Assay

1. Suspension buffer: 20 mM Tris–HCl, 20 mM NaCl, and 50 mM NH_4HCO_3, pH 7.0. Add about 80 mL water to 100 mL cylinder or a glass beaker. Weigh 0.242 g Tris and 0.117 g NaCl and transfer to the cylinder. Mix and adjust pH to 7.5 with HCl (*see* **Note 1**). Weigh 0.396 g NH_4HCO_3 and add to previously prepared solution. Mix and adjust pH to 7.0 with HCl. Add water to a volume of 100 mL, and check pH. Buffer can be stored at 4 °C for 2 weeks.
2. Reaction buffer: 20 mM Tris–HCl, 0.01 % SDS, 50 mM NH_4HCO_3, and 0.1 mM meropenem, pH 7.0. Prepare the buffer as in the previous step (except NaCl). Add 0.01 g sodium dodecyl sulfate (SDS) and mix. Weigh 3.83 mg meropenem and add to previously prepared solution. After mixing, prepare 1 mL aliquots in Eppendorf tubes and immediately freeze at −80 °C. Meropenem solution can be stored for 1 month at −80 °C (*see* **Note 2**). Avoid refreezing.
3. Matrix solution: 10 mg/mL of 2,5-dihydroxybenzoic acid in 50 % ethanol. Weigh 10 mg of 2,5-dihydroxybenzoic acid (*see* **Note 3**) in Eppendorf tube, and add 1 mL 50 % ethanol and mix. Matrix solution can be stored in the dark at 4 °C for 2 weeks.

3 Methods

1. Use a fresh culture of bacteria grown on blood or Mueller-Hinton agar (<18 h) at 35 °C (*see* **Note 4**).
2. Prepare a bacterial inoculum of 3.0 on McFarland scale in a suspension buffer.
3. Place 1 mL of bacterial suspension to an Eppendorf tube and centrifuge at 14,000 ×*g* for 3 min.
4. Remove supernatant and resuspend the pellet in 50 μl of a reaction buffer.
5. Incubate the mixture at 35 °C for 2 h.
6. Centrifuge at 14,000 ×*g* for 3 min.
7. Apply 1 μl of supernatant onto a stainless steel MALDI target (Bruker Daltonics GmbH, Bremen, Germany) plate and allow to dry at room temperature (*see* **Note 5**).
8. Cover the spot by 1 μl of DHB solution and allow to dry the spot at room temperature (*see* **Notes 5** and **6**).
9. Set up your mass spectrometer (*see* **Note 7**).
10. Calibrate the mass spectrometer (*see* **Note 8**).
11. Measure manually in at least ten different positions (*see* **Note 9**).
12. Analyze spectra using a proper software (*see* **Note 10**).
13. Interpret the results according to the criteria summarized in Table 1 (*see* **Note 11**).

Table 1
Interpretation criteria based on the peaks' presence/absence

	Carbapenemase-producing isolate	Carbapenemase-nonproducing isolate
Presence of the peaks (*m/z*) (presence of at least one peak)	358.5 (decarboxylated product) 380.5 (sodium salt of decarboxylated product)	384.5 (meropenem) 406.5 (meropenem sodium salt)
Absence of the peaks (*m/z*) (absence of all peaks)	384.5 (meropenem) 406.5 (meropenem sodium salt)	358.5 (decarboxylated product) 380.5 (sodium salt of decarboxylated product)

4 Notes

1. Concentrated HCl (12 N) can be used at first to narrow the gap from the starting pH to the pH of 7.5. From then use the diluted HCl (e.g., 0.5 N) to adjust pH to 7.0. Avoid the use of hydroxide if the pH decreased below 7.0. In that case, the buffer must be prepared once again.
2. Check the quality of meropenem solution by mass spectrometry (*see* below). The peak representing native meropenem solution 384.5 *m/z* should be dominant. Smaller peaks representing sodium salt variants (406.5 and 428.5 *m/z*) may also be presented (Fig. 2), but their intensity must not be higher than the peak representing the native molecule.
3. Other variants of dihydroxybenzoic acid (i.e., 3,4-dihydroxybenzoic acid) may not be used, because no spectra will be acquired using these molecules.
4. Cultures grown on blood, chocolate, and Mueller-Hinton agars can be used. Avoid the use of cultures from selective media (e.g., MacConkey agar) and from chromogenic media. Those cultures must be recultivated on one of the media mentioned above.
5. Placing the target with the sample/matrix to laminar flow box may hasten the drying of the spot.
6. Target with prepared spots should be measured within 20 min. If you need to postpone the measurement, the target must be stored inside the mass spectrometer under the vacuum; otherwise, meropenem will degrade.
7. If using Microflex LT mass spectrometer (Bruker Daltonics GmbH, Bremen, Germany) and FlexControl software, set up the parameters as follows: ion mode, linear positive; range, 350–460 *m/z*; ion source 1, 20 kV; ion source 2, 16.7 kV;

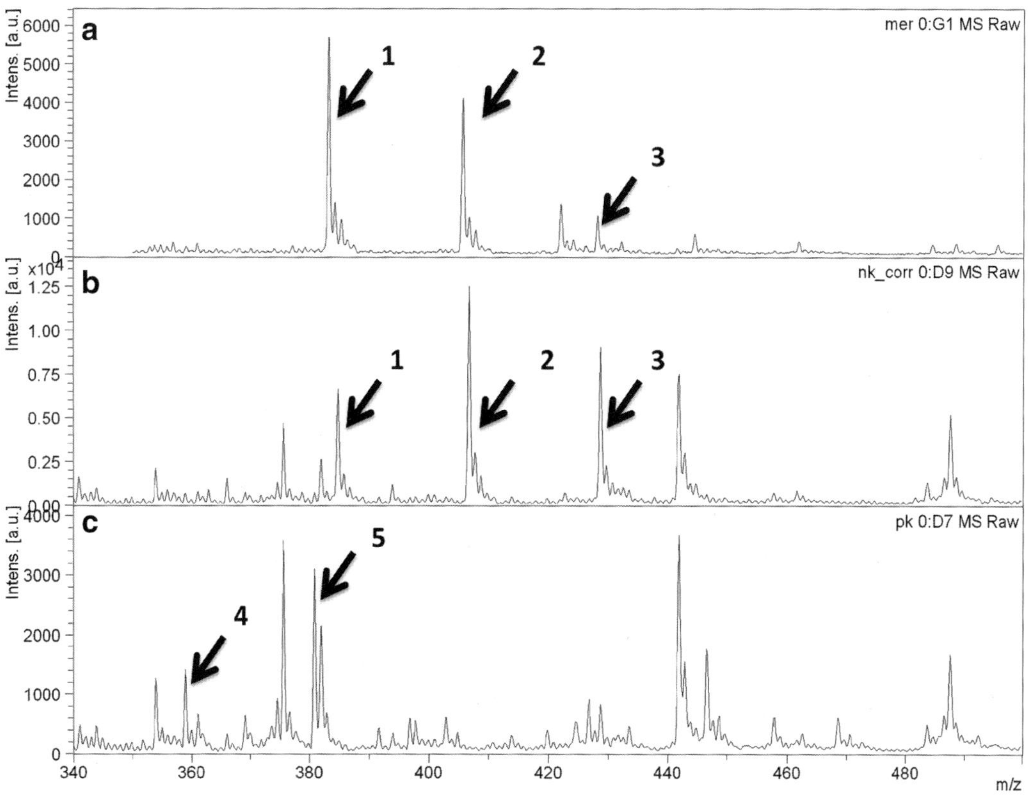

Fig. 2 Mass spectra of meropenem (**a**), carbapenemase-nonproducing isolate, negative result (**b**), and carbapenemase-producing isolate, positive result (**c**). 1, meropenem (384.5 *m/z*); 2, meropenem sodium salt (406.5 *m/z*); 3, meropenem disodium salt (428.5 *m/z*); 4, decarboxylated meropenem degradation product (358.5 *m/z*); 5, sodium salt of decarboxylated meropenem degradation product (380.5 *m/z*)

lens, 7 kV; pulsed ion extraction, 170 ns; laser frequency, 60 Hz; and digitizer trigger level, 2,500 mV. Laser intensity should be set up as described in **Note 9**.

8. Spectrometer should be calibrated using a fresh meropenem solution on the spot processed as described in **steps** 7 and **8**. Native meropenem molecule should have 384.5 *m/z*; meropenem sodium salt, 406.5 *m/z*; and meropenem disodium salt, 428.5 *m/z* (*see* Fig. 2a).

9. In every run, the meropenem solution (incubating simultaneously with the samples) as well as the positive and negative controls must be included. Using Microflex LT mass spectrometer and FlexControl software, the laser intensity should be set up to provide peaks of native meropenem in the range of 2,000–8,000 intensity.

10. We propose the use of FlexAnalysis software (Bruker Daltonics GmbH, Bremen, Germany). At the first time, the quality of the spectra of meropenem and the negative and positive controls

must be checked. If they are interpretable, other spectra can be analyzed.

11. In some producers of OXA-type carbapenemases, peaks of meropenem and its sodium salts may be presented together with degradation products. In such isolates, incubation time can be prolonged to 4 h, or other test for carbapenemase detection should be used.

Acknowledgments

This work was supported by the grant Nr. NT11032-6/2010 and by the Charles University Research Fund (Nr. P36).

References

1. Hrabak J, Chudackova E, Walkova R (2013) Matrix-assisted laser desorption ionization–time of flight (MALDI-TOF) mass spectrometry for detection of antibiotic resistance mechanisms: from research to routine diagnosis. Clin Microbiol Rev 26:103–114
2. Hrabak J, Walkova R, Studentova V, Chudackova E, Bergerova T (2011) Carbapenemase activity detection by matrix-assisted laser desorption ionization-time of flight mass spectrometry. J Clin Microbiol 49: 3222–3227
3. Burckhardt I, Zimmermann S (2011) Using matrix-assisted laser desorption ionization–time of flight mass spectrometry to detect carbapenem resistance within 1 to 2.5 hours. J Clin Microbiol 49:3321–3324
4. Hrabak J, Studentova V, Walkova R, Zemlickova H, Jakubu V, Chudackova E, Gniadkowski M, Pfeifer Y, Perry JD, Wilkinson K, Bergerová T (2012) Detection of NDM-1, VIM-1, KPC, OXA-48, and OXA-162 carbapenemases by matrix-assisted laser desorption ionization-time of flight mass spectrometry. J Clin Microbiol 50:2441–2443

Chapter 10

Molecular Detection of Antibiotic Resistance Genes from Positive Blood Cultures

Musa Y. Hindiyeh, Gill Smollan, Shiraz Gefen-Halevi, Ella Mendelson, and Nathan Keller

Abstract

Rapid detection of the bacterial causative agent causing sepsis must be coupled with rapid identification of the antibiotic resistant mechanism that the pathogen might possess. Real-time PCR (qPCR)-based assays have been extensively utilized in the clinical microbiology field as diagnostic tools for the rapid detection of specific nucleic acid (NA) targets. In this chapter, we will discuss the technical aspects of using an internally controlled qPCR assay for the rapid detection of *Klebsiella pneumoniae* carbapenemase gene (bla_{KPC}) in positive Bactec blood culture bottles. The multiplex qPCR (bla_{KPC}/RNase P) utilizes specific primers and probes for the detection of the bacterial carbapenem resistance mechanism, bla_{KPC} gene, and the internal control RNase P. The internal control of the qPCR assay is vital for detecting any inhibitors that are well known to be present in the blood culture bottles. Rapid detection of the antibiotic resistant mechanism present in the bacterial pathogen causing sepsis can help in better managing patients' infection.

Key words Blood culture bottles, bla_{KPC}, Carbapenem resistance, Multidrug resistance

1 Introduction

Physicians are currently experiencing an unprecedented challenge on how to manage septic patients in particular the ones that are infected with carbapenem-resistant members of the Enterobacteriaceae family. This was in part due to the emergence and rapid spread of plasmid-mediated carbapenem-resistant mechanism *Klebsiella pneumoniae* carbapenemase (bla_{KPC}) [1]. Carbapenems are usually the last class of antibiotics used to treat infections caused by resistant bacteria such as the ones carrying the extended-spectrum β-lactamases (ESBLs) genes. Since the description of the class A serine β-lactamases (bla_{NMC}, bla_{IMI}, bla_{SME}, bla_{GES}, and bla_{KPC}), bla_{KPC} has been the only resistance mechanism to rapidly disseminate in many parts of the world. Reports from varying parts of the world have shown that bacteria carrying the bla_{KPC} gene are usually non-susceptible to fluoroquinolones, aminoglycosides,

Nicasio Mancini (ed.), *Sepsis: Diagnostic Methods and Protocols*, Methods in Molecular Biology, vol. 1237,
DOI 10.1007/978-1-4939-1776-1_10,

and co-trimoxazole, while the majority of the isolates are susceptible to tigecycline and colistin sulfate [2, 3].

The variation in susceptibility patterns and the heterogeneous expression of the β-lactamases have complicated the detection of the bla_{KPC} resistance mechanism in the clinical laboratory [4]. The inconsistency in the performance of the standardized classical microbiology assays has led to the utilization of molecular assays for the rapid detection and identification of the carbapenem-resistant bacteria. Several highly sensitive polymerase chain reaction (PCR) assays and real-time PCR assays (qPCR) have been developed and validated for the rapid detection and identification of bla_{KPC}-positive bacteria [5–7]. However, only few assays have examined the detection of bla_{KPC}-positive bacteria directly from blood culture bottles of septic patients [8, 9].

Blood cultures are considered to be the "gold standard" for detecting the bacterial pathogen in the bloodstream of septic patients, including those that encode bla_{KPC} genes [10]. Automated blood culture systems such as the Bactec instruments (Becton, Dickinson and Company, USA), BacT/ALERT® 3D (BioMérieux, France), and VersaTREK® Instrumentation (Thermo Fisher Scientific, USA) take approximately 1–2 days, on average, to signal a positive blood culture and another 1–2 days to finalize bacterial identification and antimicrobial testing. With the advent of qPCR, the time to bacterial identification and detection of drug resistance has been reduced to 4–6 h after a positive blood culture has turned positive [11, 12]. However, the presence of PCR inhibitors in the blood culture bottles has reduced the sensitivity of the PCR assays [13].

In this manuscript we will report on our experience in performing an internally controlled qPCR (bla_{KPC}/RNase P) assay for the detection of bla_{KPC} gene in Bactec blood culture bottles. This assay can be used as a model for studying the presence of other bacterial drug resistance mechanisms in blood culture bottles.

2 Materials

Molecular testing should be performed in specifically designated molecular rooms in order to prevent nucleic acid (NA) contamination.

2.1 Specimen Processing Room for Nucleic Acid Testing (NAT)

Processing the positive blood cultures should be performed in a biological safety cabinet (BSC) in order to prevent the exposure to the pathogen inside the bottle. To obtain an aliquot from the patient's Bactec bottle, the following should be present:

1. Assorted latex powder-free gloves.
2. Absorbent bench pad.
3. 70 % isopropyl or ethanol pads.
4. Sterile syringe (2 or 5 mL).

5. Needles (18 gauge).
6. Cryovials (2.0 mL).

2.2 Specimen Extraction Room

Total NA extraction should be performed in a designated room "Specimen Extraction Room." Manual or automated extraction formats can be used to isolate total NA from the Bactec bottles.

The room should be equipped with designated tools for performing NA extraction. These include:

1. QIAamp DNA minikit for manual NA extraction from Bactec broth. Roche MagNA Pure LC instrument for automated NA extraction from Bactec broth.
2. MagNA Pure LC DNA isolation kit III (bacteria and fungi).
3. Workstation (UV Dead-Air Box).
4. Assorted latex powder-free gloves.
5. Absorbent bench pad.
6. Assorted pipettors.
7. Assorted sterile filtered tips.
8. Centrifuge for 1.5 mL and 2.0 mL tubes.
9. Eppendorf tubes (1.5 mL and 2 mL).
10. Waste containers.

2.3 Real-Time PCR Master Mix Preparation

Preparation of the reagents to run the qPCR (bla_{KPC}/RNase P) assay should be performed in a designated molecular room called "The Clean Room." The room should be equipped with designated tools for preparing the qPCR master mix. These include:

1. Disposable lab gowns.
2. Assorted latex powder-free gloves.
3. Workstation (UV Dead-Air Box).
4. Eppendorf tubes (1.5 mL and 2.0 mL).
5. Adsorbent bench pad.
6. Assorted pipettors.
7. Assorted sterile filtered tips.
8. Real-time PCR 8-well strips with caps or 96-well plates with plastic cover.
9. CoolSafe System for 0.2 mL tubes or plates, aluminum.
10. Cold plastic rack (4–8 °C) for 1.5 mL and 2.0 mL Eppendorf tubes.
11. Centrifuge for 1.5 mL and 2 mL Eppendorf tubes.
12. Centrifuge for 0.2 mL and 0.5 mL Eppendorf tubes.
13. Centrifuge for 0.2 mL 8-well strips.
14. Centrifuge for 96-well plates.

15. Molecular grade H_2O.
16. Waste containers.
17. qPCR master mix plus reaction buffer [2× (qPCR MasterMix Plus Low ROX w/o UNG (Eurogentec, Belgium)) or 2× (Thermo Scientific ABsolute Blue QPCR Low ROX Mix; Waltham, MA)]. Both mixes contain the internal reference dye 5-carboxy-X-rhodamine succinimidyl ester (ROX).
18. bla_{KPC}-specific forward primer (5′-GAT ACC ACG TTC CGT CTG G-3′) and reverse primer (5′-GCA GGT TCC GGT TTT GTC TC-3′) working stock (1.2 μM) diluted from the primers' stocks in molecular grade water. Aliquot the primers' working stocks in several well-labeled 0.5 mL Eppendorf tubes and store at −20 °C.
19. bla_{KPC}-specific probe (6-carboxyfluorescein-5′-AGC GGC AGC AGT TTG TTG ATT G-3′-6-carboxytetramethylrhodamine) working stock (0.8 μM) prepared from the probe stock in molecular grade water. Aliquot the probe working stocks in several well-labeled 0.5 mL Eppendorf tubes and store at −20 °C.
20. 10 % bovine serum albumin (BSA) Cohn fraction V at 96 % (agarose gel electrophoresis) (*see* **Note 1**). Dissolve 1 g BSA in 10 mL molecular grade water. Once dissolved, filter the 10 % BSA in 0.2 μ filter and aliquot (100 μl) in 0.5 mL Eppendorf tubes. Store BSA at −20 °C.
21. VIC-labeled RNase P internal control mix (Life Technologies, Foster City, USA) prepared as recommended by the manufacturer.

2.4 Sample Loading Room

In this location, the extracted NA from Bactec bottles will be added to the qPCR (bla_{KPC}/RNase P) master mix that was prepared in the clean room. The room should have designated molecular loading equipment:

1. Workstation (UV Dead-Air Box).
2. Assorted latex powder-free gloves.
3. Adsorbent bench pad.
4. Assorted pipettors.
5. Assorted sterile filtered tips.
6. Eppendorf tubes (2.0 mL and 1.5 mL).
7. Centrifuge.
8. Waste containers.

2.5 qPCR Equipment/Amplification Room

In this room the real-time PCR machines will be placed and connected to uninterruptible power supply (UPS). The assay described below was validated on ABI Prism 7500 sequence detection system (Life Technologies, USA). Other real-time PCR machines can be used but need to be validated.

3 Methods

3.1 Aliquoting Samples from Bactec Bottle

1. In a BSC, mix the broth inside the Bactec bottle positive for gram-negative bacilli by gentle shaking for 5–10 s.
2. Clean the rubber cap of the Bactec blood bottle with sterile 70 % isopropyl or ethanol pads.
3. Pierce the rubber cap with an 18 gauge needle fitted on 2 mL or 5 mL syringe (*see* **Note 2**).
4. Draw 1–2 mL aliquot of the broth and stow in a well-labeled sterile 2.5 mL cryovial tubes (*see* **Note 3**).
5. Store the aliquoted broth at −20 °C pending NAT.

3.2 Manual NA Extraction from Bactec Broth Using QIAamp DNA Minikit

1. Use the aliquoted broth from the Bactec bottle or thaw the stored frozen aliquoted broth positive for gram-negative bacteria at room temperature.
2. Follow the manufacturer's instructions for extracting total NA from 200 μl aliquot of the Bactec broth and from 200 μl H_2O which will be used as negative extraction control.
3. Elute extracted NA in 100 μl elution buffer.
4. Store eluted NA at −20 °C pending bla_{KPC} qPCR analysis.

3.3 Automated NA Extraction from Bactec Bottles Using Roche MagNA Pure LC Instrument (See Note 4)

1. Use the aliquoted broth from the Bactec bottle or thaw the stored frozen aliquoted broth positive for gram-negative bacteria at room temperature.
2. Follow the manufacturer guidelines to extract 100 μl Bactec broth as well as a 100 μl H_2O which will be used as negative extraction control using the MagNA Pure LC DNA isolation kit III (bacteria and fungi).
3. Make sure to perform the external bacterial lysis step to inactivate and lyse the bacteria.
4. Elute extracted DNA in 100 μl elution buffer and store it at −20 °C pending NAT analysis.

3.4 Preparation of bla_{KPC} qPCR Positive Control

1. Extract total NA using the MagNA Pure LC DNA isolation kit III or QIAamp DNA minikit from a Bactec bottle that is positive for an Enterobacteriaceae carrying the bla_{KPC} gene.
2. Transfer the extracted NA to the qPCR loading room.
3. Prepare serial logarithmic dilution (10 μl + 90 μl) of the extracted bla_{KPC} NA in sterile H_2O.
4. Run the qPCR (bla_{KPC}/RNase P) assay tin triplicate to determine the best dilution that gives a bla_{KPC} threshold cycle (C_T) value of 30.
5. Aliquot the bla_{KPC} control into several 20 μl aliquots and store at −20 °C. From each aliquot 10 μl will be used only once.

3.5 Preparation of RNase P qPCR Positive Control

1. Extract total NA using the MagNA Pure LC DNA isolation kit III or QIAamp DNA minikit from a Bactec bottle that was inoculated with patient blood.
2. Prepare serial logarithmic dilution of the NA in sterile H_2O.
3. In triplicate determine the best dilution that gives an RNase P qPCR (C_T) value of 30.
4. Aliquot the RNase P positive control into several 20 µl aliquots and store at −20 °C. From each tube, 10 µl will be used only once.

3.6 Preparing the qPCR (bla_{KPC}/RNase P) Master Mix

1. Preparation of the qPCR (bla_{KPC}/RNase P) master mix will be performed in the molecular laboratory "Clean Room."
2. Place inside the PCR workstation "UV Dead-Air Box" inside the clean room all necessary equipment and reagents for preparing the qPCR (bla_{KPC}/RNase P) master mix.
3. This includes replacing the absorbent bench pad and placing the assorted pipettors, assorted sterile filtered tips, and a waste container in the workstation.
4. A cold CoolSafe System for 0.2 mL strip/plate metal rack should also be placed inside the workstation.
5. Label a 1.5 mL or 2 mL sterile Eppendorf tube with the name of the (bla_{KPC}/RNase P) master mix that will be prepared.
6. Thaw the 2× qPCR master mix buffer (qPCR MasterMix Plus Low ROX w/o UNG) or the 2× ABsolute Blue (QPCR Low ROX Mix).
7. Thaw the bla_{KPC} primer (1.2 µM) and probe (1.2 µM) working stocks, the 10 % BSA, and the RNase P primer and probe mix on the 4 °C cold plastic rack (*see* **Note 5**).
8. After thawing the reagents, which usually takes about 5 min, gently mix the master mix buffer and the primers by gentle shaking. This should be followed by a quick spin for 10 s.
9. Pipette the reagents into the 1.5 mL labeled Eppendorf tube according to the calculations in Table 1 (*see* **Note 6**).
10. Once all the master mix ingredients are added, gently mix the reagents by vortexing for 3 s followed by spinning for 10 s.
11. Place the bla_{KPC}/RNase P master mix in the 4 °C cold plastic rack.
12. Place the appropriate number of real-time PCR 8-well strips or a real-time PCR plates in the CoolSafe System for 0.2 mL strip/plate metal rack. Table 2 can be used as a model for how to load the strips/plates.
13. With a 20 µl pipettors, pipette 15 µl master mix, and release the master mix in the appropriate designated well in the strip

Table 1
qPCR (*bla*$_{KPC}$/RNase P) master calculation table with the reagents included in the master mix, their corresponding concentrations, and volumes

Reagent	Working stock concentration	Final well concentration	Volume (μl)
PCR master mix	2×	1×	12.5
bla$_{KPC}$-forward primer	30 pmol/1.2 μM	7.5 pmol/300 nM	0.25
bla$_{KPC}$-reverse primer	30 pmol/1.2 μM	7.5 pmol/300 nM	0.25
bla$_{KPC}$-probe	20 pmol/800 nM	5 pmol/200 nM	0.25
BSA	10 %	0.5 %	1.25
RNase P mix	0.5×	0.5×	0.5
Total volume			15
DNA template			10

Table 2
Illustration for distribution of *bla*$_{KPC}$/RNase P master mix and NA template in the strips/plates

	1	2	3	4	5	6	7	8	9	10	11	12
A	*bla*$_{KPC}$ (*pos. cont.*)	Pt. 1	Pt. 8									
B	RNase P (pos. cont.)	Pt. 2	Pt. 10									
C	H_2O (neg. cont.)	Pt. 3	Pt. 11									
D		Pt. 4	Pt. 12									
E		Pt. 5	Pt. 13									
F		Pt. 6	Pt. 14									
G		Pt. 7	Pt. 15									
H	Extraction (neg. cont.)	Pt. 8	Pt. 16									

or plates (*see* **Note 7**). Table 2 illustrates a model for how the strips/plates can be loaded. Continue to add the mix to all the appropriate wells.

14. Once loading the strips/plates is completed, gently place the strip cover above the loaded wells, and move the strips/plates to the NA "Loading Room."

3.7 Loading the NA in the Plastic qPCR Strips/ Plates

1. Loading the NA template in the *bla*$_{KPC}$/RNase P master mix will be performed in the molecular laboratory "Loading Room."
2. Place inside the PCR workstation "UV Dead-Air Box" all necessary equipment and reagents for loading the NA template.

This includes replacing the absorbent bench pad and placing the assorted pipettors, assorted sterile filtered tips, and a waste container in the workstation. A cold CoolSafe System metal rack should also be placed inside the workstation.

3. Place the strips/plates that were loaded with the master mix in the cold CoolSafe System metal rack.
4. Centrifuge the patient's Eppendorf tubes containing the extracted NA at 3,000 × *g* for 1 min.
5. Load 10 μl of patient's NA in the appropriate master mix wells (*see* **Note 8**). This should be followed by loading the negative extraction control, the negative test control, and the positive (bla_{KPC} and RNase P) test controls (*see* **Note 9**).
6. After loading all the NAs, close the strips/plates with the appropriate cap/plastic cover.
7. Centrifuge the strips/plates to mix the bla_{KPC}/RNase P master mix with the NA template for 5 s in a small strip/plate centrifuge.
8. Return the loaded strips/plates into the cold CoolSafe System metal rack.
9. Move the loaded/spun strips to the real-time PCR instrument room.
10. The assay described here was validated on the Applied Biosystems (ABI) 7500. Other machines can be used but they should be validated.

3.8 Real-Time PCR Instruments Room

1. Load the strips/plates into the ABI 7500 instrument according to the manufacturer's instructions.
2. Select the FAM and VIC dyes to be read by the ABI 7500 instrument (*see* **Note 10**).
3. Perform the qPCR (bla_{KPC}/RNase P) assay under the following conditions: 2 min at 50 °C, 10 min at 95 °C, 50 cycles × (15 s at 95 °C, 1 min at 60 °C).
4. Once the run is completed, standardize the analysis of the qPCR (bla_{KPC}/RNase P) run at a specific C_T value, Fig. 1 (*see* **Note 11**).

3.9 Interpretation of the Results

Evaluate the positive (bla_{KPC} and RNase P) test control, negative (H_2O) test control, and extraction negative control results of the run:

1. The positive test control (FAM bla_{KPC} and the VIC RNase P) C_T values should be around 30 ± 1 SD (*see* **Note 12**).
2. The negative test control and extraction negative control FAM and VIC labels C_T values should be undetectable.
3. Consider a blood culture as positive for the presence of the bla_{KPC} gene if the C_T value of the FAM label is below Ct of 40

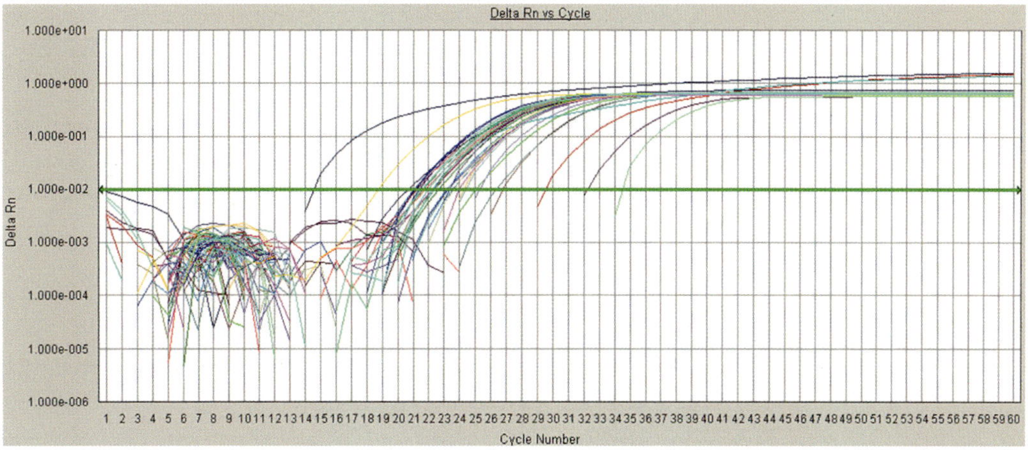

Fig. 1 Snapshot ABI Prism 7500 sequence detection system (Life Technologies, USA) amplification plot. The *X* axis represents the number of cycles that were performed, and the y axis represents the fluorescence signal detected by the CCD camera of the machine. The *different colored curves* represent a typical run of the qPCR (*bla*$_{KPC}$/RNase P) assay where the amplification of the *bla*$_{KPC}$ and the RNase P genes are shown. The *green horizontal line* represents the "threshold" of the qPCR (*bla*$_{KPC}$/RNase P) assay. In our hands, we always set the qPCR (*bla*$_{KPC}$/RNase P) threshold at 0.1

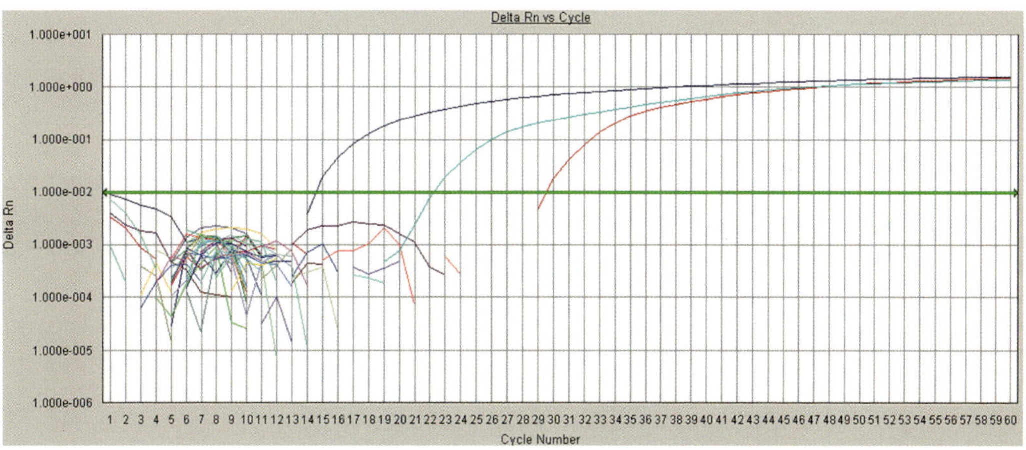

Fig. 2 Snapshot ABI Prism 7500 sequence detection system (Life Technologies, USA) amplification plot of the *bla*$_{KPC}$ (FAM) label. Two blood culture bottles (*dark blue and light blue*) were positive for the *bla*$_{KPC}$ gene. Both had C_T value less than 40. The *dark blue cure* crossed the threshold at C_T 15, while the *light blue* crossed the threshold at C_T 22. The positive *bla*$_{KPC}$ test control (*red*) gave a C_T value around 30, while both the negative test control and the extraction negative control did not cross the threshold and were reported negative

and the internal control RNase P VIC Ct value was below 45 (Fig. 2) (*see* **Note 13**).

4. Consider a blood culture as positive for the presence of the *bla*$_{KPC}$ gene if a blood culture shows a *bla*$_{KPC}$ FAM C_T value less than 40 while the VIC C_T value of RNase P was greater than 45 or negative.

5. A blood culture is considered negative if the FAM *bla*$_{KPC}$ C_T value was greater than 40 or negative and the VIC Ct value was less than 45 (*see* **Note 14**).
6. A blood culture result is considered uninterruptable if the *bla*$_{KPC}$ FAM C_T value did not cross the threshold value or greater than 40 while the VIC RNase P C_T value was greater than 45 or negative (*see* **Note 15**).

3.10 *bla*$_{KPC}$ Results Reporting

1. Report the results to the physician according to your institution guidelines.
2. Samples positive for the *bla*$_{KPC}$ gene should be immediately reported to the health-care provider and to the infection control team.

4 Notes

1. Use bovine serum albumin (BSA) Cohn fraction V at 96 % (agarose gel electrophoresis). Do not use higher purity as the qPCR C_T values will be elevated by 2–3 cycles.
2. Positive Bactec bottles might be under pressure due to the production of gas by the bacteria inside the bottle. A splash might occur upon piercing the Bactec bottle cap, thus aliquoting of samples must be performed in a BSC.
3. While adding the broth in the cryovials, make sure not to touch the top of the vial with the tip as some of the fluid might leak outside the vial once the vial screw cap is closed. This can increase the chance of contaminating the outer part of the vial and thus contaminating other vials.
4. NucliSENS® easyMAG®, a benchtop automated nucleic acid extraction machine, cannot be used since the small resin beads inside the Bactec bottles can clog the machine tubing, thus malfunctioning the machine.
5. Other qPCR master mixes from other suppliers can be used, but they need to be validated first.
6. Always add 10 % extra volume of each reagent to account for pipetting errors.
7. Load the wells with the master mix by releasing the fluid halfway through on the side of the wells.
8. Loading the strips/plates should follow a robot movement away from the wells in order to reduce the chance of any NA contamination.
9. Aliquoted controls should be used only once. Once the control is loaded, the tubes should be disposed in the waste basket.
10. It is important to select ROX dye as a background passive dye when using the ABI 7500 instrument.

11. We recommend standardizing the threshold of the bla_{KPC}/RNase P runs at 0.1. This way the bla_{KPC} and the RNase P test control values will be standardized from one run to another.
12. Add the positive control C_T values in an excel sheet or in any other statistical program in order to monitor the compliance with Westgard quality control rules.
13. The average C_T value of bla_{KPC} gene in a positive Bactec bottle should be around 18.8 (15.9–26.6), while the average C_T value of RNase P gene in a positive Bactec bottle is 27.6 (21–41.8) [9].
14. During the validation of the assay, we did not encounter a single sample with a FAM bla_{KPC} signal above 40 [9].
15. During the validation of the assay, we did not encounter a single sample with such results [9].

Acknowledgment

The authors would like to acknowledge the Chaim Sheba Medical Center laboratory team in particular the bacteriology and virology staff members. Without their help, this assay would have not been validated.

References

1. Yigit H, Queenan AM, Anderson GJ, Domenech-Sanchez A, Biddle JW, Steward CD, Alberti S, Bush K, Tenover FC (2001) Novel carbapenem-hydrolyzing beta-lactamase, KPC-1, from a carbapenem-resistant strain of Klebsiella pneumoniae. Antimicrob Agents Chemother 45(4):1151–1161. doi:10.1128/AAC.45.4.1151-1161.2001
2. Giakoupi P, Maltezou H, Polemis M, Pappa O, Saroglou G, Vatopoulos A, Greek System for the Surveillance of Antimicrobial R (2009) KPC-2-producing Klebsiella pneumoniae infections in Greek hospitals are mainly due to a hyperepidemic clone. Euro Surveill 14(21)
3. Nordmann P, Cuzon G, Naas T (2009) The real threat of Klebsiella pneumoniae carbapenemase-producing bacteria. Lancet Infect Dis 9(4):228–236. doi:10.1016/S1473-3099(09)70054-4
4. Tenover FC, Kalsi RK, Williams PP, Carey RB, Stocker S, Lonsway D, Rasheed JK, Biddle JW, McGowan JE Jr, Hanna B (2006) Carbapenem resistance in Klebsiella pneumoniae not detected by automated susceptibility testing. Emerg Infect Dis 12(8):1209–1213. doi:10.3201/eid1208.060291
5. CDC (2011) Multiplex real-time PCR detection of klebsiella pneumoniae carbapenemase (KPC) and New Delhi metallo-β-lactamase (NDM-1). CDC. http://www.cdc.gov/HAI/pdfs/labSettings/KPC-NDM-protocol-2011.pdf
6. Hindiyeh M, Smollen G, Grossman Z, Ram D, Davidson Y, Mileguir F, Vax M, Ben David D, Tal I, Rahav G, Shamiss A, Mendelson E, Keller N (2008) Rapid detection of blaKPC carbapenemase genes by real-time PCR. J Clin Microbiol 46(9):2879–2883. doi:10.1128/JCM.00661-08
7. Schechner V, Straus-Robinson K, Schwartz D, Pfeffer I, Tarabeia J, Moskovich R, Chmelnitsky I, Schwaber MJ, Carmeli Y, Navon-Venezia S (2009) Evaluation of PCR-based testing for surveillance of KPC-producing carbapenem-resistant members of the Enterobacteriaceae family. J Clin Microbiol 47(10):3261–3265. doi:10.1128/JCM.02368-08
8. Francis RO, Wu F, Della-Latta P, Shi J, Whittier S (2012) Rapid detection of Klebsiella pneumoniae carbapenemase genes in Enterobacteriaceae directly from blood culture bottles by real-time PCR. Am J Clin Pathol 137(4):627–632. doi:10.1309/AJCP9SNHJG2QGLWU

9. Hindiyeh M, Smollan G, Grossman Z, Ram D, Robinov J, Belausov N, Ben-David D, Tal I, Davidson Y, Shamiss A, Mendelson E, Keller N (2011) Rapid detection of blaKPC carbapenemase genes by internally controlled real-time PCR assay using bactec blood culture bottles. J Clin Microbiol 49(7):2480–2484. doi:10.1128/JCM.00149-11
10. Washington JA 2nd, Ilstrup DM (1986) Blood cultures: issues and controversies. Rev Infect Dis 8(5):792–802
11. Ruimy R, Dos-Santos M, Raskine L, Bert F, Masson R, Elbaz S, Bonnal C, Lucet JC, Lefort A, Fantin B, Wolff M, Hornstein M, Andremont A (2008) Accuracy and potential usefulness of triplex real-time PCR for improving antibiotic treatment of patients with blood cultures showing clustered gram-positive cocci on direct smears. J Clin Microbiol 46(6):2045–2051. doi:10.1128/JCM.02250-07
12. van Haeften R, Palladino S, Kay I, Keil T, Heath C, Waterer GW (2003) A quantitative LightCycler PCR to detect Streptococcus pneumoniae in blood and CSF. Diagn Microbiol Infect Dis 47(2):407–414
13. Gebert S, Siegel D, Wellinghausen N (2008) Rapid detection of pathogens in blood culture bottles by real-time PCR in conjunction with the pre-analytic tool MolYsis. J Infect 57(4): 307–316. doi:10.1016/j.jinf.2008.07.013

Chapter 11

Bacterial and Fungal DNA Extraction from Blood Samples: Manual Protocols

Michael G. Lorenz, Helge Mühl, and Claudia Disqué

Abstract

A critical point of molecular diagnosis of systemic infections is the method employed for the extraction of microbial DNA from blood. A DNA isolation method has to be able to fulfill several fundamental requirements for optimal performance of diagnostic assays. First of all, low- and high-molecular-weight substances of the blood inhibitory to downstream analytical reactions like PCR amplification have to be removed. This includes human DNA which is a known source of false-positive results and factor decreasing the analytical sensitivity of PCR assays by unspecific primer binding. At the same time, even extremely low amounts of microbial DNA need to be supplied to molecular diagnostic assays in order to detect low pathogen loads in the blood. Further, considering the variety of microbial etiologies of sepsis, a method should be capable of lysing Gram-positive, Gram-negative, and fungal organisms. Last, extraction buffers, reagents, and consumables have to be free of microbial DNA which leads to false-positive results. Here, we describe manual methods which allow the extraction of microbial DNA from small- and large-volume blood samples for the direct molecular analysis of pathogen.

Key words Microbial DNA from blood, Human DNA removal, MolYsis™, Pediatric blood samples, Large blood volume, Bacteremia, Fungemia

1 Introduction

A crucial step in the direct molecular diagnosis of sepsis is the isolation of microbial DNA from blood. The quantity and quality of bacterial and fungal DNA recovered from the blood contribute to the overall sensitivity and specificity of analytical systems. In-house and commercial methods are available for the removal of low and high molecular weight components of the blood, including sugars, amino acids, proteins, and heme and the elution of purified DNA at the end of the isolation procedure. For the diagnosis of systemic infections, another factor, human DNA, is known to negatively interfere with amplification-based assays [1, 2]. The generally low titers of pathogens in the blood of septic patients correspond to femtogram to picogram amounts of microbial DNA

Nicasio Mancini (ed.), *Sepsis: Diagnostic Methods and Protocols*, Methods in Molecular Biology, vol. 1237,
DOI 10.1007/978-1-4939-1776-1_11,

Table 1
Sources of potential DNA contamination[a]

Material	% False-positives (found/tested)	Origin	References
(a) Blood collection tubes	17 (31/185)	*Aspergillus* spp.	[13]
(b) Nucleic acid extraction and processing			
Zymolyase	n.d.	*Saccharomyces cerevisiae*	[14]
DNA extraction	100 (20/20)	*Burkholderia* spp., *Pseudomonas saccharophila*, *Ralstonia* spp., *Alcaligenes* spp.	[15]
	20 (4/20)	*Legionella* spp., *Aspergillus* spp.	[16, 17]
	n.d.	*Aspergillus* spp., *Candida* spp.	[18]
	n.d.	*Brucella* spp.	[19]
(c) Plastic consumables			
Pipette tips	18 (6/32)	Bacteria	[b]

[a]Samples of the same or different lots or samples from different manufacturers; signals were observed in negative PCR controls using molecular grade water; species were identified by sequencing of the amplicons and BLAST search; *n.d.* not determined

[b]Among three manufacturers, one showed severe contamination of the tips. The other products (PCR tubes, pipette tips) were continuously free of microbial DNA contamination as analyzed by 16S/18S rDNA PCR ($n = 32–320$, different lots tested)

which face microgram amounts of human DNA released from white blood cells during extraction. This 2 to 3 orders of magnitude-fold mass excess of human DNA can be the reason for unspecific amplification and loss in sensitivity, in particular when using broad-range bacterial primers for conserved genes like the 16S rRNA gene [1, 2]. Therefore, a method is desirable that removes the human DNA. Further, an ideal method would have the potential of lysing microorganisms with different cell wall structures as they are represented by Gram-positive, Gram-negative, and fungal organisms. Finally, buffers, reagents, and consumables have to be free of contaminating microbial DNA in order to prevent false-positive results (Table 1). To this topic belongs also the safe handling during DNA extraction and the organization of the laboratory to keep contamination by microbial DNA at a minimum.

In the following sections, we describe protocols of a method, MolYsis™, that are suitable for the enrichment and contamination-free extraction of microbial DNA at high recovery [3] from small- and large-volume blood samples. The methods presented have been evaluated in various clinical studies employing blood, other primary sterile body liquids, and tissues [3–12]. A large in-house evaluation showed that DNA from more than 200 bacterial and fungal species was extracted from clinical material (Table 2). The method described removes human DNA efficiently (Fig. 1) and thereby greatly enhances the sensitivity of PCR or real-time PCR assays targeting the 16S rRNA gene of bacteria and also *Staphylococcus aureus*-specific and methicillin-resistant genes [2, 7].

2 Materials

It is important to use whole blood only that is stabilized by anticoagulation agents, including EDTA, citrate, or heparin (*see* **Note 1**).

2.1 Pretreatment and Extraction of Small Blood Samples (≤1 ml) from Pediatric Patients

1. Buffers, enzymes, spin columns, and elution tubes supplied with the kit *MolYsis*™ *Complete5* (Molzym, Bremen, Germany).
2. 2-Mercaptoethanol (14.3 Mol/l).
3. Disposable gloves and laboratory coats.
4. Adjustable micropipettes (up to 20 μl, up to 200 μl, and up to 1,000 μl) (*see* **Note 2**).
5. Nucleic acid- and nuclease-free, aerosol-resistant pipette tips.
6. Sterile and DNA-free 2 ml microcentrifuge tubes.
7. Racks for tubes.
8. A cooling rack adjusted to −15 °C to −25 °C.
9. A vortex mixer.
10. A thermal shaker with a microcentrifuge tube adapter.
11. A desktop microcentrifuge (≥12,000 × *g*).

2.2 Pretreatment and Extraction of Large Blood Samples (5–10 ml) from Adult Patients

1. Buffers, enzymes, spin columns, and elution tubes supplied with the kit *MolYsis*™ *Complete10* (Molzym).
2. 2-Mercaptoethanol (14.3 Mol/l).
3. Disposable gloves and laboratory coats.
4. Disposable sterile 10 ml pipettes.
5. 50 ml Falcon tubes (*see* **Note 3**).
6. Disposable gloves and laboratory coats.
7. Adjustable micropipettes (up to 20 μl, up to 200 μl, and up to 1,000 μl) (*see* **Note 2**).
8. Nucleic acid- and nuclease-free, aerosol-resistant pipette tips.
9. Sterile and DNA-free 2 ml microcentrifuge tubes.
10. Racks for tubes.
11. A cooling rack adjusted to −15 °C to −25 °C.
12. A vortex mixer.
13. A thermal shaker with a microcentrifuge tube adapter.
14. A desktop microcentrifuge (≥12,000 × *g*).

Table 2
Species identified by sequencing and BLAST analysis (www.sepsitest-blast.net) of amplicons from PCR amplifications using DNA extracted from blood and other clinical materials following the MolYsis™ procedure

Organisms identified		
Gram-negative bacteria	*Providencia stuartii*	*Leifsonia* sp.
Acinetobacter spp.	*Pseudomonas* spp. *Pseudoxanthomonas spadix*	*Microbacterium aurum* *Micrococcus* spp.
Aeromonas veronii	*Ralstonia pickettii*	*Mycetocola* sp.
Bacteroides fragilis	*Raoultella planticola*	*Mycobacterium* spp.
Bartonella quintana	*Schlegelella aquatica* *Serratia* spp.	*Mycoplasma* sp. *Nocardia* sp.
Bordetella petrii	*Sphingomonas* sp.	*Paenibacillus* sp.
Borrelia garinii	*Spirosoma rigui*	*Parvimonas micra*
Bradyrhizobium sp.	*Shigella flexneri*	*Peptoniphilus harei*
Brevibacterium spp.	*Stenotrophomonas maltophilia*	*Peptostreptococcus stomatis*
Burkholderia fungorum	*Tepidimonas thermarum*	*Planomicrobium okeanokoites*
Campylobacter coli	*Variovorax* sp.	*Propionibacterium acnes*
Candidatus Neoehrlichia	*Veillonella* sp.	*Rothia* spp.
Citrobacter freundii	*Weeksella* sp.	*Ruminococcus productus*
Cloacibacterium normanense	*Zoogloea* sp.	*Staphylococcus* spp.
Comamonas testosteroni	Gram-positive bacteria	*Streptococcus* spp.
Coxiella burnetii	*Actinomyces* sp.	*Tropheryma whipplei*
Dialister invisus	*Aerococcus urinaeequi*	*Vagococcus carniphilus*
Edwardsiella tarda	*Anaerococcus* spp.	Fungi
Enhydrobacter aerosaccus	*Bacillus* spp.	*Aspergillus* spp.
Enterobacter spp.	*Bifidobacterium* spp.	*Candida* spp.
Escherichia spp.	*Brevibacterium* spp.	*Cladosporium cladosporioides*
Fusobacterium nucleatum	*Carnobacterium viridans*	*Cryptococcus* spp.
Haemophilus spp.	*Clostridium* spp.	*Didymella exitialis*
Helicobacter pylori	*Corynebacterium* spp.	*Davidiella tassiana*
Hyphomicrobium facile	*Dolosigranulum pigrum*	*Malassezia* spp.
Janthinobacterium lividum	*Enterococcus* spp.	*Peniophora nuda*
Klebsiella spp.	*Eremococcus coleocola*	*Saccharomyces cerevisiae*
Lautropia mirabilis	*Exiguobacterium* sp.	*Schizophyllum radiatum*

(continued)

Table 2
(continued)

Organisms identified		
Leptotrichia sp.	*Facklamia* spp.	*Sistotrema brinkmannii*
Methylobacterium sp.	*Finegoldia magna*	*Sporobolomyces* sp.
Moraxella spp.	*Gemella* spp.	*Udeniomyces pannonicus*
Morganella morganii	*Granulicatella adiacens*	Protist
Neisseria spp.	*Janibacter* sp.	*Plasmodium falciparum*
Parabacteroides distasonis	*Jeotgalicoccus pinnipedialis*	
Paracoccus aminovorans	*Kocuria* spp.	
Petrobacter sp.	*Lactobacillus* spp.	
Proteus spp.	*Lactococcus lactis*	

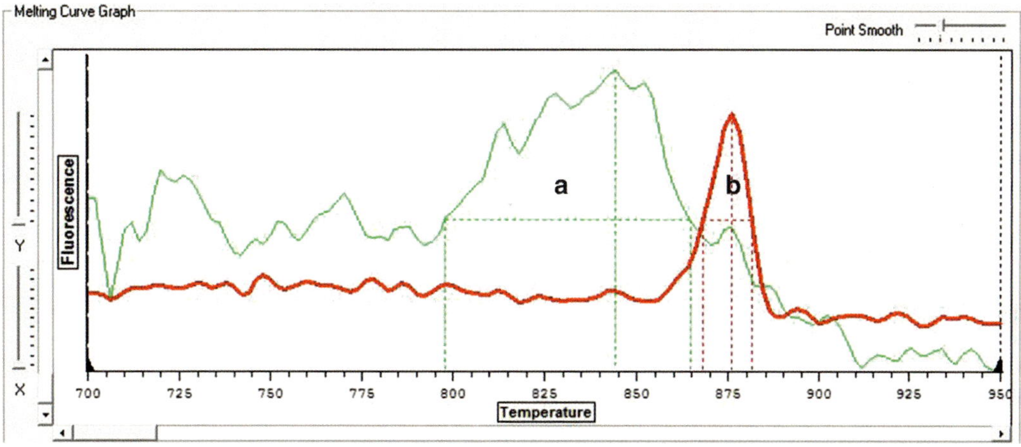

Fig. 1 Dissociation analysis of 16S rDNA real-time PCR (*Mastermix 16S*, Molzym) amplification (DNA Engine Opticon, BioRad). (**a**) analysis of a sample showing an area of unspecific amplicons from human DNA (total DNA extract); (**b**) a sample clearly showing a single bacteria-specific peak with human DNA removed following the MolYsis™ pretreatment

3 Methods

All procedures are performed at room temperature (18–25 °C), except for enzymatic reactions which are incubated as indicated below. Keep enzymes at −15 °C to −25 °C by placing the vials in a cooling rack precooled in the freezer. Take care to work at conditions to avoid DNA contamination (*see* **Note 4**). Before the setup of the DNA isolation system, follow the instructions for control of the performance of the kits (*see* **Note 5**).

3.1 Small-Volume Sample Pretreatment (Pediatric Blood Samples)

1. The protocol is designed for up to 1 ml EDTA (or citrate or heparin) blood samples which are pipetted from the blood

collection tube into a DNA-free 2 ml Eppendorf tube. If less sample volume is available, fill up with supplied buffer SU to the 1 ml mark of the 2 ml Eppendorf tube. Add 250 μl buffer CM and vortex at full speed for 15 s to mix thoroughly (*see* **Note 6**). Incubate on the bench at room temperature (18–25 °C) for 5 min to lyse the human cells (*see* **Note 7**).

2. Add 250 μl buffer DB1 and 10 μl MolDNase B. Vigorously vortex for 15 s to mix and incubate at room temperature for 15 min to degrade the human DNA (*see* **Note 8**).
3. Centrifuge the tube at full speed for 10 min to sediment human cell debris and potentially present microorganisms. Remove the supernatant by using a 1 ml pipette tip, taking care not to disturb the sediment.
4. Add 1 ml buffer RS and resuspend the sediment by vortexing (*see* **Note 9**).
5. Centrifuge the tube at full speed for 5 min. Carefully remove the supernatant with a 1 ml pipette tip. Continue with **step 1** of Subheading 3.3.

3.2 Large-Volume Sample Pretreatment (Blood Samples from Adults)

The analytical sensitivity expressed as colony-forming units (cfu)/ml detected increases with the volume of blood extracted (Table 3). The protocol described below allows the processing of 5–10 ml EDTA (or citrate or heparin) blood:

1. Pipette the sample from the blood collection tube into a 50 ml Falcon tube (*see* **Note 3**) and fill up with buffer SU to the 10 ml mark. Then add 4 ml buffer CM and vigorously vortex

Table 3
Influence of the blood volume extracted using MolYsis™ procedure on the analytical sensitivity of bacteria[a]

Strain	Blood volume (ml)	Titer (cfu/ml)	PCR result[b]
S. aureus	1	60	+++
	2	30	+++
	5	12	+++
	10	6	+++
E. coli	1	120	++−
	2	60	+++
	5	24	+++
	10	12	+++

[a]aStrains spiked at multiples of the detection limits (1 ml blood; S. aureus: 20 cfu/ml; E. coli: 40 cfu/ml) into negative blood at the final concentrations indicated. Assay: universal 16S rRNA gene real-time PCR (*Mastermix 16S*, Molzym) and dissociation analysis to detect bacteria-specific peaks (*see* Fig. 1). Sequence analysis of amplicons confirmed the identity of the spiked strains

[b]+, bacteria-specific signal; −, negative result; *n* = 3

for 10 s to mix (*see* **Note 6**). Incubate at room temperature (18–25 °C) for 5 min to lyse the human cells.

2. Add 4 ml buffer DB1 and 10 μl MolDNase B and vortex for 10 s. Incubate at room temperature for 15 min.
3. Centrifuge the Falcon tube at 9,500 × *g* for 10 min. Thereafter, carefully decant the supernatant.
4. Add 1 ml buffer RS, vigorously vortex until the sediment has been completely resuspended.
5. Transfer the lysate by pipetting into a 2 ml Eppendorf tube and centrifuge at full speed for 5 min. Carefully remove the supernatant with a 1 ml pipette tip. Continue with **step 1** of Subheading 3.3.

3.3 DNA Extraction and Purification

1. Add 80 μl buffer RL and 1.4 μl 2-mercaptoethanol to the sediment (*see* **Note 10**) in the 2 ml Eppendorf tube and resuspend by stirring with the pipette tip and pipetting in and out. Vortex at full speed for 10 s to homogenize.
2. For cell wall degradation of potentially present microorganisms, add 20 μl BugLysis solution and vortex for 10 s. Incubate tube in a thermomixer at 37 °C and 1,000 rpm for 30 min.
3. Add 150 μl buffer RP and 20 μl proteinase K to the lysate and vigorously vortex for 10 s. Incubate at 56 °C and 1,000 rpm for 10 min.
4. Add 250 μl buffer CS and vortex for 10 s to mix. Then add 250 μl binding buffer AB to the lysate and vortex for 10 s.
5. Transfer the lysate to a spin column by pipetting. Close the lid and centrifuge the column at full speed for 30–60 s. Remove the column from the collection tube, discard the flow-through by decanting and replace the column to the collection tube.
6. Wash the column by adding 400 μl buffer WB and centrifuging at full speed for 30–60 s. Decant the flow-through and replace the column to the collection tube.
7. Wash and dry the column by adding 400 μl supplied DNA-free 70 % ethanol and centrifuging at full speed for 3 min. Transfer the column to a supplied 1.5 ml elution tube, taking care to avoid splashing of ethanol to the column (*see* **Note 11**).
8. Pipette 70 °C hot supplied DNA-free deionized water (*see* **Note 12**) to the center of the membrane of the column, close the lid, and let stand for 1 min. Centrifuge at full speed for 1 min to elute the DNA. Discard the column and store the eluted DNA at 4 °C to 12 °C if analyzed at the same day or freeze at -15 °C to -25 °C until further use (*see* **Note 13**).

4 Notes

1. The blood should be processed for molecular analysis at the same day of collection. If this is not possible, the blood collection tube can be stored in the refrigerator (4–12 °C) for up to 3 days. Do not freeze the blood, because microbial cells tend to lyse as a result of freeze-thaw cycles. Because of the DNase treatment during sample pretreatment (**step 2**), breakage of cells leads to the loss of microbial DNA and thus, at the threshold of the system, to false-negative results. If samples are collected during, e.g., a retrospective study and have to be stored, the blood should be stabilized by a DNA-free cryoprotectant before freezing (for instance, *UMD Tubes*, Molzym).
2. Use only sterilized, guaranteed DNA-free disposables (e.g., *Biosphere*®, SARSTEDT, Germany; *Biopur*®, Eppendorf, Germany). It is important that only filter tips are used to avoid contamination of the pipette by aerosols. Wear protective gloves and a disposable lab coat at any handling step during DNA preparation.
3. We routinely use 50 ml *Cellstar*® tubes (Greiner Bio-One, Germany, order no. 227261). If you wish to use another brand, make sure that the tubes can be centrifuged at RCF of 9,500 × *g*.
4. In order to protect yourself from potentially present infectious agents and to keep a contamination-free environment, take care to perform extractions under a laminar flow hood that has been sterilized by UV irradiation before use. Arrange a set of equipment needed for the processing of samples in the laminar flow hood and keep there for future extractions. After working, clean the surface of the hood and the equipment with a DNA decontamination agent that is compatible with the surfaces (inspect the manuals of the products). We routinely use *DNA-ExitusPlus IF*® (AppliChem, Germany). For reasons of avoidance of DNA contamination, it is important that the place of DNA extraction is well separated (optimally in another laboratory) from places where PCR amplifications are performed and master mixes are set up. Frequently change protective gloves during handling. To avoid cross contamination of samples during pretreatment and extraction, it is important to always clear the lid of the Eppendorf tube by a pulse centrifugation after each step involving vortexing (*see* Fig. 2).
5. Perform controls for the setup of the system to test the performance of the extraction procedure. For the determination of the sensitivity of the analytical system used, prepare a dilution series of cultures of microorganisms, e.g., *S. aureus*, and spike into negative EDTA blood samples. The sensitivity threshold

Fig. 2 Blood lysate after pretreatment of 1 ml EDTA blood (**step 2**; *see* Subheading 3.1). Note that lysate contaminates the lid of the tube after vortexing. The lid is cleared by a pulse centrifugation (5 s)

depends on the strain, the blood volume extracted, and the assay used. In order to be relevant for the detection of systemic infections in septicemic patients, make sure that your analytical system is sensitive enough to detect ≤ 10 *E. coli* genome equivalents per assay. An example of a sensitive assay is given in Table 3, which shows the influence of the blood volume extracted on the analytical sensitivity. Cross contamination during extraction should be tested for by extracting negative EDTA blood samples.

6. Follow the instruction for vortexing carefully. Thorough mixing of solutions is important for optimal results. At the end (**step 5**), the lysate should appear opaque or slightly transparent (*see* Fig. 2).
7. Generally clear the lid after each vortexing (*see* Fig. 2). This avoids contamination of the performing individual by pathogens potentially present in the sample and sample cross contamination.
8. Note that also free-floating DNA from dead microbes is degraded. As a consequence, the method supplies DNA only from live microbes (growing or nongrowing). This is an aspect to be considered when analyzing blood from patients under antibiotic treatment [20].
9. The appearance of the sediments among different blood samples can vary from faintly visible to abundant (*see* Fig. 3). Vortexing may sometimes not completely resuspend the sediment. In this case resuspend by pipetting in and out for several times using a 1 ml pipette tip.
10. Take care not to inhale 2-mercaptoethanol. Practically, per sample, premix 80 μl buffer RL and 1.4 μl 2-mercaptoethanol or multiples thereof for more than one sample in a DNA-free Eppendorf tube under an extractor hood. Pipette 80 μl to the sediment of the sample tube and continue with the protocol.

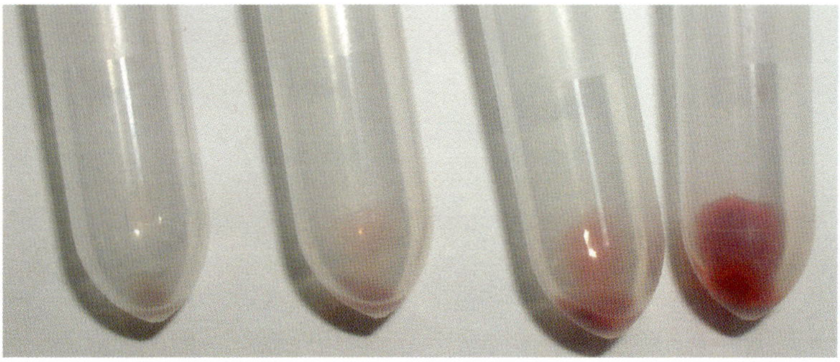

Fig. 3 Sediments of EDTA blood samples from different individuals after centrifugation of lysates (**step 5**; *see* Subheading 3.1). The sediments constitute of blood cell debris and potentially present pathogen cells. The examples demonstrate the considerable variability in the amount of cell debris that reflects the heterogeneity of septicemic blood

11. Ethanol is a strong inhibitor for polymerases employed in PCR amplification assays. If the column is contaminated by splashed ethanol, remove the column, decant the flow-through, replace the column to the tube, and centrifuge for another 1 min.
12. Pipette multiples of 0.1 ml water for sample elution into a DNA-free Eppendorf tube and heat to 70 °C before use.
13. Avoid freeze-thaw cycles which can lead to DNA breakage and loss of PCR amplification. If analyzed the same day of preparation, the eluate can be stored in the refrigerator (4–12 °C).

References

1. Handschur M, Karlic H, Hertel C et al (2010) Preanalytic removal of human DNA eliminates false signals in general 16S rDNA PCR monitoring of bacterial pathogens in blood. Comp Immunol Microbiol Infect Dis 32:207–219
2. Disqué C (2007) Einfluss der DNA-Extraktion auf die PCR-Detektion von Sepsiserregern (in German: Influence of DNA extraction on the detection of sepsis pathogens by PCR). BIOspektrum 06:627–629
3. Horz HP, Scheer S, Huenger F et al (2008) Selective isolation of bacterial DNA from human clinical specimens. J Microbiol Methods 72:98–102
4. Wellinghausen N, Siegel D, Winter J, Gebert S (2009) Rapid diagnosis of candidaemia by real-time PCR detection of *Candida* DNA in blood samples. J Med Microbiol 58:1106–1111
5. Wellinghausen N, Siegel D, Gebert S, Winter J (2009) Rapid detection of Staphylococcus aureus bacteremia and methicillin resistance by real-time PCR in whole blood samples. Eur J Clin Microbiol Infect Dis 28:1001–1005
6. Wellinghausen N, Kochem AJ, Disqué C et al (2009) Diagnosis of bacteremia in whole-blood samples by use of a commercial universal 16S rRNA gene-based PCR and sequence analysis. J Clin Microbiol 47:2759–2765
7. Hansen WLJ, Bruggeman CA, Wolffs PFG (2009) Evaluation of new preanalysis sample treatment tools and DNA isolation protocols to improve bacterial pathogen detection in whole blood. J Clin Microbiol 47:2629–2631
8. Kühn C, Disqué C, Mühl H et al (2011) Evaluation of commercial universal rRNA gene PCR plus sequencing tests for identification of bacteria and fungi associated with infectious endocarditis. J Clin Microbiol 49:2919–2923
9. Esteban J, Alonso-Rodriguez N, del-Prado G et al (2012) PCR-hybridization after sonication improves diagnosis of implant-related infection. Acta Orthop 3:299–304

10. Xu Y, Børsholt RV, Simonsen O et al (2012) Bacterial diversity in suspected prosthetic joint infections: an exploratory study using 16S rRNA gene analysis. FEMS Immunol Med Microbiol 65:291–304
11. Benítez-Páez A, Álvarez M, Belda-Ferre P et al (2013) Detection of transient bacteraemia following dental extractions by 16S rDNA pyrosequencing: a pilot study. PLoS One 8:e57782. doi:10.1371/journal.pone.0057782
12. Haag H, Locher F, Nolte O (2013) Molecular diagnosis of microbial aetiologies using SepsiTest™ in the daily routine of a diagnostic laboratory. Diagn Microbiol Infect Dis 76: 413–418
13. Harrison E, Stahlberger T, Whelan R et al (2010) *Aspergillus* DNA contamination in blood collection tubes. Diagn Microbiol Infect Dis 67:392–394
14. Loeffler J, Hebart H, Bialek R et al (1999) Contaminations occurring in fungal PCR assays. J Clin Microbiol 37:1200–1202
15. Mohammadi T, Reesink HW, Vandenbroucke-Grauls CMJE, Savelkoul PHM (2005) Removal of contaminating DNA from commercial nucleic acid extraction kit reagents. J Microbiol Methods 61:285–288
16. Evans GE, Murdoch DR, Anderson TP et al (2003) Contamination of Qiagen DNA extraction kits with *Legionella* DNA. J Clin Microbiol 41:3452–3453
17. van der Zee A, Peeters M, de Jong C et al (2002) Qiagen DNA extraction kits for sample preparation for *Legionella* PCR are not suitable for diagnostic purposes. J Clin Microbiol 40: 1126
18. Fredricks DN, Smith CS, Meier A (2005) Comparison of six DNA extraction methods for recovery of fungal DNA as assessed by quantitative PCR. J Clin Microbiol 43: 5122–5128
19. Queipo-Ortuño MI, Tena F, Colmenero JD, Morata P (2008) Comparison of seven commercial DNA extraction kits for the recovery of *Brucella* DNA from spiked human serum samples using real-time PCR. Eur J Clin Microbiol Infect Dis 27:109–114
20. Orszag P, Disqué C, Keim S et al (2013) Monitoring of patients supported by extracorporeal membrane oxygenation for systemic infections by broad-range rRNA gene PCR amplification and sequence analysis. J Clin Microbiol 52:307–311. doi:10.1128/JCM.02493-13

Chapter 12

Bacterial and Fungal DNA Extraction from Blood Samples: Automated Protocols

Michael G. Lorenz, Claudia Disqué, and Helge Mühl

Abstract

Automation in DNA isolation is a necessity for routine practice employing molecular diagnosis of infectious agents. To this end, the development of automated systems for the molecular diagnosis of microorganisms directly in blood samples is at its beginning. Important characteristics of systems demanded for routine use include high recovery of microbial DNA, DNA-free containment for the reduction of DNA contamination from exogenous sources, DNA-free reagents and consumables, ideally a walkaway system, and economical pricing of the equipment and consumables. Such full automation of DNA extraction evaluated and in use for sepsis diagnostics is yet not available. Here, we present protocols for the semiautomated isolation of microbial DNA from blood culture and low- and high-volume blood samples. The protocols include a manual pretreatment step followed by automated extraction and purification of microbial DNA.

Key words Microbial DNA extraction, Blood culture, Blood, Human DNA removal, MolYsis™, easyMAG®, SelectNA™, Bacteremia, Fungemia

1 Introduction

DNA extraction is among the critical parameters determining the sensitivity and specificity of analytical systems (see previous chapter). Most DNA isolation systems are designed for research and require adaptation to the needs of routine diagnostics of systemic infections by microorganisms. Among other parameters, the isolation of microbial DNA from only a few cells per milliliter blood is a key issue. An example for the need of optimization of standard laboratory methods is given by Podnecky et al. [1]. They evaluated manual and automated commercial DNA isolation products for DNA extraction from blood spiked with *Burkholderia pseudomallei*. It became clear that with the best extraction kit in terms of DNA recovery, still relatively high detection thresholds were observed (5.5×10^3 cfu/ml using a real-time PCR assay). In fact, the method failed to indicate *B. pseudomallei* in the blood of patients with confirmed melioidosis.

Nicasio Mancini (ed.), *Sepsis: Diagnostic Methods and Protocols*, Methods in Molecular Biology, vol. 1237,
DOI 10.1007/978-1-4939-1776-1_12, © Springer Science+Business Media New York 2015

Efforts to lower the limit of detection of sepsis pathogens in the blood by new DNA isolation approaches made great progress in the past years. Commercial methods are available for high recovery of microbial DNA combined with highly sensitive amplification-based assays, including (1) total DNA isolation and multiplex real-time PCR (SeptiFast®, Roche Diagnostics, Switzerland), (2) enrichment of microbial DNA by binding to proteins specific for non-methylated bacterial and fungal DNA followed by multiplex PCR analysis (VYOO®, Analytik Jena, Germany), and (3) degradation of human DNA combined with broad-range PCR and sequencing (SepsiTest™, Molzym, Germany) or multiplex PCR (MagicPlex®, Seegene, Korea). Sample pretreatment and the use of large blood volumes of up to 10 ml appear to enable the detection of microorganisms at loads [2] prevailing in the blood of septicemic patients (1 to 30 cfu/ml; 3, 4).

Automation of DNA extraction is crucial for its implementation as part of a diagnostic system in the routine laboratory. A broad selection of automated solutions is available for diagnostic purposes, including protocols for the isolation of DNA from particular microorganisms from a variety of clinical materials. However, to date only few automated solutions have been evaluated with the aim to be integrated into the molecular diagnostic pathway to an accurate and sensitive detection of sepsis pathogens in the blood [5–9].

In the following sections, protocols for the semiautomated extraction of microbial DNA from blood will be described. The system comprises of a manual part for the MolYsis™ (Molzym, Germany) technology-based pretreatment of samples for the removal of human DNA and enrichment of microorganisms and automated procedures for the extraction and purification of microbial DNA. The procedures supply enriched microbial DNA at quantities which allow the sensitive detection of pathogens by assays involving real-time PCR or PCR followed by microarray hybridization (Table 1). Here, three protocols are presented that enable the isolation of microbial DNA from blood culture material and from 1 ml and 5 to 10 ml whole blood.

2 Materials

2.1 Semiautomated Extraction of Blood Culture

1. Incubated blood culture bottles (BACTEC®, Becton Dickinson; BacT/ALERT®, bioMérieux).
2. Buffers and enzymes supplied with the kit *MolYsis*™ *Plus* (Molzym, Germany).
3. 2-Mercaptoethanol (14.3 Mol/l).
4. Disposable gloves and laboratory coats.
5. Adjustable micropipettes (up to 20 μl, up to 200 μl, and up to 1,000 μl) (*see* **Note 1**).

Table 1
Semiautomated isolation of microbial DNA and detection thresholds of bacteria spiked into samples

DNA isolation method	Sample	Spiked strain	Assay	Limit of detection (cfu/ml)	Reference
(a) *MolYsis*™ *Plus* kit + *NucliSens*® *easyMAG*® device	Blood culture medium (0.2 ml)	*S. aureus*	LC TaqMan (*tuf*) real-time PCR	10	[6]
(b) *SelectNA*™ *blood pathogen* kit + *SelectNA*™ device	EDTA blood (1 ml)	*E. coli*, *K. pneumoniae*, *S. aureus* (MRSA), *S. agalactiae*, *E.faecalis*, *L. monocytogenes*	PCR + microarray[a]	11–600	[8]
		E. coli, *S. aureus*	16S rDNA real-time PCR[b]	20–40	[c]
(c) *MolYsis*™ *Basic5* kit + *NucliSens*® *easyMAG*® device	EDTA blood (5 ml)	*P. aeruginosa*, *E. coli*, *S. aureus*, *E. faecium*, *K. pneumoniae* + *S. aureus*	PCR + microarray[d]	10–100	[7]

[a]Prove-it™ Bone and Joint assay (Mobidiag, Finland)
[b]Mastermix 16S Complete (Molzym, Germany)
[c]Own results
[d]In-house assay

6. Nucleic acid- and nuclease-free, aerosol-resistant pipette tips.
7. Sterile and DNA-free 2 ml microcentrifuge tubes.
8. Racks for tubes.
9. A cooling rack adjusted to −15 to −25 °C.
10. A vortex mixer.
11. A thermal shaker with a microcentrifuge tube adapter.
12. A desktop microcentrifuge (≥12,000 ×*g*).
13. Off-board *NucliSens*® lysis buffer (bioMérieux, France).
14. Magnetic silica beads (bioMérieux, France).
15. Onboard reagents of the *NucliSens*® *easyMAG*® instrument (bioMérieux, France).
16. A *NucliSens*® *easyMAG*® instrument (bioMérieux, France).

2.2 Semiautomated Extraction of Whole Blood (1 ml Samples)

1. Whole blood samples stabilized by anticoagulation agents, including EDTA, citrate, or heparin (*see* **Note 2**).
2. Off-board buffers, onboard buffer cartridges, and enzymes supplied with the kit *SelectNA*™ *Blood Pathogen* kit (Molzym).
3. 2-Mercaptoethanol (14.3 Mol/l).
4. Disposable gloves and laboratory coats.
5. Adjustable micropipettes (up to 20 μl, up to 200 μl, and up to 1,000 μl) (*see* **Note 1**).
6. Nucleic acid- and nuclease-free, aerosol-resistant pipette tips.
7. Sterile and DNA-free 2 ml microcentrifuge tubes.
8. Racks for tubes.
9. A cooling rack adjusted to −15 to −25 °C.
10. A vortex mixer.
11. A thermal shaker with a microcentrifuge tube adapter.
12. A desktop microcentrifuge (≥12,000 ×*g*).
13. A *SelectNA*™ instrument (Molzym).

2.3 Semiautomated Extraction of Whole Blood (5–10 ml Samples)

1. Whole blood samples stabilized by anticoagulation agents, including EDTA, citrate, or heparin (*see* **Note 2**).
2. Buffers and enzymes supplied with the kit MolYsis™ Basic10 (Molzym).
3. 2-Mercaptoethanol (14.3 Mol/l).
4. Disposable gloves and laboratory coats.
5. Adjustable micropipettes (up to 20 μl, up to 200 μl, and up to 1,000 μl) (*see* **Note 1**).
6. Nucleic acid- and nuclease-free, aerosol-resistant pipette tips.
7. Sterile and DNA-free 2 ml microcentrifuge tubes.

8. Racks for tubes.
9. A cooling rack adjusted to −15 to −25 °C.
10. A vortex mixer.
11. A thermal shaker with a microcentrifuge tube adapter.
12. A desktop microcentrifuge (≥12,000 × *g*).
13. A high-speed centrifuge with fixed angle rotor for 50 ml tubes (≥9,500 × *g*).
14. Disposable sterile 10 ml pipettes.
15. Sterile 50 ml Falcon tubes (Cellstar tubes, order no. 227261, Greiner Bio-One).
16. A *NucliSens*® *easyMAG*® instrument.

3 Methods

All procedures of the sample pretreatment are performed at room temperature (18–25 °C), except for enzymatic reactions as indicated. Keep enzymes at −15 to −25 °C by placing the vials in a cooling rack. Follow the instruction for working at conditions to avoid DNA contamination (*see* **Note 3**). For the setup of the DNA isolation system, follow the instructions for controlling the performance of the kit (*see* **Note 4**).

3.1 Blood Culture Sample Pretreatment and Automated Extraction by NucliSens® easyMAG®

1. The protocol is outlined for 0.2 ml blood culture from BACTEC® or BacT/ALERT® bottles. Pipette a liquid sample from an incubated blood culture bottle into a DNA-free 2 ml microcentrifuge tube. Use the components of the *MolYsis*™ *Plus* kit (Molzym) for the following **steps 1** through 7. Add 50 μl buffer CM and vortex at full speed for 10 s to mix thoroughly (*see* **Note 5**). Incubate on the bench for 5 min to lyse the human cells (*see* **Note 6**).
2. Add 50 μl buffer DB1 and 10 μl MolDNase A. Vigorously vortex for 10 s to mix and incubate for 15 min.
3. Centrifuge the tube at full speed for 10 min to sediment human cell debris and potentially present microorganisms. Remove the supernatant by using a 1 ml pipette tip, taking care not to disturb the sediment.
4. Add 1 ml buffer RS and resuspend the sediment by vigorous vortexing for 10 s. Alternatively, stir the sediment with the pipette tip and resuspend by pipetting in and out (*see* **Note** 7).
5. Centrifuge the tube at full speed for 5 min. Carefully remove the supernatant with a 1 ml pipette tip.
6. Add 80 μl buffer RL and 1.4 μl 2-mercaptoethanol to the sediment and resuspend by pipetting in and out (*see* **Note 8**). Finally, vortex for 10 s to homogenize.

7. Add 20 μl BugLysis solution and vortex for 10 s. Incubate tube in a thermal shaker at 37 °C under constant vigorous shaking (1,000 rpm) for 30 min.
8. Add the pretreated sample (100 μl) to a *NucliSens® Lysis* tube adjusted to room temperature. Mix by full-speed vortexing for 10 s. Add 140 μl magnetic silica (bioMérieux) and incubate at room temperature for 10 min.
9. Add the lysed sample into the vial row and transfer to the *NucliSens® easyMAG®* device. Select the specific B extraction protocol 2.0.1 and set the elution volume to 100 μl.

3.2 Whole Blood Sample Pretreatment and Automated Extraction by SelectNA™ Instrument: 1 ml Protocol

Use *SelectNA™ Blood Pathogen* Kit for sample pretreatment. Prepare the *SelectNA™* instrument as advised in the manual to be ready for the extraction of pretreated samples (see protocol below). Make sure that the instrument has been run through the UV decontamination program before the extraction of the samples.

1. Pipette 1 ml EDTA (or citrate or heparin) stabilized blood sample into a supplied DNA-free sample tube. Add 250 μl buffer CM and vortex at full speed for 10 s to mix (*see* **Note 7**). Incubate on the bench for 5 min to lyse human cells (*see* **Note 5**).
2. Add 250 μl buffer DB1 and 10 μl MolDNase B to the lysate. Vortex for 10 s and incubate on the bench for 15 min to degrade released human DNA.
3. Centrifuge the tube at full speed for 10 min to sediment human cell debris and potentially present microorganisms. Remove the supernatant by pipetting, taking care not to disturb the sediment.
4. Add 1 ml buffer RS and resuspend the sediment by full-speed vortexing (*see* **Note 7**). Alternatively, stir the sediment with the pipette tip and resuspend by pipetting in and out.
5. Centrifuge the tube at full speed for 5 min. Remove the supernatant by pipetting.
6. Add 80 μl buffer RL and 1.4 μl 2-mercaptoethanol to the sediment (*see* **Note 8**) in the 2 ml sample tube and resuspend by pipetting in and out. Vortex for 10 s to homogenize.
7. Transfer the sample tube to the *SelectNA™* instrument and follow the instructions of the manual.

3.3 Whole Blood Sample Pretreatment and Automated Extraction by NucliSens® easyMAG®: 5–10 ml Protocol

1. Use the *MolYsis™ Basic10* kit (Molzym) for sample pretreatment. For this, pipette the contents of a blood collection tube (EDTA, citrate, or heparin blood) into a sterile 50 ml Falcon tube and fill up by pipetting buffer SU to the 10 ml mark. Add 4.0 ml buffer CM, vortex at full speed for 10 s, and let stand for 5 min.
2. Add 4.0 ml buffer DB1 and 10 μl MolDNase B to the lysate and vortex at full speed for 10 s. Incubate on the bench for 15 min to degrade released human DNA.

3. Centrifuge the closed Falcon tube in a high-speed centrifuge at 9,500 × *g* for 10 min. Thereafter, carefully decant the supernatant.
4. Add 1 ml buffer RS and vortex until the sediment has been resuspended (*see* **Note 7**). Transfer the suspension by pipetting (1 ml pipette tip) into a sterile 2 ml microcentrifuge tube.
5. Continue with **step 5** of the protocol in Subheading 3.1.

4 Notes

1. Use only sterilized, guaranteed DNA-free disposables (e. g., Biosphere®, SARSTEDT, Germany, or Biopur®, Eppendorf, Germany). It is important that only filter tips are used to avoid contamination of the pipette by aerosols. Wear protective gloves and a disposable lab coat at any handling step during DNA preparation.
2. The stabilized blood should be processed for molecular analysis at the same day of collection. If this is not possible, the blood collection tube can be stored in the refrigerator (4–12 °C) for up to 3 days. Do not freeze the blood, because microbial cells tend to lyse as a result of freeze-thaw cycles. This in turn leads to the loss of microbial DNA because of the DNase treatment during sample pretreatment. If, for instance, samples are collected during a retrospective study and have to be stored for longer periods until extraction, the blood should be stabilized by a DNA-free cryoprotectant before freezing (for instance, UMD Tubes, Molzym).
3. In order to protect yourself from potentially present infectious agents and to keep a contamination-free environment, take care to perform extractions under a class II safety cabinet that has been sterilized by UV irradiation before use. Arrange a set of equipment needed for the processing of samples in the cabinet and keep there for future extractions. After working, clean the surface of the cabinet and the equipment with a DNA decontamination agent that is compatible with the surfaces (inspect the manual of the product). We routinely use DNA-ExitusPlus IF® (AppliChem, Germany). For reasons of avoidance of DNA contamination, it is important that the place of DNA extraction is well separated from places where PCR amplifications are performed and master mixes are set up. Ideally, these three places should be in separate laboratories. Frequently change protective gloves during handling.
4. Perform controls for the setup of the system to test the performance of the extraction procedure. For the determination of the detection threshold, prepare a dilution series of cultures of microorganisms, e. g., *S. aureus*, and spike into negative EDTA blood samples. The sensitivity threshold depends on

the strain, the blood volume extracted, and the assay used. In order to obtain results that are relevant for the detection of systemic infections in septicemic patients, make sure that your analytical system is sensitive enough to detect ≤10 *E. coli* genome equivalents per assay (approx. 50 fg *E. coli* genomic DNA). Examples of sensitive assays and thresholds reached are given in Table 1. Cross contamination during extraction should be tested by extracting negative EDTA blood samples.

5. Vortexing is crucial for the optimal performance of the procedure as a whole. In particular, quantitative lysis of human cells and degradation of released DNA is only achieved when buffer CM (**step 1**) and DB1-MolDNase B (**step 2**) are homogenously mixed with the blood sample.
6. To avoid cross contamination of samples during pretreatment and extraction, it is important to always clear the lid of the microcentrifuge tube by a pulse centrifugation after each step involving vortexing (for an image, *see* Fig. 2 in previous chapter).
7. Depending on the blood sample, the sediment may be rigid and resuspension may take some time. Resuspension is important to wash away components inhibitory to subsequent enzymatic processes, including cell wall hydrolysis and protein degradation.
8. Take care not to inhale 2-mercaptoethanol. Practically, premix 80 μl buffer RL and 1.4 μl 2-mercaptoethanol or multiples of the components for more than one sample in a DNA-free microcentrifuge tube under an extractor hood. Pipette 80 μl to the sediment of the sample tube and continue with the protocol.

References

1. Podnecky NL, Elrod MG, Newton BR et al (2013) Comparison of DNA extraction kits for detection of *Burkholderia pseudomallei* in spiked human whole blood using real-time PCR. PLoS One 8:58032. doi:10.1371/journal.pone.0058032
2. Hansen WLJ, Bruggeman CA, Wolffs PFG (2009) Evaluation of new preanalysis sample treatment tools and DNA isolation protocols to improve bacterial pathogen detection in whole blood. J Clin Microbiol 47:2629–2631
3. Loonen AJM, Jansz AR, Kreeftenberg H et al (2011) Acceleration of the direct identification of *Staphylococcus aureus* versus coagulase-negative staphylococci from blood culture material: a comparison of six bacterial DNA extraction methods. Eur J Clin Microbiol Infect Dis 30:337–342
4. Wiesinger-Mayr H, Jordana-Lluch E, Martró E et al (2011) Establishment of a semi-automated pathogen DNA isolation from whole blood and comparison with commercially available kits. J Microbiol Methods 85:206–213
5. Laakso S, Mäki M (2013) Assessment of a semi-automated protocol for multiplex analysis of sepsis-causing bacteria with spiked whole blood samples. Microbiologyopen 2:284–292
6. Hansen WLJ, Bruggeman CA, Wolffs PFG (2013) Pre-analytical sample treatment and DNA extraction protocols for the detection of bacterial pathogens from whole blood. Methods Mol Biol 943:81–90
7. Loonen AJM, Bos MP, van Meerbergen B et al (2013) Comparison of pathogen DNA isolation methods from large volumes of whole blood to improve molecular diagnosis of bloodstream infections. PLoS One 8:e72349. doi:10.1371/journal.pone.0072349
8. Kellogg JA, Manzella JP, Bankert DA (2000) Frequency of low-level bacteremia in children from birth to fifteen years of age. J Clin Microbiol 38:2181–2185
9. Yagupsky P, Nolte FS (1990) Quantitative aspects of septicemia. Clin Microbiol Rev 3: 269–279

Chapter 13

Broad-Range PCR for the Identification of Bacterial and Fungal Pathogens from Blood: A Sequencing Approach

Eva Leitner and Harald H. Kessler

Abstract

Broad-range PCR has become a valuable tool for the identification of microorganisms in the clinical laboratory over the last years. It was primarily used to identify slow-growing and fastidious microorganisms with poor biochemical activity. Nowadays, it is also used to identify microorganisms directly from clinical samples such as blood or punctuates from primarily sterile body sites. In these specimens, the usage of broad-range PCR is challenging regarding contamination and standardization. To overcome these problems, a new test system, the SepsiTest™, was introduced recently employing broad-range PCR for the identification of microorganisms in septic patients. In this chapter, the test system is described and the equipment necessary listed.

Key words Sepsis, Broad-range PCR, Sequencing, SepsiTest, Identification, Bacteria, Fungi

1 Introduction

Broad-range gene polymerase chain reaction (PCR) including sequencing of the amplification product developed originally in the environmental microbiology has become a valuable diagnostic tool in clinical microbiology over the last years [1–3]. For prokaryotes, the ribosomal RNA (rRNA) genes (5S, 16S, and 23S) and intragenic regions are commonly used for taxonomic purposes. The most favored gene for broad-range PCR has become the 16S rRNA gene which is about 1,550 bp long including conserved and variable regions [1]. The 16S rRNA gene is not only present in all living organisms but also in several copies distributed over the genome, thereby increasing the sensitivity for this target [4].

Eukaryotic ribosomal genes are similarly designed with conserved and variable regions with the operon including four ribosomal genes (5S, 5.8S, 18S, and 28S). The internal transcribed spacer (ITS) and parts of the 28S rRNA gene are most widely used for fungal identification [3, 5].

Nicasio Mancini (ed.), *Sepsis: Diagnostic Methods and Protocols*, Methods in Molecular Biology, vol. 1237,
DOI 10.1007/978-1-4939-1776-1_13,

For broad-range DNA sequencing, a universal PCR using primers in the regions mentioned above is run, followed by sequencing of the amplification product. Advances during the past years lead to improved quality of the sequences and made sequencing technology (Sanger sequencing) available even in small laboratories [1, 2]. For identification of the microorganism, the sequence obtained is matched with sequences provided by databases. For the Basic Local Alignment Search Tool (BLAST) search, several open-access DNA sequence databases are available such as CMR (http://cmr.jcvi.org/tigr-scripts/CMR/CmrHomePage.cgi), EMBL (http://www.ebi.ac.uk/ena/), GenBank (http://www.ncbi.nlm.nih.gov/genbank/), and RDP (http://rdp.cme.msu.edu/). These databases may differ in quality of the sequences and should be selected with the understanding of their strengths and limitations [1, 5].

Recently, a new test system was introduced employing broad-range PCR for the identification of microorganisms in septic patients [6, 7]. The SepsiTest™ (Molzym, Bremen, Germany) is an in vitro diagnostic (IVD)/Conformité Européene (CE)-labeled test system consisting of a DNA extraction and amplification/detection assay. The DNA extraction assay includes spins, enzymes, and buffers for enzymatic lysis of cells and extraction of bacterial and fungal DNA. The amplification/detection assay includes three PCR mixes (for bacteria, yeasts, and the external control), the DNA polymerase, the DNA staining solution, and sequencing primers for bacteria and fungi. The SepsiTest™ fulfills the majority of recommendations regarding DNA-free reagents and controls and contamination-free workflow. Furthermore, DNA extraction can be performed on an automated platform [8]. In this chapter, the protocol employing manual DNA extraction is described.

2 Materials

2.1 DNA Extraction

2.1.1 Instruments Required (Manual DNA Extraction)

1. Laminar flow.
2. Vortex.
3. Microcentrifuge.
4. Thermomixer.

2.1.2 SepsiTest™ Reagents as Provided by the Manufacturer (Molzym)

1. DNA extraction assay.
2. Amplification/detection assay.

2.2 DNA Amplification/Detection

2.2.1 Instruments Required

1. UV workstation.
2. Vortex.
3. Microcentrifuge.
4. Real-time PCR instrument.

2.2.2 SepsiTest™ Reagents (Molzym)

1. Master mix components.
2. PCR positive control DNA.

2.3 Sequencing

2.3.1 Instruments Required

1. Vortex.
2. Microcentrifuge.
3. Thermocycler.
4. Sequencer for Sanger sequencing.

2.3.2 Reagents

1. PCR purification kit.
2. SepsiTest™ sequencing primers (Molzym).
3. Cycle sequencing kit.

In all workstations, precise pipettes up to 10 μl, up to 20 μl, up to 200 μl, and up to 1,000 μl with compatible disposable filter-tip pipette tips and cooling racks must be available. Additionally, sterile disposables including 0.2-ml microtubes and 1.5-ml tubes with cap, real-time amplification vials, gloves, and sleeves are necessary. Except of the SepsiTest™ test system, suppliers are not mentioned here because additional equipment and disposables are selectable up to user.

3 Methods

The workup must be done according to the guidelines for molecular diagnostics, e.g., separated workstations for DNA extraction, master mix preparation, and PCR amplification and detection. The workstations must be decontaminated from DNA carefully. DNA-free disposables must be used during the whole procedure. DNA-free environment during opening of the microtubes and reagent containers as well as master mix handling must be warranted to avoid laboratory contamination.

3.1 Sample Preparation Employing the Manual DNA Extraction Protocol

Put the vials containing the specific buffers required for extraction into a reagent rack according to the sequence of steps (*CM*→ *DB1* → *RS*→ *RL*→ *RP*→ *CS*→ *AB*→ *WB*→ *WS*→ *ES* buffers).

Take a cooling rack for MolDNase B, BugLysis, and proteinase K.

Heat the thermomixer to 37 °C.

1. Use a sample preparation rack and label two blood sample (BS) tubes per specimen for duplicate extraction and transfer 1.0 ml of fresh whole EDTA blood into each tube.
2. Add 250 μl of buffer *CM* to each BS tube. After vortexing at 12,000 × *g* for 15 s, keep at room temperature for 5 min. (*CM is a chaotropic buffer that lyses blood cells. Note: CM buffer is irritating.*)
3. Spin down and add 250 μl of buffer *DB1* to the BS tubes.

4. Add 10 µl of MolDNase B (must be stored and used at −20 °C) to the lysate, vortex immediately for 15 s, and keep the BS tubes at room temperature for 15 min.
5. Centrifuge the BS tubes at 12,000 × *g* for 10 min.
6. Remove the supernatant carefully and discharge.
7. Add 1 ml of buffer *RS* to the pellet and resuspend it by stirring with the pipette tip by pipetting up and down (remaining material on the tip must be stripped off), and finally vortex for homogenizing.
8. Centrifuge the BS tubes at 12,000 × *g* for 5 min.
9. Remove the supernatant carefully and discharge.

After this step, the procedure can be interrupted by freezing the sample at −20 °C. For further processing, thaw the sample to room temperature and proceed with the next step.

10. Add 80 µl of buffer *RL* to the BS tubes and resuspend the pellet (*see* **step** 7).
11. Spin down and add 20 µl of BugLysis (must be stored and used at −20 °C).
12. Add 1.4 µl of ß-mercaptoethanol (*Note: ß-mercaptoethanol is toxic*).
13. Vortex the tubes for 15 s.

After this step, you may continue with automated DNA extraction if a suitable platform is available.

14. Incubate in the thermomixer at 37 °C and 1,000 rpm for 30 min.
15. After incubation, set the thermomixer at 56 °C (required for the next step).
16. After spinning the BS tubes briefly, add 150 µl of buffer *RP* and 20 µl of proteinase K (should be stored and used at −20 °C).
17. After vortexing for 15 s, incubate in the thermomixer at 56 °C and 1,000 rpm for 10 min.
18. After incubation, set the thermomixer at 70 °C (required for the elution step) and put the vial *ES buffer* into the thermomixer.
19. After spinning the BS tubes briefly, add 250 µl of buffer *CS* to each of the BS tubes and vortex for 15 s at full speed.
20. After spinning the BS tubes briefly, add 250 µl of buffer *AB* to each of the BS tubes and vortex for 15 s at full speed.
21. After spinning the BS tubes briefly, transfer each lysate into the spin columns (SC) that had been put into 2.0-ml collection tubes (CT).

22. Close the cap of the CT and centrifuge at 12,000 × *g* for 60 s.
23. Remove the SC from the centrifuge. Open the lid and remove the SC to place it in a newly supplied 2.0-ml CT. Discharge the CT containing the fluid.
24. Add 400 μl of buffer *WB* to the SC and repeat **step 22**.
25. Add 400 μl of buffer *WS* to the SC and centrifuge at 12,000 × *g* for 3 min.
26. Remove the closed CT from the centrifuge carefully! Remove the SC from the CT and place it into a supplied sterile 1.5-ml elution tube (ET). Discharge the CT containing the fluid.
27. Add 100 μl of the preheated buffer *ES* into the center of the SC. Close the cap of the ET and incubate at room temperature for 1 min.
28. Centrifuge at 12,000 × *g* for 1 min to elute the DNA.
29. Remove the SC from the ET and close the cap. Discard the SC.
30. Store the ET containing the eluate at 4 °C if analyzed within 24 h or freeze at −20 °C until further use.

3.2 Real-Time PCR Employing the LightCycler 2.0 (Roche Diagnostics GmbH, Mannheim, Germany)

1. Thaw the vials *H_2O* (DNA-free water), *MA Bac* (assay bacteria), *MA Yeasts* (assay fungi), *MA IC* (external control), and *DS* (DNA staining solution, SYBR® Green 1) to room temperature.
2. Vortex the vials for a few seconds and spin briefly.
3. Place the vial containing the DNA polymerase (*MolTaq 16S*) in a cooling rack (−20 °C).
4. Put the reagents according to the sequence of steps required as well as the glass capillaries into the adapter block (4 °C).
5. For each master mix (bacteria, fungi, and external control), use a 1.5-ml microtube and prepare master mixes according to Table 1.

Table 1
Bacteria, fungi, and external control PCR mix for one reaction

Reagent	Volume (μl)
H_2O	6.0
MA[a]	8.0
DS	2.0
MolTaq 16S	0.8
Total volume	16.8

[a]MA: bacteria (16S), yeasts (18S), external control

Table 2
Real-time PCR program

Program step	Cycles	AM	TT (°C)	IT (s)	TTR (°C/s)	STT	SS	SD	AM
Denaturation	1	None	95	60	20.00	0	0	0	None
PCR	40	Quantification	95	5	20.00	0	0	0	None
			55	5	20.00	0	0	0	None
			72	25	20.00	0	0	0	Single
Melting curve	1	Melting curves	95	0	20.00	0	0	0	None
			65	15	20.00	0	0	0	None
			95	0	0.05	0	0	0	Continuous
Cooling	1		40	5	20.00	0	0	0	None

AM analysis mode, *TT* target temperature, *IT* incubation of time, *TTR* temperature transition rate, *STT* secondary target temperature, *SS* step size, *SD* step delay, *AM* acquisition on mode

6. After the preparation of the master mixes, mix them gently by pipetting up and down. Spin briefly.
7. Pipet 16 μl each of the masters mixes into the the 20-μl LightCycler capillaries. Start with pipetting 5 μl of the negative control, seal it, and continue with 5 μl of the eluates and 5 μl of the positive control.
8. For the real-time PCR, the protocol is shown in Table 2.

The PCR analysis is performed using the absolute quantification mode and the T_m calling mode in the fluorimeter channel 530 with color compensation to evaluate the crossing points (Cp) and the T_m peak, respectively. A true positive result is defined as follows: the external control appears at the expected value at a Cp 18 ± 2, the sample shows a positive result in the melting curve analysis, and the negative control does not show any signal (except of primer dimers at T_m between 78 and 82 °C) (Figs. 1 and 2). A low Cp value of the sample indicates a high pathogen titer in blood, and vice versa.

3.3 Sequencing Employing the Applied Biosystems Sequencer 3130 Genetic Analyzer (Life Technologies Corporation, Carlsbad, CA, USA)

Sequencing is performed if a true positive result is obtained with the real-time PCR:

1. After real-time PCR amplification, the PCR mix is transferred into a 1.5-ml microtube by inverted brief spinning of the capillary in a microcentrifuge.
2. Subsequently, amplification products are purified with the QIAquick® PCR Purification Kit (QIAGEN, Hilden, Germany).
3. Add 100 μl of buffer *PB* to the PCR mix.

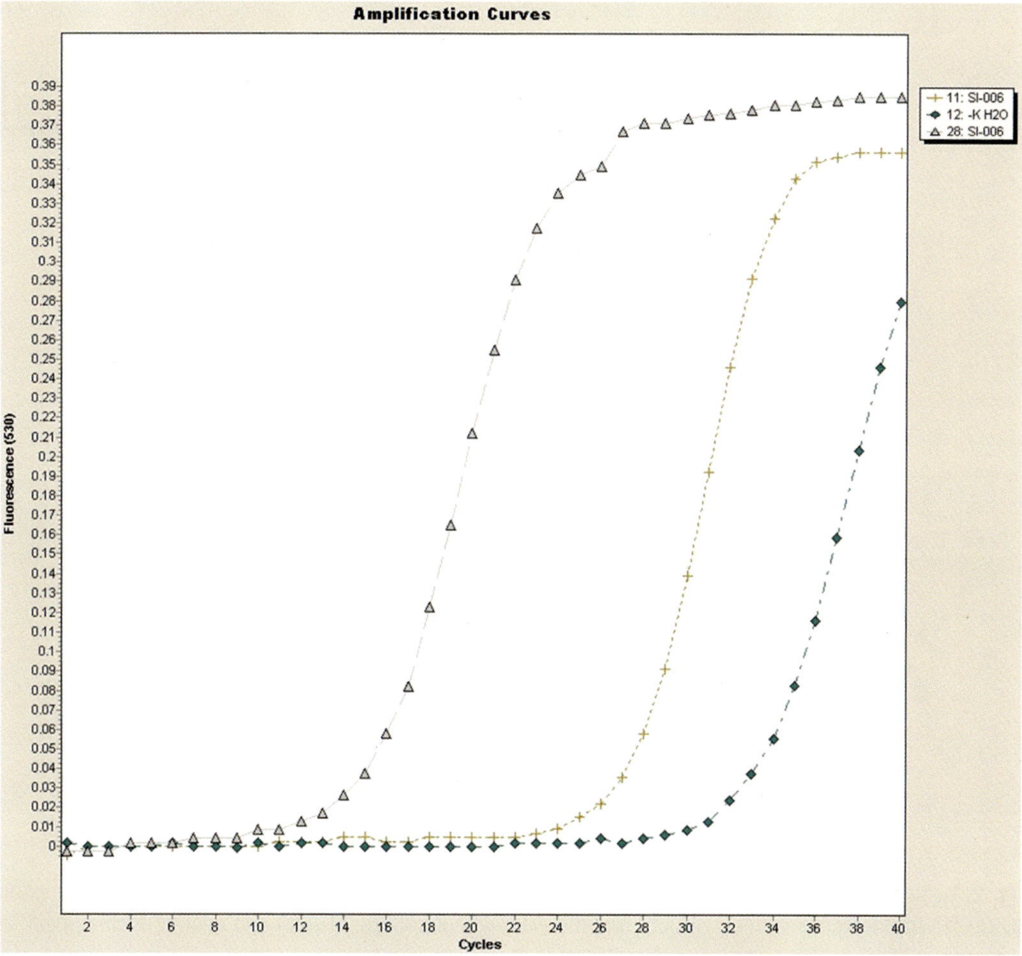

Fig. 1 Amplification curves in the qualitative detection mode (channel 530) showing positive results for the specimen (11: SI-006) and the external control (28: SI-006) and a negative result for the negative control (12: -K H_2O) with a slope due to the formation of primer dimers

4. Place the QIAquick SC in the 2-ml CT provided and add the mixed sample.
5. Centrifuge at 12,000 × *g* for 60 s.
6. Discharge the fluid and place the SC back into the same CT.
7. Add 750 μl of buffer *PE* to the SC and centrifuge at 12,000 × *g* for 60 s.
8. Discard the fluid and place the SC back in the same CT.
9. Centrifuge again for 60 s to remove residual buffer.
10. Place the SC in a sterile 1.5-ml tube and add 30 μl of buffer *EB* to the center of the SC.
11. Leave at room temperature for 60 s and then centrifuge at 12,000 × *g* for 60 s.

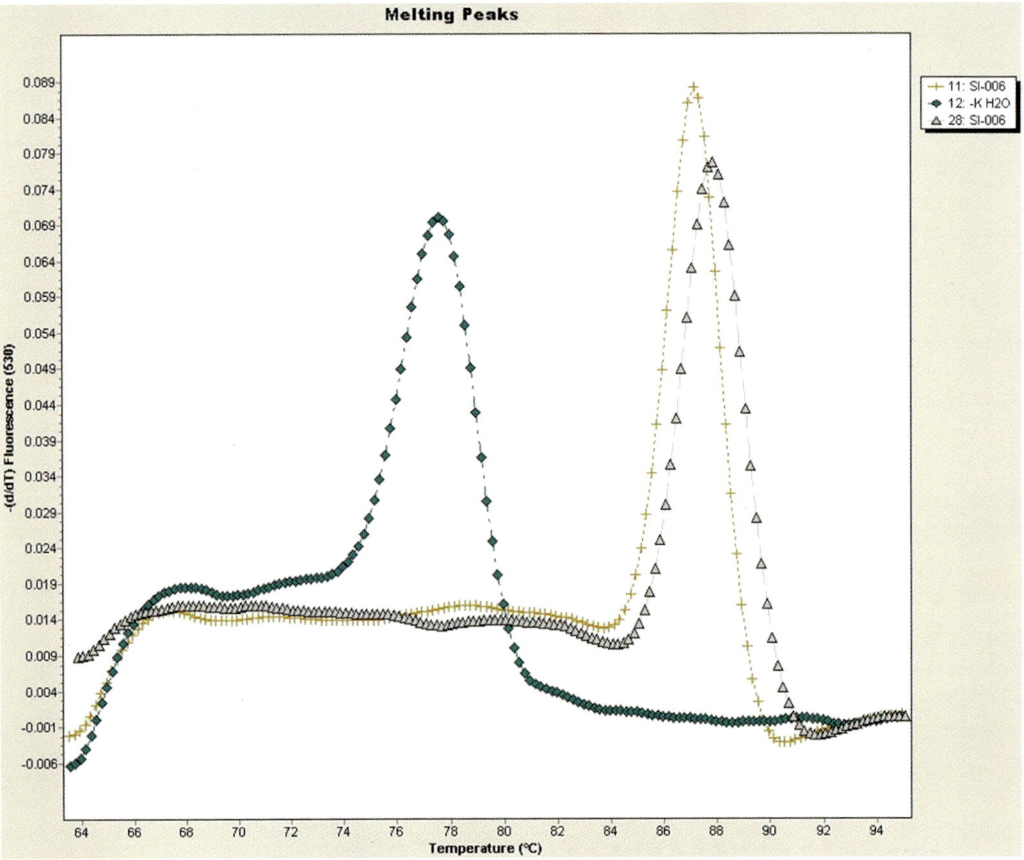

Fig. 2 Melting curves in the T_m calling mode (channel 530) showing melting peaks as expected (within 86–90 °C) for the specimen (11: SI-006) and the external control (28: SI-006) and another melting peak as expected (<82 °C) for the negative control (12: -K H_2O)

For the chain terminator reaction, the BigDye Terminator v3.1 and the sequencing primers from the SepsiTest™ are used. The BigDye Terminator v3.1 Cycle Sequencing Kit (Life Technologies Corporation) includes reagent components for the chain terminator reaction in a ready, premixed format. The only manual step is to provide the template and the template-specific primers in order to perform a fluorescent-based chain terminator reaction.

The SepsiTest™ Kit includes forward sequencing primers for gram-positive bacteria (SeqGP16), gram-negative bacteria (SeqGN16), and fungi (SeqYeast18). Table 3 shows the PCR mix for the chain terminator reaction.

The PCR program for the generation of DNA fragments consists of the following steps: denaturation at 96 °C for 60 s, followed by 30 cycles of 15 s at 96 °C, 30 s at 53 °C, and 90 s at 60 °C, and a final cool down withholding at 4 °C.

Table 3
PCR mix for the chain terminator reaction

Reagent	Volume (μl)
BigDye v3.1	1.8
5× Sequencing buffer	2.0
Primer (10 pmol)	1.0
DNA template	3.0
ddH_2O	2.2
Total volume	10.0

For purification after the chain terminator reaction, ethanol precipitation is used. After the addition of Hi-Di™ Formamide, the sample is sequenced with the Applied Biosystems Sequencer 3130 Genetic Analyzer according to the manufacturer's instructions.

Sequences obtained are analyzed with the Sequencing Analysis Software v5.2 and a BLAST search using the Molzym SepsiTest™ database (http://www.sepsitest-blast.de/de/index.html).

4 Notes

1. Clinicians need to collect the sample carefully according to the existing guidelines to avoid skin contaminations. If combined with blood culture collection, the specimen for blood culture must be collected prior to the whole blood specimen for the SepsiTest™.
2. To avoid contamination during the DNA extraction process, special attention needs to be paid to DNA-free workstations, consumables, and workflow. Each bacterial or fungal DNA present in the neighborhood is detectable with this test system!
3. Weak positive samples with indeterminate sequencing result due to the insufficient amount of DNA for the sequencing reaction were usually judged negative in different studies, although they should have been interpreted as positive without identification result.
4. The SepsiTest™ database is offered for BLAST search by the manufacturer. We recommend using a second BLAST search tool, especially in samples with poor sequencing result. In those samples, interpretation with the SepsiTest™ database may not be possible due to the too short sequence.

5. Limitations of the test systems include:
   (a) Neither reagent control nor internal control is implemented in the test system from the beginning onward. The use of a reagent control is recommended, although it may increase cost.
   (b) Only forward sequencing primers are offered. Reverse primers would increase the quality of sequences significantly.
6. With the use of broad-range PCR, especially when used directly in clinical samples, challenges regarding contamination and standardization of this method have been reported [2, 9, 10]. To overcome this, different recommendations regarding sample collection, controls, and correct interpretation of results are available [2, 5].

References

1. Clarridge JE 3rd (2004) Impact of 16S rRNA gene sequence analysis for identification of bacteria on clinical microbiology and infectious diseases. Clin Microbiol Rev 17:840–862, table of contents
2. Sontakke S, Cadenas MB, Maggi RG, Diniz PP, Breitschwerdt EB (2009) Use of broad range16S rDNA PCR in clinical microbiology. J Microbiol Methods 76:217–225
3. Atkins SD, Clark IM (2004) Fungal molecular diagnostics: a mini review. J Appl Genet 45: 3–15
4. Klappenbach JA, Saxman PR, Cole JR, Schmidt TM (2001) rrndb: the ribosomal RNA operon copy number database. Nucleic Acids Res 29:181–184
5. CLSI, Clinical and Laboratory Standards Institute (2008) Interpretive criteria for identification of bacteria and fungi by DNA target sequencing; approved guideline. CLSI dokument MM18-A (ISBN 1-56238-664-6)
6. Wellinghausen N, Kochem AJ, Disque C, Muhl H, Gebert S, Winter J, Matten J, Sakka SG (2009) Diagnosis of bacteremia in whole-blood samples by use of a commercial universal 16S rRNA gene-based PCR and sequence analysis. J Clin Microbiol 47:2759–2765
7. Leitner E, Kessler HH, Spindelboeck W, Hoenigl M, Putz-Bankuti C, Stadlbauer-Kollner V, Krause R, Grisold AJ, Feierl G, Stauber RE (2013) Comparison of two molecular assays with conventional blood culture for diagnosis of sepsis. J Microbiol Methods 92:253–255
8. Wiesinger-Mayr H, Jordana-Lluch E, Martro E, Schoenthaler S, Noehammer C (2011) Establishment of a semi-automated pathogen DNA isolation from whole blood and comparison with commercially available kits. J Microbiol Methods 85:206–213
9. Sleigh J, Cursons R, La Pine M (2001) Detection of bacteraemia in critically ill patients using 16S rDNA polymerase chain reaction and DNA sequencing. Intensive Care Med 27:1269–1273
10. Fenollar F, Raoult D (2007) Molecular diagnosis of bloodstream infections caused by non-cultivable bacteria. Int J Antimicrob Agents 30(Suppl 1):S7–S15

Chapter 14

Real-Time PCR-Based Identification of Bacterial and Fungal Pathogens from Blood Samples

Madeleine Mai, Iris Müller, Daniela Maneg, Benedikt Lohr, Achim Haecker, Gerd Haberhausen, and Klaus-Peter Hunfeld

Abstract

Latest major contributions in the field of sepsis diagnostics result from advances in PCR technologies permitting new standards in speed and quality, given the fact that a timely diagnosis is the decisive factor to the survival of patients with bloodstream infections.

Multiplex real-time PCR is a quantitative method for simultaneous amplification and detection of different targeted DNA molecules within hours. Nevertheless, various studies have shown a number of technical shortcomings as well as a high heterogeneity in sensitivity.

The present method allows the standardized and rapid detection and identification of 25 common bacteria and fungi responsible for bloodstream infections from whole blood samples by using LightCycler® SeptiFast (LC-SF) test, based on real-time PCR.

Key words Multiplex real-time PCR, Polymerase chain reaction, LightCycler SeptiFast test, Bloodstream infection, Whole blood, Rapid diagnosis, Sepsis, MagNA pure compact instrument

1 Introduction

The survival rate of septic patients mainly depends on a rapid and reliable diagnosis, since in cases of severe sepsis there is an average 7.6 % decrease in survival rate per hour from the onset of hypotension without effective antimicrobial treatment [1, 2]. Nevertheless, as the gold standard for identification of bloodstream pathogens so far, blood culture takes typically ≥24 h to obtain positive results with a case-dependent sensitivity as being incapable to detect certain pathogens at all or only after a certain time period [3, 4]. Early-stage treatment with antibiotics or the presence of fastidious pathogens may result in negative blood cultures and low sensitivity [3, 5]. Molecular biological techniques represent promising options for a rapid detection and identification of bloodstream pathogens and their inactivated fragments. Multiplex real-time PCR allows speeding up the

Nicasio Mancini (ed.), *Sepsis: Diagnostic Methods and Protocols*, Methods in Molecular Biology, vol. 1237,
DOI 10.1007/978-1-4939-1776-1_14, © Springer Science+Business Media New York 2015

Table 1
Master list of detectable *microorganisms* (Table modified from [7])

Gram negative	Gram positive	Fungi
Escherichia coli	*Staphylococcus aureus*	*Candida albicans*
Klebsiella (*pneumoniae/ oxytoca)*	*Coagulase negative Staphylococci* (include *S.epidermidis, S. haemolyticus*)	*Candida tropicalis*
Serratia marcescens	*Streptococcus pneumoniae*	*Candida parapsilosis*
Enterobacter (cloacae/ aerogenes)	Streptococcus *spp.* (include *S. pyogenes, S. agalactiae, S. mitis*)	*Candida krusei*
Proteus mirabilis	*Enterococcus* faecium	*Candida glabrata*
Pseudomonas aeruginosa	*Enterococcus* faecalis	*Aspergillus fumigatus*
Acinetobacter baumannii		
Stenotrophomonas maltophilia		

diagnosis of sepsis with a sensitivity up to 80 % and a specificity up to 95 % in cases of bacteremia, and a sensitivity up to 61 % and a specificity up to 99 % in cases of fungemia [6].

In this chapter, a protocol is described for the LightCycler® SeptiFast (LC-SF) test based on multiplex real-time PCR for detection and identification of 25 pathogens, including gram-positive and gram-negative bacteria and fungi. This protocol follows the protocol described by Lehmann et al. [7] with an improvement concerning the automation of the sample preparation. The automation by using a MagNA Pure Compact Instrument reduces the turnaround time (TAT) from 6 to 3.5–5 h without impairing sensitivity and specificity [8].

The following three steps will be described:

Step 1. First, the blood samples are prepared for automated nucleic acid isolation by using the MagNA Lyser (15 min). DNA extraction and purification are automatically executed using MagNA Pure Compact extraction instrument with a TAT of 30 min [8].

Step 2. The LC-SF is performed with three parallel PCR mixes for gram-positive bacteria, gram-negative bacteria, and fungi using a hot start Taq polymerase for amplification [7]. According to the 25 species of the master list (Table 1), either universal or specific primers amplify internal transcribed spacer regions (ITS) of the ribosomal DNA. The target sequences are located between the 16S and the 23S ribosomal DNA sequences of gram-negative and gram-positive bacteria, and between the 18S and 5.8S ribosomal sequences of fungi. During the amplification, the increase of the specific real-time PCR products was determined by using sequence-specific DNA probes labeled with four different dyes and by automated measurement of the resulting fluorescence [7].

Step 3. The identification of species and controls is performed automatically by melting curve analysis of the hybridization probes. The probes reliably discriminate between the different species by use of specific melting temperatures of the corresponding amplicons. The melting temperature depends on fragment length, composition of sequence, and degree of homology between the hybridization probe and the target DNA [7].

The three steps can be performed in 3.5–5 h, depending on the number of samples.

2 Materials

2.1 Sample

1. 1.5 ml minimum sample volume of non-centrifuged whole blood.
2. Sterile ethylenediaminetetraacetic acid (EDTA)-containing monovettes.

2.2 PCR Kits and Buffers

All used PCR reagents are commercially available from Roche Diagnostics:

1. LightCycler® Septi*Fast* Kit MGRADE (reference number (RN) 04 469 046 001), including internal control (IC) and negative control, detection mix: primers and probes (DM G+; DM G−; DM F), deoxynucleoside triphosphates, hot start *Taq* polymerase (RM 1a and RM 1b), adenosine triphosphates (ATP), buffers, Mg^{++}, reagent controls (RC G+,RC G−, RC F), AmpErase (uracil-*N*-glycosylase).
2. Septi*Fast* Lys Kit MGRADE (RN 04 404 432 001).
3. MagNA Pure Compact Nucleic Acid Isolation Kit 1 (RN 03 730 972 001), including cartridges, tip trays, 300 μl protease and chaotropic lysis buffer, magnetic glass particles (MGP), wash buffer and elution buffer.
4. Phosphate-buffered saline (PBS).
5. LightCycler® Multicolor Compensation Set.

2.3 Instrumentation

1. Septi*Fast* Cooling Block.
2. LightCycler® Capillaries MGRADE (100 μl).
3. Micro tubes 2.0 ml Type H/MagNA Pure Compact sample tubes.
4. MagNA Lyser Instrument.
5. MagNA Pure Compact Instrument.
6. LightCycler® 2.0 Instrument.
7. LC Carousel Centrifuge 2.0.
8. SeptiFast Software Set V2.0.

3 Methods

3.1 Multiplex PCR Assay Scheme

The assay scheme is a three-step procedure:

1. Lysis/purification of DNA.
2. Real-time PCR amplification and detection of PCR products.
3. Identification of species and controls.

3.2 Lysis/Purification of DNA

3.2.1 Preparation

1. Carry out all procedures after cleaning the bench and performing the cleaning procedure of the MagNA Pure Compact to decontaminate instruments from bacterial DNA including exposure with ultraviolet light and by using decontamination reagents according to laboratory standards (*see* **Note 1**).
2. Allow refrigerated or frozen patient samples (*see* **Notes 2–5**), internal control (IC) (*see* **Note 6**), and negative control (NC) to thaw at room temperature.
3. Mix the IC gently to ensure homogeneity and centrifuge the homogenate.
4. Place all Master Mix tubes on the Septi*Fast* Cooling Block for thawing at room temperature.
5. Ensure that an actual color compensation run was performed (*see* **Note 7**).

3.2.2 MagNA Lyser®

1. Transfer 1,500 μl EDTA whole blood samples and the NC into the SeptiFast Lys Kit vials.
2. Transfer the vials into the MagNA Lyser Instrument (*see* **Note 8**).
3. Store the lysed samples for 10 min at room temperature to allow settling of ceramic beads and separation of cell debris.

3.2.3 MagNA Pure Compact

1. Transfer 500 μl of the lysed patient samples into the MagNA Pure Compact sample tubes.
2. To prepare the NC, place 400 μl of the PBS buffer into the MagNA Pure Compact sample tube; add 200 μl of the lysed NC, gently vortex the mix, and centrifuge it. You have to discard 100 μl to get the needed 500 μl of the prepared NC.
3. Add 4 μl of the IC to the patient sample tubes and the NC.
4. Load all tubes after brief vortexing, for centrifugation onto the MagNA Pure Compact Instrument (*see* **Note 9**).
5. Insert the necessary equipment (elution tubes, tip trays, reagent cartridges) from the MagNA Pure Compact Nucleic Acid Isolation Kit 1 into the instrument.
6. Open the tubes (*see* **Note 10**).

7. Choose the Bacteria Protocol DNA Bacteria V3.2 (sample 400 μl, eluate 200 μl, IC none) and follow the instructions according to the manual (*see* **Note 11**).
8. After the run, close the eluate tubes and place them on the clean bench.

3.3 Real-Time PCR Amplification and Detection of PCR Products

1. Carefully decontaminate the Septi*Fast* Cooling Block (*see* **Note 12**) before placing it onto the clean bench.
2. Vortex and centrifuge the detection mix tubes (except the RM 1a) before fitting them directly into the provided recesses on the Septi*Fast* Cooling Block.
3. Open RM 1b, RM 1a, DM G+, DM G−, and DM F.
4. Pipette 200 μl of the assembled RM 1a into DM G+, DM G−, and DM F to get the Master Mix MM G(+), MM G(−), and MM F (*see* **Note 13**). A homogenous mixture will be achieved by gently and frequently pipetting up and down.
5. Place the capillaries into the provided recesses: three capillaries for the RCs (position 1,2,3 on the Cooling Block), three capillaries for the NC eluate (position 4,5,6), and three capillaries for the patient eluates (position 7,8,9).
6. Pipette 50 μl MM G(+) in the capillaries labeled with G(+).
7. Pipette 50 μl MM G(−) in the capillaries labeled with G(−).
8. Pipette 50 μl MM F in the capillaries labeled with F.

   Prepare three PCR mixes for each eluted sample: gram-positive bacteria, gram-negative bacteria, and fungi. Pipette the patient eluate into the three capillaries labeled G(+), G(−), and F (*see* **Note 14**).
9. Mix the content of each sample by carefully pipetting up and down and use a separate pipette or pipette tip for each sample.
10. Close the capillaries with the Capping Tool.
11. After repeating **steps 9–10** for all eluates, place the NC eluate in the capillaries (position 4,5,6).
12. Transfer the capillaries into the LightCycler 2.0 Instrument after centrifugation in the LC Carousel Centrifuge 2.0.
13. Load the capillaries which can start now the run through the LightCycler 2.0 Instrument.

3.4 Identification of Species and Controls

1. After the end of the run, mark manually the melting curves with vertical sliders.
2. After setting a slider, the melting point (T_m) value and the corresponding peak height are automatically calculated (Fig. 1).
3. The finished file containing the amplicons resulting from PCR reactions in specimens and controls can be edited in the Septi*Fast* Identification Software (SIS) for analyzing and interpretation (*see* **Note 15**).

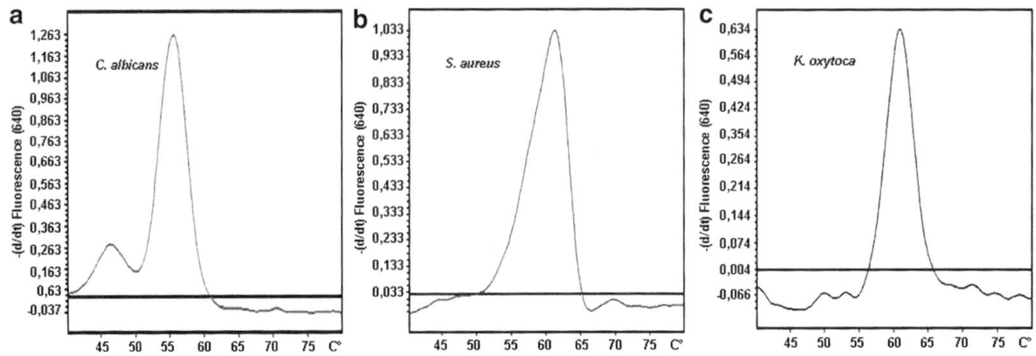

Fig. 1 Examples of characteristic melting peaks and melting curve registrations with microorganism identification by respecting melting temperatures (internal control: T_m = 46 °C). (**a**) *C. albicans* (T_m = 55 °C), (**b**) *S. aureus* (T_m = 62 °C), and (**c**) *K. oxytoca* (T_m = 61 °C). Figure modified from [7]

4 Notes

The following limitations of the LC-SF should always be taken into account:

(a) There are only 25 different pathogens detectable which are defined in the Septi*Fast* Test master list (Table 1). Although a number of about 20–25 species cover over 90 % of sepsis-causing pathogens [9], it has to be noted that there are other pathogens which cannot be detected by this method.

(b) An actual meta-analysis [5] has shown an overall sensitivity of 0.75 (95 % CI, 0.65–0.83) with a specificity of 0.92 (95 % CI, 0.90–0.95) for LC-SF to detect bacteremia or fungemia. Because of the sensitivity of the test system as given by the manufacturer, it is possible to obtain a valid result only if at least about 100 CFU/ml of microorganisms are present in the samples (*Candida glabrata*, *Streptococcus* spp., and coagulase-negative *Staphylococcus* spp.) respectively 30 CFU/ml (for all other specified organisms (*see* Table 1)). In some cases (in particular in respect of patients with a suspected diagnosis of endocarditis), it could be more reasonable to raise the amount of bacteria or fungi in the blood by culturing the blood specimens 48 h before starting the PCR run. By this additional culture, a considerably higher sensitivity of 0.95 (95 % CI, 0.94–0.96) can be achieved, but at the cost of a—by this period—extended TAT of the test [10]. Another possibility that might compensate for this problem could be the broad-range PCR amplification of conserved bacterial DNA sequences [10]. Based on the slightly limited sensitivity with highly preserved specificity, the LC-SF test seems to have higher rule-in than rule-out diagnostic value [6]. A unique advantage of PCR is the ability to detect inactivated bacterial cells. Due to advanced approaches, molecular methods are nowadays even able to distinguish viable from inactivated bacterial cells [11, 12]. The LC-SF has especially proven its worth as an adjunct to blood culture for neutropenic [13, 14], pediatric [15], intensive care, and general medicine [16] patients [17]. A further benefit is the reduction of the risk of contamination by using LC-SF (amplification and detection in a single-tube format) [17].

1. Several necessary measures should be taken to avoid PCR contamination with exogenous DNA sequences. Buffers have to be stored in small aliquots and discarded after use. Exogenous DNA has to be inactivated by UV irradiation at 254 nm wavelength for at least 10 min. After each usage the bench should be cleaned with a decontamination solution (e.g., LTK-008™ from Biodelta) followed by wiping all surfaces with DEPC (diethylpyrocarbonate)-treated H_2O.
2. Specimens should be stored at 2–8 °C and assayed within 72 h after collection. Storage of the blood samples at ambient temperature (15–25 °C) requires the initiation of the analysis no later than 4 h following the time of collection of the samples.
3. Eluates after the analysis can be stored for 30 days (−15 to −25 °C), 8 days (2–8 °C), or 4 h at ambient temperature (15–25 °C).
4. We highly recommend performing the test from whole blood samples collected with K-EDTA tubes to minimize the chance of obtaining false-positive or false-negative reactions.
5. If blood samples with leukocytes over 30,000/μl or under 1,000/μl are used, the negative results are not reliable.
6. The LC Septi*Fast* Kit includes an IC consisting of synthetic double-stranded DNA molecules with primer binding sites identical to those of the target sequences but differing in their HybProbe binding sites.
7. Color compensation should be performed at least every 6 months, preferably every time a new lot of probes is used. It helps to compensate device- and channel-specific interferences of emission spectra of the differently labeled DNA probes. You need the LightCycler® Multicolor Compensation Set containing fluorescent dyes for the five different channels. The color compensation run requires 20 μl of every component from the set and is running in the LightCycler® 2.0 Instrument.
8. The vials contain ceramic beads for mechanical lysis of both blood cells and pathogens by vigorously shaking, using up to 7,000 rpm for 70 s. They should be tightly closed before transferring them into the MagNA Lyser® Instrument.
9. The specimens and the NC need each one cartridge and one tip tray from the MagNA Pure Compact Nucleic Acid Isolation Kit 1.
10. Change gloves just before opening the eluate tubes.

11. The MagNA Pure Compact System automatically performs all steps for isolation and purification. Magnetic glass particles (MGP) that can bind the isolated DNA to their surfaces are added to the lysed samples. Afterward, several washing steps remove unbound substances, and the purified DNA is then eluted. In case manual extraction is used, the Septi*Fast* Prep Kit can be applied to get the eluates following the instructions according to literature [7]. Incubate the lysed specimens at 56 °C for 15 min with gentle mixing with a protease and chaotropic lysis buffer as well as an IC. The buffer releases the nucleic acid and protects the released DNA from DNAses in the whole blood. Transfer the mixtures after addition of binding buffer to a spin column with a glass fiber insert (1,900 × *g*; 3 min). Wash for 2 min with 1,800 μl of inhibitor removal buffer (at 4,200 × *g*) to remove the unbound substances (salts, proteins, cellular fragments) and in a second washing step for 10 min (4,200 × *g*) with 1,600 μl of wash buffer. Add 100 μl of heated elution buffer (70 °C), incubate for 5 min, and centrifuge for 2 min at 4,200 × *g* to elute adsorbed nucleic acids from the column.
12. For decontamination you can use, e.g., LTK-008™ from Biodelta.
13. You can store the Master Mixes (MM G(+), MM G(−), and MM F) at minus 20 °C.
14. Ensure that each PCR mix in a 100 μl capillary consists of 50 μl Master Mix and 50 μl eluted sample.
15. The Septi*Fast* Identification Software (SIS) is highly recommended for analyzing and interpretation of your samples.

References

1. Degoricija V, Sharma M, Legac A et al (2006) Survival analysis of 314 episodes of sepsis in medical intensive care unit in university hospital: Impact of intensive care unit performance and antimicrobial therapy. Croat Med J 47: 385–397
2. Kumar A, Roberts D, Wood KE et al (2006) Duration of hypotension before initiation of effective antimicrobial therapy is the critical determinant of survival in human septic shock. Crit Care Med 34:1589–1596
3. Book M, Lehmann LE, Zhang XH et al (2013) Monitoring infection: from blood culture to polymerase chain reaction (PCR). Best Pract Res Clin Anaesthesiol 27:279–288
4. Daniels R (2011) Surviving the first hours in sepsis: getting the basics right (an intensivist's perspective). J Antimicrob Chemother 66 (Suppl 2):ii11–ii23
5. Lehmann LE, Hunfeld K-P, Steinbrucker M et al (2010) Improved detection of blood stream pathogens by real-time PCR in severe sepsis. Intensive Care Med 36(1):49–56
6. Chang S-S, Hsieh W-H, Liu T-S et al (2013) Multiplex PCR system for rapid detection of pathogens in patients with presumed sepsis: a systemic review and meta-analysis. PLoS One 8(5): e62323. doi:10.1371/journal.pone.0062323
7. Lehmann LE, Hunfeld K-P, Emrich T et al (2008) A multiplex real-time PCR assay for rapid detection and differentiation of 25 bacterial and fungal pathogens from whole blood samples. Med Microbiol Immunol 197(3): 313–324
8. Regueiro B, Varela E, Martinez-Lamas M et al (2008) Automation of Septifast® for molecular diagnosis of infection in septic patients. Crit Care 12(Suppl 5):P12

9. Wolk DM, Fiorello AB (2010) Code sepsis: rapid methods to diagnose sepsis and detect hematopathogens. Part II: challenges to the laboratory diagnosis of sepsis. Clin Microbiol Newslett 32(6):42–49
10. Tissari P, Zumla A, Tarkka E et al (2010) Accurate and rapid identification of bacterial species from positive blood cultures with a DNA-based microarray platform: an observational study. Lancet 375:224–230
11. Kobayashi H, Oethinger M, Tuohy MJ et al (2009) Limiting false-positive polymerase chain reaction results: detection of DNA and mRNA to differentiate viable from dead bacteria. Diagn Microbiol Infect Dis 64:445–447
12. Weigel KM, Jones KL, Do JS et al (2013) Molecular viability testing of bacterial pathogens from a complex human sample matrix. PLoS One 8:e54886. doi:10.1371/journal.pone.0054886
13. Mancini N, Clerici D, Diotti R et al (2008) Molecular diagnosis of sepsis in neutropenic patients with haematological malignancies. J Med Microbiol 57:601–604
14. von Lilienfeld-Toal M, Lehmann LE, Raadts AD et al (2009) Utility of a commercially available multiplex real-time PCR assay to detect bacterial and fungal pathogens in febrile neutropenia. J Clin Microbiol 47:2405–2410
15. Mussap M, Molinari MP, Senno E et al (2007) New diagnostic tools for neonatal sepsis: the role of a real-time polymerase chain reaction for the early detection and identification of bacterial and fungal species in blood samples. J Chemother 19(Suppl 2):31–34
16. Louie RF, Tang Z, Albertson (2008) Multiplex polymerase chain reaction detection enhancement of bacteremia and fungemia. Crit Care Med 36:1487–1492
17. Mancini N, Carletti S, Ghidoli N et al (2010) The era of molecular and other non-culture-based methods in diagnosis of sepsis. Clin Microbiol Rev 23(1):235–251

Chapter 15

Host Response Biomarkers in the Diagnosis of Sepsis: A General Overview

Marianna Parlato and Jean-Marc Cavaillon

Abstract

Critically ill patients who display a systemic inflammatory response syndrome (SIRS) are prone to develop nosocomial infections. The challenge remains to distinguish as early as possible among SIRS patients those who are developing sepsis. Following a sterile insult, damage-associated molecular patterns (DAMPs) released by damaged tissues and necrotic cells initiate an inflammatory response close to that observed during sepsis. During sepsis, pathogen-associated molecular patterns (PAMPs) trigger the release of host mediators involved in innate immunity and inflammation through identical receptors as DAMPs. In both clinical settings, a compensatory anti-inflammatory response syndrome (CARS) is concomitantly initiated. The exacerbated production of pro- or anti-inflammatory mediators allows their detection in biological fluids and particularly within the bloodstream. Some of these mediators can be used as biomarkers to decipher among the patients those who developed sepsis, and eventually they can be used as prognosis markers. In addition to plasma biomarkers, the analysis of some surface markers on circulating leukocytes or the study of mRNA and miRNA can be helpful. While there is no magic marker, a combination of few biomarkers might offer a high accuracy for diagnosis.

Key words Diagnosis, Acute phase proteins, Cytokines

1 Introduction

Any severe insults (burns, trauma, pancreatitis, severe surgery, resuscitated cardiac arrest) are accompanied by a "systemic inflammatory response syndrome" (SIRS) [1]. Almost concomitantly, mechanisms aimed to dampen the inflammatory process are initiated. The consequences of this regulatory step could be an alteration of the immune status, also known as "compensatory anti-inflammatory response syndrome" (CARS) [2]. A sterile inflammation associated with SIRS is initiated by damage-associated molecular patterns (DAMPs) or alarmins released by cells after their necrosis, by injured tissues, or by activated cells. Interestingly, most of the DAMPs activate immune cells through similar sensors as the pathogen-associated molecular patterns (PAMPs) harbored

Nicasio Mancini (ed.), *Sepsis: Diagnostic Methods and Protocols*, Methods in Molecular Biology, vol. 1237,
DOI 10.1007/978-1-4939-1776-1_15,

or released by bacteria during sepsis do. As a consequence, since DAMPs and PAMPs initiate the inflammatory cascade through similar receptors, most of the generated mediators are identical. Accordingly, the challenge remains to define biomarkers from the host that would be able to distinguish patients with sepsis from those with non-infectious SIRS. A Canadian physician, William Osler (1849–1919), nicely defined the consequences of the host response in sepsis: "except on few occasions, the patient appears to die from the body's response to infection rather than from it." Thus, this response is the consequence of an exacerbated inflammatory process that can be monitored by the presence in biological fluids of both pro- and anti-inflammatory mediators. It is worth mentioning that their excessive production allowing their detection can be considered as the "tip of the iceberg" [3], meaning that the failure to detect them, for example, in the bloodstream, does not mean that they are not playing any role since they can be trapped by their specific high-affinity receptors on target cells. In a recent review on sepsis biomarkers [4], 178 different biomarkers were retrieved. Since then, every year new ones are reported. While this chapter will not address all of them, it will discuss the most interesting ones and will mainly address the biomarkers as a tool for diagnosis. Despite in a large number of studies mentioned in this chapter, authors have reported highly significant differences between the levels of a given biomarker measured in sepsis versus SIRS patients, this is not sufficient to guarantee a marker of interest. Indeed, the overlapping in the concentrations reported in both groups of patients should be minimum.

Often, biomarkers also appear of interest in terms of prognosis ("prognosticator"). Quite often high levels of biomarkers are associated with enhanced severity and poor outcome. However, this information remains of limited interest since a large number of available clinical data are sufficient for the physicians to apprehend the prognosis, and clinical scores remain often the most reliable information in terms of prognostic. In contrast, biomarkers that would allow distinguishing the occurrence of sepsis among SIRS patients would be of great interest to initiate as early as possible the appropriate use of antibiotics. Indeed, it is known that any delay in the initiation of antibiotic therapy has major consequences in terms of survival [5]. So far, no biomarkers have shown sufficient specificity and sensitivity to be validated [6], and most probably only a combination of biomarkers will allow to reach sufficient efficacy for diagnosis [7]. In addition, host biomarkers can also be useful for patient stratification in clinical studies or to define patients who could benefit of a given treatment (e.g., interleukin-6 (IL-6) [8]) or to ascertain the efficiency of a given treatment such as a successful antibiotherapy (e.g., procalcitonin, PCT [6]). Table 1 summarizes the different uses of biomarkers.

Table 1
Uses of biomarkers

Screening
To identify patients at increased risk of adverse outcome to inform a prophylactic intervention or further diagnostic test
Diagnosis
To establish a diagnosis to inform a treatment decision and to do so more reliably, more rapidly, or more inexpensively than available methods
Risk stratification
To identify subgroups of patients within a particular diagnostic group who may experience greater benefit or harm with therapeutic intervention
Monitoring
To measure response to intervention to permit the titration of dose or duration of treatment
Surrogate end point
To provide a more sensitive measure of the consequences of treatment that can substitute for a direct measure of a patient-centered outcome

From Marshall JC and Reinhart K. Biomarkers of sepsis Critical Care Medicine 2009, 37, 2290–22

2 Plasma Biomarkers

In this review, we will address the host plasma markers as well as the biomarkers that can be studied at the leukocyte levels either as cell surface (Table 2) or after analysis of cellular mRNA or miRNA. We will only focus on biomarkers identified in humans.

2.1 Acute Phase Proteins

Acute phase proteins are produced by the liver in response to numerous inflammatory cytokines (e.g., interleukin-1 (IL-1), IL-11, IL-22, tumor necrosis factor (TNF), leukemia inhibitory factor (LIF), transforming growth factor-β (TGF-β)), but IL-6, named initially as "hepatocyte-stimulating factor," is the main activator. Their role is mainly to favor immune defense; to contribute to the elimination of microbial products, cellular debris, and released products (e.g., hemoglobin); and to neutralize some inflammatory mediators such as free radicals or proteases. C-reactive protein and serum amyloid A protein display the largest discrepancy between their levels at homeostasis and after an inflammatory insult. Haptoglobin, α1-glycoprotein acid, α1 antitrypsin, α1 antichymotrypsin, and fibrinogen are other acute phase proteins significantly enhanced during inflammation, whereas ceruloplasmin is less increased. In contrast, to maintain oncotic pressure, the levels of other proteins are diminished (e.g., albumin, transferrin, fibronectin). The kinetics of appearance and disappearance of each acute phase proteins are different and

Table 2
Main biomarkers of interest

Acute phase proteins
C-reactive protein
Serum amyloid A
LPS Binding protein
Pentraxin 3
Procalcitonin
Tissue injury biomarkers
Lactate
Hyaluronan
Pancreatic stone protein
Heat shock proteins
Alarmins (DAMPs)
DNA
HMGB-1
S100A8/9
Galectin-3
Cytokines
Interleukin-1
Interleukin-1 receptor antagonist
Interleukin-6
Interleukin-10
Interleukin-13
Interleukin-18
Interleukin-27
Tumor necrosis factor
Macrophage migration inhibitory factor
Chemokines
Interleukin-8 (CXCL8)
IP-10 (CXCL10)
Monocyte chemotactic factor-1 (CCL2)
Macrophage inflammatory protein-1α/β (CCL3; CCL4)
RANTES (CCL5)
Hormones
Leptin
Testosterone/oestradiol
Vasopressin/copeptin
Natriuretic peptides
Apoptosis-related biomarkers
Fas and FasL
CK18
Soluble receptors
Soluble CD14
Soluble MD2
Soluble ST2
Soluble TREM-1
Soluble TNF R
Soluble IL-2R (sCD25)
Soluble CD163
Soluble decoy receptor 3
Soluble urokinase-type plasminogen activator receptor
Enzymes
Elastase
Metaloproteinase
Dipeptidylpeptidase
Phospholipase A2
YKL-40
Granzyme A
Coagulation biomarkers
Antithrombin
Protein C
Thrombomodulin
Plasminogen activator inhibitor
von Willebrand factor
Vascular endothelial biomarkers
Soluble ICAM1
Soluble E-selectin
Soluble L-selectin
Soluble VCAM-1
Soluble-ELAM-1
Angiopoietin
Vascular endothelial growth factor
Endothelin
Endocan
Adrenomodullin
Heparin-binding protein
Growth arrest-specific 6
Miscellaneous
Fibronectin
Selenium
Morphine
Gelsolin
Osteopontin
C3a
Cell-surface biomarkers
HLA-DR
TLR4
CD14
CD25
CD40
CD48
CD64
CD69
CD80
TREM1
CX3CR1

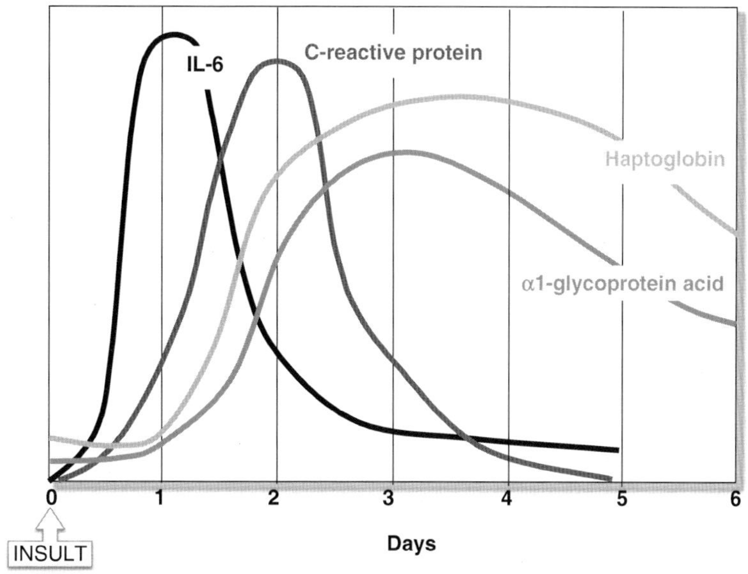

Fig. 1 Kinetics of IL-6 and main acute phase proteins after an inflammatory insult on day 0

allow defining the timing within the inflammatory process (Fig. 1). In addition to the classical acute phase proteins, a new list of molecules also produced by the liver has been established such as the lipopolysaccharide (LPS) binding protein (LBP), the soluble CD14, or the IL-1 receptor antagonist (IL-1Ra).

2.1.1 C-Reactive Protein

C-reactive protein (CRP) was discovered by Tillett and Francis in 1930 [9] as a serum entity present in rabbits with pneumonia able to bind a polysaccharide fraction C prepared from pneumococci and absent in normal sera. Then, it was reported to be found independently of the type of infection. The protein was crystallized in 1947 [10], and its structure was revealed in 1977 as a pentameric molecule, composed of five identical subunits arranged in cyclic symmetry [11]. Present at homeostasis (<5 μg/mL), it can reach levels higher than 500 μg/mL during inflammation. Numerous studies have reported significantly higher levels of CRP in sepsis patients as compared to critically ill adult patients with SIRS [12–15] independently of the clinical score [16]. Also in newborns, CRP levels were independently predictive of positive blood culture [17]. Interestingly, the combination of CRP and temperature increased the specificity for infection diagnosis to 100 % among critically ill patients [18]. Among patients admitted in an emergency department for suspected sepsis, procalcitonin (PCT) best predicted septicemia, but CRP better identified clinical infection [19]. In terms of prognosis, the highest levels of CRP 2 and 3 days after the onset of postoperative sepsis did not distinguish between survivors and non-survivors, whereas on day 7, non-survivors had significantly higher levels of CRP than survivors [20]. A similar investigation confirmed in patients with severe sepsis that on day 7, CRP had a predictive value in terms of outcome with an accuracy higher than IL-6, PCT, and acute physiology and chronic health evaluation (APACHE II) score [21]. CRP levels can also be a useful tool to monitor the efficiency of initial antimicrobial therapy: CRP levels decreased more rapidly and to a greater degree in sepsis patients with a favorable response to initial antibiotics [22]. In contrast, an increase in CRP of at least 2.2 mg/dl in the first 48 h was associated with ineffective initial treatment. CRP was shown to be a reliable diagnostic marker of neonatal sepsis with the same diagnostic accuracy as PCT [23]. Several meta-analyses illustrated that the usefulness of CRP depends on the type of patients for whom an infection is suspected. In hospitalized patients, CRP had a lower diagnostic accuracy than PCT for suspected infection [24]. In contrast, CRP had a higher accuracy than PCT for suspected acute appendicitis [25], in diagnosing infective endocarditis [26], in detecting infectious complication after hematopoietic stem cell transplantation [27], and in detecting bacterial infection in SIRS patients (area under the curve (AUC) of the receiver operating characteristic (ROC) = 0.81 for CRP versus 0.75 for PCT) [28]. Depending on the studies, the sensitivity of CRP varies from 30 to 97.2 % and its specificity from 67 to 100 % in adult and pediatric sepsis [29].

2.1.2 Serum Amyloid A

Serum amyloid A (SAA) is another major acute phase reactant [30]. Present at homeostasis (1–5 μg/mL), it can reach levels higher than 1 mg/mL during inflammation. SAA induces extracellular matrix-degrading enzymes, acts as a chemoattractant for monocytes and neutrophils, inhibits the oxidative burst response, and prevents platelet aggregation. During inflammation, SAA becomes the major high-density lipoprotein (HDL)-associated apolipoprotein and binds to cholesterol either facilitating its delivery to cells or its removal from sites of tissue damage. In patients with septic shock, its levels correlate with those of CRP [31]. SAA displayed a better capacity than CRP to differentiate between infectious and non-infectious febrile episodes in neutropenic patients [32]. SAA was reported to be an early and accurate marker of neonatal early-onset sepsis, better than CRP [33], whereas contradictory data have been reported for its use as an indicator of neonatal late-onset sepsis [34, 35]. When associated with other measurements (CRP, sICAM-1, and sE-selectin), it led to enhanced diagnostic performance for the diagnosis of neonatal infection [36]. In adult sepsis, the interindividual variability is considerable, independently of body weight or age [37, 38].

2.1.3 LPS Binding Protein

LPS binding protein (LBP) binds to the endotoxin of Gram-negative bacteria and acts as a catalyzer, favoring LPS bioactivities. It is constitutively present in the serum (5–10 μg/mL) and increased during inflammation (up to 200 μg/mL). Serum LBP levels have been regularly reported to be higher in sepsis patients than in SIRS patients [39]. Among surgical intensive care unit (ICU) patients, LBP moderately discriminated sepsis from SIRS and had a lower accuracy than IL-6 or PCT [40]. LBP levels on admission were similar in patients with Gram-negative or Gram-positive infection and in survivors and non-survivors. In an emergency department, LBP performed similarly to IL-6 and CRP to distinguish between SIRS and sepsis (AUC = 0.86 versus 0.87 and 0.84, respectively) and was superior to PCT as a diagnostic marker for infection (AUC = 0.74) [41]. A strong correlation was found between LBP and CRP ($r = 0.84$) and lower with IL-6 ($r = 0.57$). While LBP was of no use to identify infective endocarditis [42], LBP was the only factor independently associated with severe bacterial infection in a multivariate analysis in cirrhotic patients with ascites [43]. Measurements of LBP in pleural fluid was as efficient as CRP and soluble TREM-1 (sTREM-1) to identify patients with infectious effusions (AUC = 0.87 versus 0.87 and 0.86, respectively) and better than PCT (AUC = 0.57) [44].

2.1.4 Pentraxin 3

Pentraxin 3 (PTX3) is not an acute phase protein *per se*, since it is not produced by liver cells, but by endothelial cells, fibroblasts, epithelial cells, adipocytes, monocytes/macrophages, and neutrophils during inflammation in response to PAMPs and inflammatory cytokines such as IL-1 and TNF. PTX3 shares with CRP a similar pentameric structure. PTX3 can bind to complement factor C1q favoring the complement activation and to bacteria favoring their opsonization. PTX3 is elevated in critically ill patients with a gradient from SIRS to sepsis to severe sepsis to septic shock [45]. PTX3 is an early predictor of bacteremia and septic shock in hematological patients after intensive chemotherapy [46]. In emergency room, high levels of PTX3 were found in patients with severe sepsis (AUC = 073) [47]. In addition, elevated levels of PTX3 during the first days after diagnosis of bacteremia and sepsis are independent prognostic biomarkers of mortality [48, 49].

2.1.5 Procalcitonin

See Chapter 18.

2.2 Tissue Injury Biomarkers

2.2.1 Lactate

Lactate was first described in 1780 by Karl Wilhelm Scheele, a Swedish chemist, as a substance in sour milk. Its presence in human blood was first demonstrated in 1843 by Joseph Scherer, a German chemist and physician, in women who died of puerperal fever [50]. Lactate is converted from pyruvate, the end product of glycolysis, by the action of the enzyme lactate dehydrogenase. Lactate formation is favored by tissue hypoperfusion, increased aerobic glycolysis, mitochondrial dysfunction, or impaired activity of pyruvate dehydrogenase. In critically ill patients, lactate is produced in tissues other than those producing lactate at homeostasis (i.e., muscle, skin, brain, intestine). In sepsis patients with multiple organ failure (MOF), it was reported that lactate is secreted from the most severely affected organs [51]. It is released at the sites of infection and inflammation and is thought to be related to the augmented glycolysis in the recruited and activated leukocytes at the sites of infection [50]. Both an increased production and a decreased clearance lead to the enhanced levels of circulating lactate observed in patients, although injection of LPS in human volunteers suggested that enhanced lactate levels reflect an increased lactate production in other places than the muscles. Its levels correlate with anaphylatoxins C3a and C4a, elastase, and IL-6 [52]. High lactate levels in sepsis have been regularly correlated with poor outcome [53, 54], and its early clearance is associated with improved outcome [55, 56]. Enhanced lactate levels have also been regularly described in any SIRS patients such as burn [57], trauma [58], and surgery [59]. In many cases, lactate levels, particularly associated with poor clearance, were predictor of mortality. In a study involving emergency department older patients with ($n=777$) or without ($n=665$), lactate values were associated with mortality regardless of the presence or absence of infection [60].

2.2.2 Hyaluronan

Hyaluronan is a glycosaminoglycan present in the extracellular matrix and in the vascular glycocalyx layer. Tissue injury is associated with hyaluronan release, and enhanced levels in plasma are detected in sepsis patients [61]. The concentration of plasma hyaluronan in sepsis patients is ninefold that observed in healthy controls, but the size of the fragment (50–500 kDa) did not differ between control and sepsis patients [62]. Although hyaluronan levels are higher in sepsis as compared to SIRS patients, hyaluronan has a lower predictive value as compared to CRP or PCT [63].

2.2.3 Pancreatic Stone Protein

Pancreatic stone protein (PSP) is a lectin binding protein initially identified in patients with pancreatitis. PSP is produced in the pancreas in response to stress and also by Paneth cells of the small intestine and the fundic cells of the stomach. In trauma patients, PSP levels were higher in those who developed local infection or sepsis [64]. Levels were higher in patients with septic shock as compared to patients with severe sepsis and were more accurate than CRP, PCT, IL-6, and IL-8 in predicting in-hospital mortality [65]. At the time of admission in ICU, PSP displays the best performance as a diagnostic marker of sepsis (as compared to PCT, sCD25, IL-1β, IL-6, and IL-8; AUC=0.93). The addition of sCD25 or PCT to PSP improved its diagnostic accuracy [66]. In patients with peritonitis, PSP was the best predictor for death in the ICU [67].

2.2.4 Heat Shock Proteins

Heat shock proteins (HSP) are released by cells during stress and injury, and peripheral blood mononuclear cells from sepsis patients expressed significantly higher level of HSP70 than cells from healthy controls [68]. In SIRS and in septic patients, body temperature affects the gene expression of several HSP and heat shock transcription factors (HSF) [69]. Indeed, fever and Toll-like receptor (TLR) agonists synergize to induce HSP70 release [70]. There are three genes in the HSP70 family. A single gene polymorphism of HSP70-2 was associated with an increased risk to develop septic shock among patients admitted with community-acquired pneumonia [71]. In severe trauma patients, elevated HSP72 serum levels were associated with survival [72]; in contrast, in children with septic shock, the highest levels of HSP70 were found among the non-survivors [73]. The levels of HSP70 were shown to correlate with the prooxidant status of the serum of sepsis patients, and the highest expression of HSP70 was also monitored in non-surviving adult patients [74]. Plasma levels of HSP70 in patients with severe sepsis were higher than the levels measured in brain-dead patients or in resuscitated cardiac arrest patients, although those who ultimately died from shock also had high levels [75]. In this study, high and significant correlations were found between HSP70 and IL-8 ($r=0.9$) sTREM-1 ($r=0.87$), sTNF R ($r=0.68$), and IL-6 ($r=0.66$). Finally, in children with septic shock, levels of HSP60 were significantly higher than in critically ill children without sepsis [76].

2.3 Damage-Associated Molecular Patterns

Damage-associated molecular patterns (DAMPs) behave as endogenous danger signals. They are mainly derived from the necrosis of cells and the release of their intracellular content.

2.3.1 DNA

The presence of increased plasma levels of DNA has been regularly reported in patients with trauma, myocardial infarction, cancer, and stroke. In critically ill patients, plasma levels of DNA were higher in those who were diagnosed with severe sepsis or septic shock [77]. The same study revealed that higher levels were also measured in patients who subsequently died in the ICU or in the hospital. Among patients with severe sepsis or septic shock, the value of cell-free plasma DNA at admission was also found to be predictive for outcome [78]. A quantification method has been set up to specifically measure the DNA from neutrophil extracellular traps (NETs). Initially, high circulating free DNA/NETs levels with recurrent increased values between days 5 and 9 were associated with subsequent sepsis, MOF, and death [79]. Cell-free DNA concentrations in plasma of patients with fever of unknown origin help to diagnose infection and sepsis and were increased according to the severity of the infection [80]. Plasma cell-free DNA concentration proved to be a specific independent prognostic biomarker in bacteremia and sepsis [81, 82]. Circulating mitochondrial DNA has also been reported in sepsis patients of whom levels were higher than in healthy controls [83] but similar to the levels detected in emergency department patients [84]. Controversial conclusions have been reached regarding the concentration of mitochondrial DNA on day 1 and its potential use for predicting outcome in sepsis patients [83, 85].

2.3.2 High-Mobility Group Box 1

High-mobility group box 1 (HMGB1) is a nuclear factor bound to DNA. It is secreted by activated monocytes and macrophages and passively released by necrotic or damaged cells, and it further triggers inflammation [86]. HMGB1 was shown to be a late mediator of endotoxin-induced lethality in mice and to be present in plasma of sepsis patients in larger amounts in the non-survivors [87]. Similarly, higher plasma levels of HMGB1 were found in non-surviving patients with community-acquired pneumonia [88]. In the majority of sepsis patients, HMGB1 levels remain high up to 1 week [89]. Large concentrations of HMGB1 can also be detected in abdominal fluids from patients with peritonitis and in bronchoalveolar lavages of patients with pneumonia [90]. While HMGB1 was found in different groups of patients with SIRS (severe trauma, hemorrhagic shock, burn, stroke), no significant differences were observed between sepsis, severe sepsis, and septic shock [91], and the use of HMGB1 to dissociate SIRS and sepsis has not been fully addressed.

2.3.3 S100A8/S100A9

S100A8 (an 8 kDa protein, also called calgranulin A or myeloid-related protein-8 (MRP8)) and S10019 (a 14 kDa protein, also called calgranulin B or MRP14) are considered as DAMPs or alarmins, although they are released by activated cells independently of any cell death [92]. They are present in any biological fluids bathing inflamed tissues (synovial fluid of patients with arthritis) or in the serum of patients with local (chronic inflammatory bowel disease) or systemic disorders (systemic lupus erythematosus). It is also present in the serum of patients with severe sepsis and in abdominal fluid from patients with peritonitis [93]. In human volunteers receiving LPS, the peak of 100A8/100A9 was observed 5 h after the injection. S100A8/100A9 complex (also called calprotectin) displayed a greater diagnostic accuracy than CRP in identifying neonatal sepsis [94]. S100A9 mRNA expression measured on days 7–10 was significantly higher in patients who were about to contract hospital-acquired infection compared with those who were not [95]. To our knowledge, no study has compared the levels of S100A8 and/or S100A9 in non-infectious SIRS patients with those measured in sepsis patients.

2.3.4 Galectin-3

Galectin-3 is a 30 kDa intracellular lectin with a broad biological functionality. It can be passively released from damaged cells. Circulating galectin-3 concentrations are increased in patients with sepsis as compared to healthy controls or patients with pancreatitis. Levels of galectin-3 are higher in patients with septic shock than in patients with sepsis and among non-survivors as compared to survivors [96].

2.4 Apoptosis-Related Biomarkers

Apoptosis is a hallmark of sepsis [97]. It particularly affects lymphocytes, NK cells and dendritic cells, as well as endothelial and epithelial cells. In contrast, apoptosis of neutrophils is decreased. Both the presence of proapoptotic activity on renal tubular cells [98] and antiapoptotic activity on neutrophils [99] have been reported for sera derived from SIRS patients.

2.4.1 Soluble Fas and FasL

Fas ligand (FasL) (sCD178) is a soluble homotrimeric molecule cleaved from the cell surface, which belongs to the TNF superfamily. Its binding with its Fas receptor (CD95) induces apoptosis. The Fas receptor can be shed from the cell surface and found as a soluble receptor (sFas), behaving as an inhibitor of apoptosis. Removal of sFas from septic patient sera diminished its antiapoptotic effect [100]. Indeed, the levels of sFas are increased in sepsis adults [100, 101] and pediatric [102] patients. In the latter study, sFas expression was significantly increased as compared to critically ill children only on day 3 but was higher in sepsis non-survivors as compared to survivors on day 1 and day 3. In contrast, a study performed in 132 adult patients with bacteremia failed to associate enhanced levels of sFas with clinical score [103]. In both groups, FasL ligand was not increased in sepsis patients as compared to healthy controls.

2.4.2 CK18

During apoptosis of epithelial cells, activated caspases 3, 6, and 7 cleave cytokeratin 18 (CK18) into proteolytic fragments, which diffuse into the serum. CK18 is an intermediate filament protein of which cleavage leads to neoepitope recognized by specific antibodies. Both full length and cleaved fragments can be found in the circulation of sepsis patients [104]. Levels of CK18 fragments were higher in the serum of sepsis patients as compared to trauma patients [105] and to patients who underwent major abdominal surgery [106]. CK18 fragments appeared to be an early predictor of survival in sepsis patients [106].

2.5 Cytokines

The first reports on the presence of circulating cytokines and chemokines in sepsis patients were depending on appropriate measurements. Their analysis was first achieved thanks either to their bioactivity or to radioimmunoassays before enzyme-linked immunosorbent assay (ELISA) became available. Nowadays, the use of multiplex cytokine analysis technologies allows the measurement of a large number of analytes within a very small volume of biological samples. Usually below detection limit in healthy subjects, their presence in the plasma or serum illustrates the cytokine storm associated with sepsis (Table 3). Some are rarely found or only detected in very low amounts and will not be further discussed.

Table 3
First reports on the increased levels of circulating cytokines and chemokines in human sepsis

Cytokine	Year	References
Tumor necrosis factor (TNF)	1986	Waage et al. Scand J Immunol 24: 739 [507]
Interleukin-1β (IL-1β)	1988	Girardin et al. N Engl J Med 319: 397 [166]
Gamma interferon (IFNγ)	1988	Girardin et al. N Engl J Med 319: 397 [166]
Interleukin-6 (IL-6)	1989	Waage et al. J Exp Med 169: 33 [127] Hack et al. Blood 74: 1704 [121]
Interleukin-8 (IL-8; CXCL8)	1992	Hack et al. Infect Immun 60: 2835 [192] Friedland et al. Infect Immun 60: 2402 [191]
Leukemia inhibitory factor (LIF)	1992	Waring et al. J Clin Invest 90: 2031 [182]
Granulocyte colony-stimulating factor (G-CSF)	1993	Gessler et al. Blood 82 : 3177 [175]
Interleukin-10 (IL-10)	1994	Marchant et al. Lancet 343: 707 [508]
Interleukin-1 receptor agonist (IL-1Ra)	1994	Rogy et al. J Am Coll Surg 178: 132 [111] Van Deuren et al. J Infect Dis 169: 157 [112]
Monocyte chemoattractant protein-1 and protein-2 (MCP-1 and MCP-2; CCL2 & CCL8)	1995	Bossink et al. Blood 86: 3841 [193]
M-CSF and GM-CSF	1995	Waring et al. Clin Exp Immunol 102: 501 [509]
Interleukin-4	1995	DiPiro et al. Arch. Surg. 130: 1159 [510] Zeni et al. J Infect Dis 172: 1171 [511]
Transforming growth factor-β (TGFβ)	1996	Marie et al. Ann Intern Med 125: 520 [512]
Lymphotoxin-α (Ltα)	1996	Sriskandan et al. Lancet 348: 1315 [513]
Macrophage inflammatory protein-1α and -1β (MIP-1α and MIP-1β); CCL3 & CCL4)	1996	Fujishima et al. Intens. Care Med. 22, 1169 [196]
Interleukin-12 (IL-12)	1997	Presterl et al. Am J Respir Crit Care Med 156: 825 [153]
Interleukin-15 (IL-15)	1999	Lauw et al. J Infect Dis 180: 1878 [160]
Interleukin-18 (IL-18)	1999	Lauw et al. J Infect Dis 180: 1878 [160]
Macrophage migration inhibitory factor (MIF)	2000	Calandra et al. Nat Med. 6: 164 [170]
RANTES (CCL5)[a]	2000	Carrol et al. J.Infect. Dis. 182, 363 [217]
Interleukin-13 (IL-13)	2004	Collighan et al. Br. J. Surg. 91, 762 [185]
Interleukin-22 (IL-22)	2010	Bingold et al. Shock 34: 337 [514]
Interleukin-27 (IL-27)	2012	Wong et al. Crit. Care 16 : R213 [186]

[a]A rare case for which the levels are decreased as compared to healthy controls

2.5.1 Interleukin-1β

IL-1β is produced through the activation of the inflammasome, the cleavage of its precursor form by caspase-1, and its release requires other signal including ATP. Accordingly, accumulation of intracellular IL-1 could be an accurate mean to assess cellular activation [107]. Surprisingly, in sepsis the presence of cell-associated IL-1β was only found in 50 % of the sepsis patients, and its detection in the plasma by radioimmunoassay did not exceed 50 % of the patients [108]. The analysis by ELISA did not increase its frequency, and multiplex analysis revealed very low mean levels [109]. These results illustrate the discrepancy that may exist between biomarkers and actors, since all animal models revealed a clear role of IL-1 in sepsis.

2.5.2 Interleukin-1 Receptor Antagonist

IL-1Ra is an anti-inflammatory cytokine that occupies the IL-1 receptor and counteracts the effects of IL-1. IL-1Ra was first reported in critically ill patients and in human volunteers receiving LPS injection [110] and then regularly found in sepsis patients [111, 112], correlating with severity score [113]. IL-1Ra was found to be of interest for early diagnosis of neonatal sepsis with high sensitivity and specificity [114]. However, in adults, similar levels of IL-1Ra were found in sepsis patients and non-infected critically ill patients [115] or in successfully resuscitated patients after cardiac arrest [116]. The levels of circulating IL-1Ra are >1000-fold higher than those of IL-1β [113, 117] and correlate with the levels of circulating inflammatory cytokines (Table 4).

2.5.3 Interleukin-6

IL-6 has been regularly reported as a sensitive marker whose levels are proportional to the intensity of the insult such as illustrated after surgery of different severities [118], in severe versus mild pancreatitis [119], in long-term hemodialyzed versus uremic patients [120], and in septic shock versus sepsis [121, 122]. Lower levels were also observed in non-infectious SIRS as compared to sepsis [123]. Its levels correlate with many other plasma markers such as lactate, TNF, IL-8, monocyte chemoattractant protein-1 (MCP-1), and macrophage inflammatory protein 1β (MIP-1β) and most interestingly with anti-inflammatory cytokines such as IL-10 and IL-1Ra [121, 124, 125], as well as with length of fever, length of hospital stay clinical score, worsening organ dysfunction, or failure of organ dysfunction to improve on day 3 [109, 113, 126]. IL-6 has been also regularly reported as a prognosis marker either in terms of levels measured at admission or in terms of maintained high levels during the survey [108, 127–131]. IL-6 has been reported to be an ideal marker for detecting early diagnosis of neonatal sepsis [132, 133] or as a predictive marker of occult bacteremia in febrile children [134, 135]. Combining IL-6 levels with CRP levels or with absolute neutrophil counts enhanced its accuracy for diagnosis of infection [133, 134]. In adult patients with fever [136], after cardiac surgery [137], increased IL-6 levels could

Table 4
Correlations of plasma/serum levels of anti-inflammatory IL-10 and IL-1Ra with inflammatory cytokines in sepsis patients (data are expressed as *r* values)

	TNF	IL-6	IL-8	References
IL-10	0.66	0.71	0.85	Van Deuren et al. J. Infect. Dis. 1995, 172, 433 [147]
	0.57	0.87	0.70	Lehmann et al. Infect. Immun. 1995,63, 2109 [143]
	0.76	0.68	0.61	Gomez-Jimenez et al. J. Infect. Dis. 1995, 171, 472 [148]
	0.66	0.59	nd	Riordan et al. Arch Dis. Child 1996, 45, 453 [149]
	0.70	0.77	0.77	Kasai et al. Res. Com. Mol. Path. Pharm. 1997, 98, 34 [515]
	0.79	nd	nd	Rodriguez-Gaspar et al. Cytokine 2001, 15, 232 [123]
	nd	0.97	0.85	Cavaillon et al. Scand. J. Infect. Dis. 2003, 35, 535 [125]
	nd	0.79	0.79	Vedrine et al. Cytometry B Clin. Cytom 2004, 60B, 14 [154]
	nd	0.75	0.55	Tamayo et al. Eur Cytok. Netw. 2011, 22, 82 [150]
IL-1Ra	nd	0.79	nd	Fischer et al. Blood 1992, 79, 2196 [110]
	nd	0.45	nd	Rogy et al. J. Am. Coll Surg. 1994, 178, 132 [111]
	0.52	0.72	0.72	Van Deuren et al. J. Infect. Dis. 1995, 172, 433 [147]
	0.60	0.74	0.71	Kasai et al. Res. Com. Mol. Path. Pharm. 1997, 98, 34 [515]
	nd	0.74	0.80	Cavaillon et al. Scand. J. Infect. Dis. 2003, 35, 535 [125]

predict the occurrence of infection, but this was neither observed after trauma [138] nor for differentiating among patients with hematological malignancies fever of unknown origin from sepsis [139]. In SIRS patients, PCT appeared better than IL-6 for the diagnosis of bacterial infection, and IL-6 exhibited a better kinetics for monitoring the effectiveness of antibiotic treatment and outcome [140].

2.5.4 Interleukin-10

IL-10 is a potent anti-inflammatory cytokine. When IL-10 was reported for the first time in patients with sepsis, it was noticed that patients with septic shock had higher levels than those without shock [141]. IL-10 levels correlate with clinical scores [123, 142]. Cases involving fatalities in patients with meningococcal disease had high levels [143]. A similar observation was made in patients with sepsis [123, 144] or with fever [145]. When IL-10 and IL-6 levels are combined, it further improves their prognosis accuracy [146]. Most interestingly, it has been regularly reported that IL-10 levels correlate with the levels of other inflammatory cytokines [123, 125, 143, 147–150] (Table 4). This is of premium importance since it further illustrates that both pro- and anti-inflammatory responses are concomitant [151]. No clear evidence suggests that

IL-10 can help to discriminate between non-infectious SIRS and sepsis [150], although significantly higher levels of IL-10 were found in bacteremic SIRS patients than in non-bacteremic ones [152]. IL-10 was shown to have a high predictive value of outcome (AUC = 0.79) [131]

2.5.5 Interleukin-12

IL-12 is a heterodimeric cytokine made by two chains, p35 and p40, shared with IL-35 and IL-23, respectively. The whole molecule is sometimes called IL-12p70. Although IL-12 has been regularly mentioned in reviews on biomarkers of sepsis, IL-12 is not such a biomarker. In fact, many authors mentioned that they failed to detect significant amounts of IL-12 in the plasma of patients with sepsis [109, 153–155]. In fact, when authors succeeded to detect measurable amounts of IL-12, they reported lower levels in sepsis patients than in healthy controls [156, 157]. A similar observation was reported for IL-12p40 in neonatal sepsis [158]. However, in a report on IL-12p40, a significant higher level of IL-12p40 was observed in adult patients with severe sepsis as compared to healthy controls, but surprisingly, this was not the case for patients with septic shock [159].

2.5.6 Interleukin-18

IL-18 is a member of the IL-1 superfamily, and its precursor form is similarly processed as the IL-1β by caspase-1 following the inflammasome activation. It belongs to the family of cytokines that favors the production of IFN-γ. IL-18 has been regularly detected in the bloodstream of sepsis patients [160, 161]. IL-18 levels are higher in sepsis patients than in trauma patients, in patients with shock than in those without, and possibly in Gram-positive sepsis patients than in Gram-negative sepsis patients [162]. Levels of IL-18 were also found to be higher in non-surviving sepsis patients than in survivors [156]. IL-18 levels correlate with the development of sepsis in surgical patients [163]. In post-traumatic SIRS, IL-18 concentration over days 3–6 was significantly increased among patients who developed sepsis. These increases were noted to be apparent 2–3 days before the clinical diagnosis of sepsis [164].

2.5.7 Tumor Necrosis Factor

Tumor necrosis factor (TNF) is with IL-1, the main orchestrator of inflammation. It was the first cytokine to be identified in the blood of sepsis patients (Table 3). Its presence in the plasma of patients with meningococcal sepsis was first identified using bioassay [165] and radioimmunoassay [166] before the use of ELISA became available [167]. A correlation between high levels of TNF and poor outcome was regularly reported [166, 168, 169]. TNF levels often correlated with clinical scores [113, 168]. TNF levels have a lightly weaker predicting value for positive blood culture than PCT (AUC = 0.67 versus 0.69) [131]. It is worth noting that most recent reports mentioned levels lower than 10 pg/mL. Knowing that TNF is highly sensitive to freezing/thawing, TNF does not appear as an easy tool to differentiate between SIRS and sepsis.

2.5.8 Macrophage Migration Inhibitory Factor

Macrophage migration inhibitory factor (MIF) is a pro-inflammatory cytokine released by immune cells and the pituitary gland of which plasma levels are enhanced in sepsis and septic shock patients as compared to healthy controls [170]. MIF levels are correlated with severity scores, lactate, and outcome [171]. Interestingly, MIF is present within peripheral blood mononuclear cells (PBMC), and its concentration was found to be increased in sepsis patients [172]. Levels of free MIF found in sepsis patients do not correlate with the levels of IL-6, IL-8, or IL-10 [173]. In samples collected 48 h after admission in ICU of patients after severe burn injury, plasma levels of patients who developed sepsis were far higher than in the other groups of patients [174].

2.5.9 Colony-Stimulating Factors

Granulocyte colony-stimulating factor (G-CSF) allows the specific differentiation of neutrophils. It is present in the plasma of sepsis neonates [175] or adult patients [176]. G-CSF has been reported to be helpful for the diagnosis of sepsis in children in pediatric ICU [177] and in adult patients with trauma [178]. G-CSF was also higher in patients with bacteremic pneumococcal pneumonia as compared to other types of pneumonia [179]. Plasma levels of G-CSF were higher in patients with sepsis or septic shock than in patients with non-infectious shock [180]. Levels of granulocyte-macrophage CSF (GM-CSF) are also enhanced in SIRS and sepsis patients, but no specific difference was noticed [181], although patients with septic shock had significantly higher levels than SIRS patients [150].

2.5.10 Other Cytokines

Other cytokines have been detected in the plasma or serum of sepsis patients. This is the case, for example, for leukemia inhibitory factor (LIF) [182, 183], oncostatin M [184], IL-13 [185], IL-15 [160], and IL-27 [186], pre-B-cell colony-enhancing factor [187], gamma interferon (IFN-γ) [188], and vascular endothelial growth factor (VEGF) [189]. However, the levels have rarely been compared to those in SIRS patients, except for IL-13 [190] and IL-27 [186], of which levels were higher in sepsis and septic shock than in non-infectious SIRS patients and IFN-γ of which levels were similar between bacteremic and non-bacteremic SIRS patients [152].

2.6 Chemokines

2.6.1 Interleukin-8 (CXCL8)

IL-8, member of the CXCL family, is one of the main chemokines that recruits neutrophils toward inflamed tissues. Since its first observation in the plasma of sepsis patients in 1992 concomitantly by two different teams [191, 192], IL-8 has been regularly found in this type of patients with levels that correlate with IL-6, lactate, anaphylatoxin C3a, elastase, and other chemokines [125, 192–195]. High levels of IL-8 have been regularly associated with higher mortality in adults [195–197] as well as in children [198]. Similarly, higher levels of IL-8 are predictor of infection in adults [199–201], children [135, 202], and neonates [203]. In addition to its presence as a free plasma molecule, IL-8 can be found in large amounts associated with circulating leukocytes [204]. Such a measurement offers a better AUC to predict sepsis in post-operative patients than plasma IL-8 [205]. Of note, among patients with MOF, significantly higher levels of IL-8 were measured in septic patients than in non-septic ones [195]. Similarly, higher levels of IL-8 were found in patients with septic shock as compared to SIRS patients [150]. As well, higher levels of IL-8 were found in febrile patients with bacterial infection as compared to febrile patients without bacterial infection [206]. However, another study failed to find any differences between sepsis and SIRS [123]. Accuracy of IL-8 for diagnosis can be improved by combining values with those of G-CSF [177] or with CRP and soluble CD25 [207]. When associated with CRP, the follow-up of IL-8 could help to reduce unnecessary antibiotic therapy in newborns [208].

2.6.2 IP-10 (CXCL10)

IP-10 is another member of the CXCL family, which displays some defensin-like antimicrobial activity. It is with MIG (CXCL9) and I-TAC (CXCL11), a chemokine induced by interferons. Accordingly, it is widely used in viral infections, and it may not be the best marker to use if viral infection could affect the studied population such as sepsis patients who may reactivate asymptomatic viral infections. In a study aimed to identify late-onset bacterial infection in preterm infants, IP-10 was shown among 11 tested chemokines and cytokines to provide the highest overall sensitivity (93 %) and specificity (89 %) with a cutoff value ≥1.25 ng/mL [209]. However, IP-10 was found in only some of the patients with urosepsis [210]. Furthermore, similar levels of IP-10 were reported in patients with pyelonephritis with negative or positive blood cultures. Accordingly, further studies are needed to fully appreciate the interest of IP-10.

2.6.3 Monocyte Chemotactic Factor-1 (CCL2)

Since its description in the plasma of sepsis patients [193], MCP-1 levels were found to be positively correlated with the levels of circulating endotoxin in patients with meningococcemia and to be far higher in patients with fulminant meningococcal sepsis than in patients with mild infection [211] and higher in non-survivors than in surviving children with *Neisseria meningitidis* sepsis [212]. A similar observation linking the levels of MCP-1 with outcome was reported for adult sepsis [213]. Among febrile neutropenic children, MCP-1 levels were significantly higher in those with a documented clinical sepsis and/or a local infection than in the group with unexplained fever [214]. However, no significant differences were noticed between patients with SIRS and patients with septic shock [150] or between febrile patients with or without bacterial infection [206].

2.6.4 Macrophage Inflammatory Protein-1α/β (CCL3. CCL4)

MIP-1α levels in meningococcal sepsis led to similar conclusion as MCP-1 in terms of severity and outcome [211, 212]. In other sepsis adult patients, neither MIP-1α nor MIP-1β levels correlate with mortality [196, 213], although MIP-1β levels predict outcome in pediatric sepsis [215]. In human volunteers injected with LPS, cyclooxygenase inhibitors enhanced MIP-1α levels. Slightly higher levels of MIP-1β levels were reported in septic shock than in SIRS [150], but neither MIP-1α nor MIP1β could help to discriminate between febrile patients with or without bacterial infection [206].

2.6.5 RANTES (CCL5)

RANTES acronym is supposed to correspond to "regulated upon activation, normal T cell expressed and (presumably) secreted"; however, it was given by its discoverers [216] to recall the name of a character seen in an Argentine science fiction movie called "Man Facing Southeast." RANTES behaves exactly as the opposite to any other inflammatory mediators. Its levels are lower in septic shock than in sepsis, in non-survivors than in survivors, and are inversely correlated with clinical score [125, 217]. Reduced levels of RANTES are similarly found in invasive fungal infection [218], in cerebral malaria [219], in septic neonates [220], and in patients with hematological malignancy undergoing chemotherapy [221]. Its levels are far lower in sepsis patients with the most severe thrombocytopenia (unpublished observation). Altogether, it appears that most circulating RANTES is derived from platelets and is associated with their reduced number. Data are still missing to appreciate the usefulness of RANTES to help to discriminate between SIRS and sepsis adult patients. In preterm infants, significantly lower levels of RANTES were reported in infected neonates as compared to the noninfected group [222].

2.7 Soluble Receptors

2.7.1 Soluble CD14 (sCD14)

CD14 is part of the LPS receptor that shuttles the endotoxin to the MD2 component associated with TLR4. Soluble CD14 is produced by hepatocytes and can also be considered as an acute phase protein [223]. The levels of sCD14 in the plasma of septic patients are higher in patients with MOF than in those without MOF [224] and are associated with mortality in patients with Gram-negative septic shock [225] or Gram-positive sepsis [226]. Soluble CD14 is similarly enhanced in neonatal sepsis [227] and could distinguish between Gram-positive and Gram-negative infections [228]. The levels are higher in critically ill neonates with sepsis than without, but this is not the case in children [229]. In adult, the levels are higher in those with bacterial infection than those with viral infection [230]. In trauma patients, levels of sCD14 are slightly higher in patients with sepsis than in non-septic ones [231]. More recently, a sCD14 subtype (sCD14-ST, presepsin) was identified, and its levels were shown to be much higher in sepsis than in subjects with SIRS [232]. However, this sCD14-ST/presepsin was unable to discriminate pediatric patients with fever of unknown origin and with negative blood culture from patients with bacteremia and sepsis [233].

2.7.2 Soluble MD2 (sMD2)

MD2 is associated with the TLR4 molecule and binds LPS. Its soluble form is present in the plasma of sepsis patients but not in those of healthy controls or patients with chronic inflammation [234]. Soluble MD2 is an acute phase protein [235]. To date no data are available comparing levels in sepsis and non-infectious SIRS.

2.7.3 Soluble ST2 (sST2)

ST2 is part of the IL-33 receptor, also made with the IL-1 receptor accessory protein (IL-1RAP). High levels of soluble ST2 have been reported in sepsis patients, higher than in trauma and abdominal surgery patients [236]. In sepsis, its levels weakly correlates with the clinical scores and IL-8 and IL-10 levels [237]. Non-survivors displayed elevated sST2 compared with survivors within the ICU [237].

2.7.4 Soluble CD25 (sCD25)

The IL-2 receptor is composed by three chains. The α-chain (CD25) is involved in the binding of IL-2 and is released from activated lymphocytes. High levels of sCD25 are reported in the plasma of sepsis patients, and higher concentrations are found in patients with organ failure [238]. In neutropenic cancer children with febrile episodes, sCD25 could not help to distinguish between fever of unknown origin and patients with infection [239]. Similarly, in emergency department, sCD25 measurements in febrile patients could not discriminate between patients with or without bacterial infection [206]. However, these observations were not confirmed in another study that reported higher levels in bacteremic/sepsis patients versus patients with fever of unknown origin [233]. Plasma levels of sCD25 were significantly higher in sepsis patients, compared to the levels in SIRS patients [240] or in non-septic patients admitted in an emergency department [206]. At admission of patients with SIRS, sCD25 levels were higher in bacteremic patients and in non-survivors [152]. Its AUC is close to that of PCT (0.81 versus 0.80, respectively) as well as its sensitivity and specificity (87.5 and 75.0 versus 91.3 and 62.5, respectively).

2.7.5 Soluble CD163 (sCD163)

CD163 is the receptor for the haptoglobin-hemoglobin complexes and acts as a scavenger receptor for hemoglobin. It is mainly shed from monocytes/macrophages in a metalloproteinase-dependent fashion. Levels of sCD163 are significantly increased in sepsis patients as compared to healthy controls and are higher in non-survivors. However, this later observation was not made in patients older than 75 [241]. Levels were not significantly different between sepsis, severe sepsis, and septic shock and weakly correlated with IL-6 and IL-10 [91], and sCD163 did not discriminate between infected and non-infected patients [242]. Of note, far higher levels of sCD163 were found in patients with hemophagocytic syndrome [243]. However, in a recent study, sCD163 was found significantly higher in sepsis than in SIRS and also significantly higher in severe sepsis than in moderate sepsis [244]. In this study, the AUC of sCD163 to distinguish between SIRS and sepsis was better than PCT (0.86 versus 0.63, respectively).

2.7.6 Soluble TNF Receptors (sTNF Rs)

There are two TNF receptors (p55 TNF R/TNF R-I/CD120a and p75 TNF R/TNF R-II/CD120b). Both receptors can be shed from the cell surface and behave as inhibitors of TNF. However, the half-life of the sTNF receptors complexed with their ligands is greatly different [245]. Enhanced levels of sTNF R-I and sTNF R-II were found in patients with sepsis as compared to healthy controls, and higher levels were found in non-survivors [246, 247]. There is a strong correlation between both soluble receptors [248, 249] and clinical scores [247] and the occurrence of organ failure [250]. In post-cardiac surgical patients, TNF R-I was significantly higher in patients who had high risk of sepsis [251], but in critically burned patients, on day 1, no significant differences for both receptors were found between sepsis and non-sepsis patients [252]. In fact, sTNF R was shown to be a predictive marker for the development of SIRS in patients after cardiopulmonary bypass [253] or trauma [254]. Altogether, these studies do not suggest that sTNF Rs could be useful markers to dissociate sepsis form SIRS.

2.7.7 Soluble Decoy Receptor 3

DcR3 is a decoy receptor in the TNF receptor superfamily. It binds particularly FasL and "homologous to lymphotoxins, inducible expression, competes with HSV glycoprotein D for HVEM, a receptor expressed on T-lymphocytes" (LIGHT) and plays a regulatory role in preventing their capacity to induce apoptosis. Its soluble form has been shown to discriminate between SIRS and sepsis patients, the latter having the highest plasma levels [255]. In sepsis patients, it correlated with the APACHE II score ($r = 0.56$). In patients with acute respiratory distress syndrome (ARDS), levels of soluble DcR3 discriminated the survivors and the non-survivors, and the occurrence of septic shock was more frequent among those who had the highest levels of DcR3 [256].

2.7.8 Soluble TREM-1

See Chapter 19.

2.7.9 Soluble Urokinase-Type Plasminogen Activator Receptor (suPAR)

See Chapter 20.

2.8 Vascular Endothelial-Related Biomarkers

2.8.1 Soluble Adhesion Molecules

Cell recruitment is initiated following the adhesion of circulating leukocytes to activated endothelial cells that neo-express adhesion molecules. E-selectin (also known as endothelial-leukocyte adhesion molecule 1 (ELAM-1/CD62E) allows the rolling of the cells on the endothelium, whereas intercellular adhesion molecule-1 (ICAM-1/CD54) and vascular cell adhesion molecule 1

(VCAM-1/CD106) allow a firm adhesion before migration. The corresponding ligands are sialyl-LewisX, lymphocyte function-associated antigen 1 (LFA1/CD11a/CD18), and α4β1 integrin, respectively. These membrane bound adhesion molecules can undergo proteolytic cleavage generating soluble forms. The plasma levels of all three soluble forms have been shown to be enhanced in patients with sepsis or SIRS and to be higher in non-survivors [257–260]. In patients with sepsis, the levels of all three soluble adhesion molecules correspond to the severity of the illness [261, 262]. A relative modest ($r = 0.48$) but significant correlation was reported between sE-selectin and sICAM-1 [262]. Different studies mentioned a higher levels of sE-selectin, sVCAM-1 and sICAM-1 in sepsis than in patients with trauma [263, 264], critically ill medical ICU patients [265, 266], patients with hypotension [267], and patients classified as noninfectious SIRS [268]. Similarly, higher levels of sE-selectin and sICAM-1 were reported in neonatal sepsis when compared to non-infected infants [36, 269].

However, comparing the circulating levels in sepsis and SIRS did not always end with significant differences. Indeed, even if significant p-values were obtained, there was often a large overlapping of the values. No difference was reported in ICU patients for sICAM-1 [261] and sE-selectin [270]. In patients after major abdominal surgery, sVCAM-1 could not discriminate between post-operative ($n = 28$) and sepsis ($n = 101$) groups, whereas sICAM-1 was significantly higher in sepsis patients [106]. In another study that also included a reasonable number of patients to lead to trustable data, those were the opposite with sVCAM-1 higher in sepsis patients ($n = 162$) versus non-infectious SIRS ($n = 162$), while sICAM-1 was not associated with sepsis. There is presently no clear explanation to understand these discrepancies.

2.8.2 Angiopoietin-2

Angiopoietin-1, produced by pericytes, acts on the Tie2-receptor on endothelial cells and favors their cell cycle. This process is antagonized by angiopoietin-2 (Ang-2), produced by endothelial cells, which inhibits angiogenesis. Ang-2 is recognized as an autocrine regulator of endothelial cell inflammatory response, favoring the action of TNF on these cells [271].

Levels of Ang-2 were first reported to be enhanced in patients with severe sepsis as compared to mild sepsis or controls [272]. This was further confirmed, and the levels of Ang-2 in severe sepsis were shown significantly higher in patients with severe sepsis than in patients with non-infectious SIRS [273–275]. Ang-2 levels were found to be significantly correlated with those of TNF ($r = 0.65$) [273], IL-6 ($r = 0.65$) [274], and sICAM-1 ($r = 0.58$) [276] and with clinical score (APACHE II: $r = 0.53$; sequential organ failure assessment (SOFA) :$r = 0.79$) [277]. Similarly, Ang-2 levels were higher in children with septic shock as compared to critically ill children and those with sepsis [278]. The enhanced level of Ang-2 is predictive of acute lung injury [279]. The proportion of surviving patients with levels below the median value of Ang-2 was higher than that of patients with levels above the median [274, 277]. Similarly, in patients with trauma, Ang-2 levels were higher in patients with worse clinical outcome [280]. Monocytes of sepsis patients were shown to be a source of Ang-2 [281]. Levels of soluble Tie2 are not significantly different between sepsis and non-sepsis patients [282]. Angiopoietin-1 levels are either similar or even lower in sepsis patients as compared to healthy controls [282, 283].

2.8.3 Vascular Endothelial Growth Factor (VEGF)

Vascular endothelial growth factor (VEGF) promotes proliferation, migration, and survival of endothelial cells, but it also favors endothelial permeability, induces the expression of cell adhesion molecules, and upregulates procoagulant activity. VEGF levels were shown to be enhanced in severe sepsis as compared to healthy controls and to correlate with the multiple organ dysfunction score [189] and cytokine levels (IL-1β, IL-10, IL-12) [284]. Surprisingly, lower levels are associated to hematological and renal dysfunction and poorer outcome [285]. But, opposite results were also reported, showing higher levels in septic shock non-survivors as compared to septic shock survivors [286]. Levels of VEGF were higher in sepsis patients than in mechanically ventilated critically ill patients [282] or than in hematological patients with neutropenic fever without infection [287].

2.8.4 Endothelin-1

Endothelin (Et-1) is a 21-amino acid peptide produced by endothelial cells with potent vasoconstricting properties derived from a large precursor (around 200 a.a.). There are two other isoforms produced by different genes. Levels of Et-1 are significantly more elevated in sepsis patients than in cardiac surgery patients and healthy controls [288]. Its levels were higher in non-surviving sepsis patients and correlate with those of thrombomodulin [289]. In burned patients, Et-1 concentrations were significantly higher in the patients who developed sepsis than in those who did not [290]. Its precursor form (big Et-1, 38 a.a.) is also present in enhanced concentration in sepsis patients and correlates with outcome [291]. Another precursor form, C-terminal-proEt-1 (45 a.a.) was also higher in sepsis and septic shock than in SIRS [292]. Newborns with positive hemoculture and severe sepsis had significantly higher levels of Et-1 as compared to other newborns with mild or moderate sepsis [293].

2.8.5 Adrenomedullin

Adrenomedullin (AM) is a hypotensive peptide acting locally as a vasorelaxant and as a systemic vasodilator, produced by vascular smooth muscle cells and endothelial cells. The gene encodes a 185-amino acid preprohormone, which after cleavage generates pro-AM, a 164-amino acid peptide. This prohormone is further processed into two biologically active peptides: AM (a.a. #95 to #146.) and pro-adrenomedullin NH2-terminal 20 peptide (PAMP, a.a. #22 to #41). A third fragment derived from pro-AM, the mid-regional fragment of pro-adrenomedullin (MR-proADM) (a.a. #45-92), is the more stable part of adrenomedullin.

Levels of adrenomedullin are enhanced in sepsis patients [294] and correlate significantly with decreases in diastolic blood pressure, systemic vascular resistance index, and pulmonary vascular resistance index values [295] as well as with levels of CRP ($r = 0.63$) [296]. AM levels are also enhanced in SIRS patients (burns, pancreatitis, trauma), although at lower concentrations than in patients with traumatic shock and severe sepsis [297]. Patients with septic shock displayed the highest levels, but whether there is a significant difference between survivors and non-survivors remains controversial [297, 298].

Pro-AM has been proposed as prognostic marker in neonatal sepsis [299] and in adult sepsis [300]. In the latter case, although there was a statistical difference between SIRS and sepsis patients, the overlapping of the values between both groups was such that pro-AM per se cannot be useful to predict infection in patients at onset of fever.

MR-proADM levels were reported to be higher among patients with sepsis, severe sepsis, or septic shock as compared to SIRS patients and to correlate with simplified acute physiology score (SAPS II) score ($r = 0.5$), IL-6 ($r = 0.53$), and PCT ($r = 0.65$) and to be higher among non-surviving patients [301]. Similar observations were published later on [292, 302].

2.8.6 Endocan

Endocan is a dermatan sulfate proteoglycan expressed by lung and kidney endothelial cells. The presence of its soluble form was found in higher quantities in patients with septic shock as compared to severe sepsis and sepsis [303]. As well, levels in sepsis were higher than in SIRS patients. Its highest correlations were observed with the levels of IL-10 ($r = 0.59$) and that of von Willebrand factor ($r = 0.60$). It could also discriminate between survivors and non-survivors. With a cutoff of 1.2 ng/mL, it could help to distinguish between SIRS and sepsis with a sensitivity of 0.825 % and a sensitivity of 100 %. While a cathepsin G-generated 14 kDa fragment is also found in the plasma of sepsis patients, there was no correlation noticed between this fragment and the whole molecule [304].

2.8.7 Heparin-Binding Protein

Heparin-binding protein (HBP), contained within the secretory and azurophilic granules of human neutrophils, induces cytoskeletal rearrangement of endothelial cells leading to vascular leakage. Plasma levels of HBP are significantly higher in patients with severe sepsis and septic shock as compared to patients with non-infectious SIRS or sepsis [305]. HBP levels could also help to discriminate between bacterial and viral infections (AUC = 0.84) [230].

2.8.8 Growth Arrest-Specific Protein 6

Growth arrest-specific protein 6 (GAS6) is a 75 kDa antiapoptotic vitamin K-dependent protein released by different cells including endothelial cells. GAS6 is a regulator of the vascular system and mediates endothelial cell survival. GAS6 promotes and accelerates the sequestration of circulating platelets and leukocytes on activated endothelium. It is also a negative regulator of the inflammatory response acting on immune cells but favors atherosclerosis and thrombus formation. GAS6 plasma concentrations in patients with severe sepsis are higher than in healthy controls and slightly higher than non-septic patients with organ failure [306, 307].

2.9 Enzymes

2.9.1 Elastase

Neutrophil elastase (also called elastase 2) is a protease that can break down certain bacterial compounds, but it also acts as a matrix-degrading protease. Elastase activity is neutralized by the acute phase protein, α1-antitrypsin (also known as α1-proteinase inhibitor). Enhanced levels of elastase were reported in patients soon after undergoing abdominal surgery, but in patients who developed sepsis, the levels were further increased [308]. In fact, the neutrophil elastase has been recovered associated with its inhibitor. The level of the complex was found to remain enhanced over a week in patients with septic shock complicated by MOF, whereas it returns to normal in patients who recovered from a hemorrhagic shock within 24 h of blood transfusion or surgery [309]. Levels of elastase and elastase complexed with its inhibitor were higher in non-surviving sepsis patients [310, 311]. The levels of the complex were significantly higher in patients with non-infectious SIRS than in patients with SIRS and microbial infection [311], whereas it was not the case when elastase was measured by its own [312]. In addition, in children with septic shock, elastase levels were predictive of the occurrence of acute kidney injury [313]. Interestingly, a significant correlation ($r = 0.69$) was reported in patients with septic shock between elastase and IL-8, a potent neutrophil activator [194].

2.9.2 Metalloproteinases

Metalloproteinases (MMPs) are a family of around 23 zinc-containing endoproteinases that participate in matrix degradation and remodeling. Levels of MMP9 (gelatinase B) but not MMP2 (gelatinase A) are higher in septic patients compared to healthy controls. A similar observation was made for their inhibitors, known as tissue inhibitors of matrix-metalloproteinases (TIMP)-1 and TIMP-2. Levels of TIMP-1 were higher in non-survivors [314]. Only a trend was observed for increased levels of MMP9 when comparing healthy controls and sepsis patients, while a significant enhancement was found for MMP10 (stromelysin 2) [315]. In another study, the authors succeeded to find a significant difference between healthy controls and patients with severe sepsis regarding the levels of MMP2 [316]. The same study also reported enhanced levels in sepsis for MMP8 (neutrophil collagenase). In a study that investigated few metalloproteinases, the authors failed to find significant difference between severe sepsis and healthy controls for MMP2, but reported significant increased levels for MMP3 (stromelysin 1) and MMP7 (matrilysin), and confirmed previous reports on MMP8 and MMP9 [317]. They also showed enhanced levels of TIMP-1, TIMP-2, and TIMP-4. Levels of MMP2 and TIMP-1, but not MMP9, were shown to be higher in non-surviving sepsis patients than in survivors [318]. Finally, MMP1 levels in sepsis patients were markedly elevated and correlated with death [319]. Most surprisingly, in a study that investigated both SIRS and sepsis patients, the authors reported an extremely important capacity of MMPs1, 2, 7, and 13 to distinguish between the two groups of patients (AUC$\geq$096), but in contradiction with other studies, they reported a decreased concentration of all four MMPs in sepsis patients as compared to healthy controls [320].

2.9.3 Phospholipase A2

Phospholipase A2 (PLA2) is present in many tissues and allows the generation of arachidonic acid from membrane phospholipids. This is the first requested step for the production of prostaglandins, leukotrienes, and platelet-activating factor. The first demonstration that PLA2 levels were elevated in the plasma of patients with septic shock was reported in 1984 [321]. The concentration of PLA2 was higher in patients with sepsis or non-septic bacterial infection than in those with viral infection [322]. The concentration of PLA2 correlated well with the concentration of CRP (r=0.61) [322], thrombomodulin (r=0.76) [323], and MOF score (r=0.67) [324]. PLA2 levels measured in trauma patients showed that levels were far higher in septic shock patients [325]. However, among patients with hematological malignancies, levels of PLA2 could not discriminate patients with fever of unknown origin from patients with sepsis [139]. Plasma levels were significantly higher in the patients who died of sepsis than in those who survived the illness [326]. The levels in patients with sepsis or septic shock were far higher at admission in ICU than levels measured in patients with multiple injuries [324] or patients with burns [327]. Similarly, in newborns admitted in neonatal ICU, levels of PLA2 were higher in neonates with documented sepsis [328]. Finally, in an emergency department, PLA2 was a better diagnosis marker of sepsis than CRP [329].

2.9.4 YKL-40

YKL-40 was named based on its three NH2-terminal amino acids tyrosine (Y), lysine (K), and leucine (L) and its molecular weight of 40 kDa. It is a member of the "mammalian chitinase-like proteins." It has some role in extracellular remodeling and angiogenesis. Circulating levels of YKL-40 were first reported to be increased in cancer and type 2 diabetes. Interestingly, a proteomic analysis of sera of sepsis patients identified YKL-40 as a new biomarker of sepsis [330]. The authors showed that the mean level in healthy controls is around 100 ng/mL and is higher than 1 μg/mL in patients with severe sepsis or septic shock. Off-pump coronary artery bypass grafting patients had YKL-40 levels similar to healthy controls. In patients with *S. pneumoniae* bacteremia, YKL-40 levels correlated with those of the soluble form of urokinase-type plasminogen activator (suPAR) [331]. A recent study revealed that levels of YKL-40 in SIRS and sepsis patients were similar, whereas patients with severe sepsis and particularly patients with septic shock had higher levels of YKL-40. The lowest levels at time of ICU admission were associated with better survival. These levels were under the influence of a single nucleotide polymorphism (SNP) of the YKL-40 gene [332].

2.9.5 Granzyme A

Granzyme A is a serine protease constitutively present in the intracytoplasmic granules of cytotoxic T cells, NK cells, and NKT cells. Plasma granzyme A was significantly decreased in septic rather than in non-septic burn patients and healthy controls [333]. However, this study came in contrast with previous ones which reported enhanced levels of granzyme A in severe sepsis patients as compared to healthy controls [159, 334].

2.10 Coagulation, Fibrinolysis, and Hemostasis Biomarkers

There is an interplay between inflammation and coagulation. Inflammatory cytokines favor an enhanced expression of tissue factor on monocytes and endothelial cells leading to increased production of prothrombin that is converted to thrombin and that in turn generates fibrin from fibrinogen. Similarly, inflammatory cytokines increase the levels of the plasminogen-activator inhibitor-1 (PAI-1) resulting in an impaired production of plasmin and thus a failure of normal fibrinolytic mechanisms. The net result is enhanced formation of fibrin clots in the microvasculature, leading to impaired tissue oxygenation and cell damage [335]. Reciprocally, factors generated during coagulation such as thrombin or factor Xa display pro-inflammatory properties. Accordingly, sepsis is associated with an altered hemostatic balance between procoagulant and anticoagulant mechanisms, and there is a relationship between severity of coagulation and inflammation abnormalities and mortality in sepsis patients.

2.10.1 Antithrombin

Antithrombin (AT) is a 432-amino acid single-chain glycoprotein of 58 kDa. It is a serine protease inhibitor that inactivates several enzymes of the coagulation cascade. In surgical patients, low or falling levels of AT were associated with a development of sepsis [336]. Reduced levels of AT were regularly reported to be associated with the occurrence of sepsis in ICU patients [337, 338] and trauma patients [339]. Among patients with sepsis, the levels of AT were lower in patients with organ dysfunction [340] and in non-survivors as compared to survivors over a week period [341].

Indeed, the levels of AT were shown to correlate with outcome of critically ill patients with suspected sepsis, although the discriminative power was poor as compared to clinical scores such as logistic organ dysfunction score (LODS) or APACHE score [342]. Furthermore, AT levels in patients with disseminated intravascular coagulation (DIC) display a rather low correlation with the SOFA score ($r = 0.37$) [343]. In patients with community-acquired pneumonia, AT levels were lower in men than in women [344]. Finally, AT levels were also shown to be reduced in newborns with late-onset sepsis and to discriminate between survivors and non-survivors [345].

2.10.2 Protein C

Activated protein C (APC) is generated following the cleavage of its precursor by thrombomodulin. Activated protein C is an inhibitor of the coagulation cascade and displays certain anti-inflammatory properties. Recombinant human activated protein C has been available for a decade to treat patients with severe sepsis, before being withdrawn from the market for lack of efficacy. Injection of endotoxin in human volunteers leads to a modest but significant decreased level of protein C [346]. Protein C deficiency is prevalent in the majority of septic patients and is associated with increased morbidity and mortality in patients with severe sepsis and septic shock [347, 348]. Protein C levels are significantly less in patients with sepsis as compared to patients with pneumonia [349]. Interestingly, the occurrence of DIC does not modify the levels of protein C found in sepsis patients [350], although protein C levels in patients with severe sepsis correlate with the international normalized ratio (INR) used to measure the extrinsic pathway of coagulation ($r = 0.7$) [346]. Among patients with organ failure, the levels of protein C were similar in septic and non-septic patients [351], but protein C deficiency was seen to be associated with subsequent pulmonary, renal, and hematologic organ failure [352]. Protein C levels in severe sepsis patients modestly correlated with the SOFA score ($r = 0.5$) [346]. In emergency department patients, levels of protein C associated with those of neutrophil gelatinase-associated lipocalin and IL-1Ra were predictive of severe sepsis and septic shock [353].

2.10.3 Thrombomodulin

Thrombomodulin (TM) is a glycoprotein expressed on the surface of endothelial cell, mesothelial cell, monocyte, and a subset of dendritic cell that acts as a receptor of thrombin and neutralizes its clotting activity. The soluble form is considered as a marker of endothelial cell injury. Its levels were higher in sepsis patients who developed organ failure [354], in non-surviving sepsis patients and correlate with those of endothelin-1 ($r = 0.63$) [289] and type II phospholipase A2 ($r = 0.76$) [323]. TM plasma levels in patients with sepsis were significantly higher than in non-septic critically ill and trauma patients [355, 356] and were suggested to be a predictive marker of sepsis and MOF in trauma patients [357]. Serum concentrations of

TM were higher in sepsis patients with DIC or MOF [358]. Of note, TM expression on monocytes was similarly enhanced in sepsis and trauma patients as compared to healthy controls [351]. Interestingly, elevated levels of TM in the most severe patients seem to be a marker rather than a deleterious mediator since treatment by recombinant TM of sepsis patients with DIC led to improved mortality [359].

2.10.4 Plasminogen Activator Inhibitor Types 1 and 2

Altered fibrinolysis is a hallmark of sepsis, illustrated by an enhanced level of plasminogen activator inhibitor types 1 and 2 (PAI-1 and PAI-2) that counteract the action of tissue plasminogen activator (tPA) and urokinase that act upstream of fibrinolysis by converting plasminogen into plasmin. PAI-1 is mainly produced by endothelial cells. Patients with septic shock have significantly enhanced levels of PAI-1, which have a strong predictive value for MOF, DIC, and mortality [360–362]. Similar observations were reported in trauma patients [363]. In sepsis patients, after the start of antibiotic treatment, high concentrations of PAI-1 persisted in the non-survivors in contrast to decreasing concentrations in most of the survivors [364]. PAI-1 levels in patients with septic shock correlate with IL-6 ($r = 0.53$) [365] and with the SOFA score. The single base-pair insertion/deletion (4G allele) within the PAI-1 promoter polymorphism is associated with high concentrations of PAI-1 in the plasma and a poor survival rate after severe trauma [366]. It also confers an increase in the risk of mortality in patients with septic shock due to a greater organ failure [367]. Regarding PAI-2, it was mainly detected in non-surviving sepsis patients [368].

2.10.5 von Willebrand Factor

von Willebrand factor (VWF) mediates the adherence of platelets to one another and to sites of vascular damage. This action is important in the formation of blood clot. VWF also acts as a carrier for factor VIII in the circulation. It can form very large multimers. Present in normal plasma (10 μg/mL), it is produced by bone marrow cells and endothelial cells. Plasma levels of VWF were found higher in septic shock patients than in patients after traumatic shock [369]. In sepsis patients, VWF levels were predictive for the development of acute lung injury [370] and were significantly higher in non-survivors [371]. Levels of VWF in patients with severe sepsis were far higher than in patients with uncomplicated SIRS after cardiopulmonary bypass [372]. In parallel among sepsis patients, the proteolytic activity of VWF inactivating protease, ADAMTS13, stepwise declined with the severity of inflammation.

2.11 Hormones

2.11.1 Sex Steroid Hormones

There is a profound alteration of the levels of sex steroid hormones in sepsis patients as compared to non-septic shock: estrone and estradiol are dramatically enhanced in female patients with sepsis and septic shock and in male patients with septic shock. In contrast, in male patients with sepsis, the levels of testosterone are significantly decreased [373, 374]. The low testosterone concentrations in the severely ill male patients correlated inversely with the APACHE score [375]. In critically ill trauma and surgical patients, estradiol but not testosterone levels were significantly higher in non-survivors [376]. In patients in a surgical ICU who had severe life-threatening illnesses with or without sepsis, prolactin levels were normal, although hypoprolactinemia was reported in 50 % of children with nosocomial sepsis and MOF [377].

2.11.2 Leptin

Leptin is a 16 kDa hormone made by adipocytes that acts on the hypothalamus to regulate food intake. Mean plasma leptin levels were threefold higher in critically ill patients than controls. The controls exhibited a nyctohemeral fluctuation in plasma leptin levels with peak levels at 11 pm; in contrast, septic patients had no nocturnal rise of leptin. Mean leptin levels were threefold higher in patients who survived a septic episode [378]. Leptin levels were found to be increased in patients with sepsis and even more in patients with septic shock. Survivors had higher levels, and leptin levels correlated with those of IL-1Ra ($r = 0.82$), IL-10 ($r = 0.79$), and sTNF R-I ($r = 0.66$) and to a lesser degree with IL-6 ($r = 0.55$) [379]. Levels of leptin were significantly higher in non-infected critically ill patients than in sepsis [380]. While in this later study, sepsis patients had lower levels than healthy controls; other studies failed to report any significant difference between healthy controls and critically ill patients and between ICU sepsis and non-sepsis patients [381, 382]. This absence of difference was confirmed in critically ill patients on admission in ICU, but on the second day of admission, leptin levels were significantly enhanced in SIRS and sepsis patients as compared to non-SIRS patients [383].

2.11.3 Vasopressin and Copeptin

Vasopressin is a 9 a.a. antidiuretic hormone derived from a pre-pro-vasopressin precursor of 164 a.a. from which copeptin, a 38 a.a. peptide, is also derived. The levels of vasopressin and copeptin are enhanced in ICU patients as compared to healthy controls. Their levels are higher in non-infected SIRS patients and in patients after cardiac surgery than in patients with sepsis and correlate to each other ($r = 0.73$) [384]. In contrast, levels of vasopressin and copeptin are increased in pediatric sepsis and even more in children with septic shock [385]. Similarly, in adult, copeptin levels increase progressively with the severity of sepsis and were an independent predictor of mortality in ventilator-associated pneumonia [386]. A similar association with outcome was reported in community-acquired pneumonia [387]. In patients with febrile neutropenia, no correlation was observed between copeptin levels and disease severity, and median levels were similar between patients without bacteremia and those with positive blood culture [388].

2.11.4 Natriuretic Peptides

Natriuretic peptides comprise a family of 3 structurally related molecules: atrial natriuretic peptide (ANP, 28 a.a.), brain natriuretic peptide (BNP, 32 a.a.), and C-type natriuretic peptide (CNP, 22 a.a.). ANP and BNP, but not CNP, have hormonal natriuretic and vasodilating activity. ANP and BNP are mainly secreted from the heart. CNP is produced by vascular endothelial cells. BNP is a sensitive diagnostic marker for heart failure.

Among septic patients, ANP levels had a better predictive value for outcome than CRP, PCT, IL-6, and even the APACHE II score [389], although this remains controversial [390]. Nevertheless, ANP levels correlated well with IL-6 levels on the day of diagnosis of septic shock (r=0.73) [391], and the N-terminal prohormone form of ANP does not seem to be predictive of outcome [392].

BNP levels in patients with septic shock reflect left ventricular dysfunction and inversely correlated to cardiac index (r=0.56) [391], and high plasma levels are associated with poor outcome [390, 392, 393]. Similarly, the N-terminal-proBNP displays predictive values for the outcome of patients with sepsis or septic shock [394]. The association between BNP levels and cardiovascular dysfunction has also been noticed in SIRS patents and to reflect the severity of the SIRS [395]. Nevertheless, in patients with sepsis or septic shock, BNP concentrations were increased regardless of the presence or absence of cardiac dysfunction [396], although some reports consider BNP as a reliable marker for identification of patients developing sepsis-induced myocardial depression [397]. These discrepancies may reflect the time of the analysis of BNP after admission and diagnosis. BNP levels correlate with APACHE II score (r=0.58) [398] or SOFA score (from r=0.58 [390] to r=0.86 [399]). Higher levels of BNP were reported in emergency department patients with SIRS than with sepsis [400]. However, another study reported that in an emergency department patients who had the highest levels of BNP had the greater risk of development of severe sepsis or septic shock [401].

CNP levels are also increased in sepsis patients as compared with patients with congestive heart failure or chronic renal failure [402]. The N-terminal fragment of CNP allows to distinguish among multiple-traumatized patients without traumatic brain injury those who will develop sepsis [403] and is higher in ICU patients at admission and on day 3 among those who will not survive [404].

2.12 Miscellaneous

2.12.1 Fibronectin

Fibronectin is a 440 kDa glycoprotein present in normal human plasma (≈300 μg/mL), produced by hepatocytes that favor the clearance of particulate debris. In the 1980s, numerous papers reported that fibronectin levels were significantly reduced in sepsis patients either in adults or neonates. Similar observations were reported in SIRS patients (burns, major surgery, trauma). Among ICU patients after major elective surgery or trauma, levels of fibronectin were lower among patients with sepsis [405]. A similar difference was reported in burns, but only from the 6th postburn day [406]. In DIC patients, there is a significant negative correlation between protein C and fibronectin [407]. In patients with fever, only those with sepsis had a significantly reduced level of fibronectin [408].

2.12.2 Others

Still many other plasma biomarkers have been mentioned in literature. In this paragraph, we will only focus on those for which a different expression has been reported in sepsis and SIRS patients. This is the case of endogenous *morphine* shown to be in part secreted by neutrophils during sepsis. The levels of serum levels of morphine were significantly higher in patients with sepsis, severe sepsis, and septic shock as compared to SIRS patients over 3 days of monitoring [409]. *Selenium* is an essential trace element. During ICU stay, there was a significant decrease in plasma selenium concentration. Patients with severe sepsis or septic shock had the lowest plasma selenium at study inclusion as compared to SIRS patients [410]. *Osteopontin* is involved in bone remodeling. Osteopontin concentration in the serum was tenfold higher in SIRS than in healthy controls and was even higher in patients with severe sepsis and septic shock and seemed to be associated with the clinical outcome [411]. *Gelsolin*, a 84 kDa protein, which exists in a cytoplasmic as well as an excreted form, is a key regulator of actin filament assembly and disassembly. At admission, plasma gelsolin levels were significantly lower in patients with severe sepsis than in non-septic critically ill ICU patients and healthy control individuals. It did not significantly differ between surviving and non-surviving [412]. Finally, the *anaphylatoxin C3a* is also enhanced in sepsis patients. Plasma levels in patients with shock and in normotensive septic patients were found similar but significantly higher than in SIRS patients [413]. C3a was found a better predictor biomarker of sepsis (AUC = 0.90) than PCT (AUC = 0.82). The AUC of a score comprising PCT and C3a values was 0.93 [312].

3 Leukocyte Biomarkers

3.1 Cell-Surface Biomarkers

3.1.1 HLA-DR

Human leukocyte antigen (HLA-DR) is a member of the family of major histocompatibility class II molecules that are expressed on the surface of macrophages and other antigen-presenting cells (APCs). Monocytes constitutively and strongly express HLA-DR. In *in vitro* experiments aimed to induce endotoxin tolerance, there is a downregulation of HLA-DR on monocytes, which correlates with impaired LPS-induced production of proinflammatory cytokines and a reduced capacity to generate reactive oxygen species [414]. It is worth mentioning that endotoxin tolerance partially mimics the reprogramming of monocytes observed in SIRS and sepsis patients [415]. The decrease of HLA-DR expression onto monocytes was first reported in trauma patients [416]. Those who developed sepsis had a further decreased expression, which was even more pronounced in those who ultimately died. In a survey of patients undergoing abdominal vascular surgery, it was shown that the expression of HLA-DR on CD14^{high} monocytes decreased rapidly during surgery, faster than on the CD14^{low} subsets [417]. IL-10 and glucocorticoids contribute to this observation [417–419]. In septic patients, the decreased cell-surface expression of HLA-DR has regularly been observed on circulating monocytes [420–422] and is now considered a reliable indicator of reprogramming in critically ill patients. Decreased HLA-DR expression in ICU patients has been shown to be predictive of infections after trauma [423], surgery [424], transplantation [425], pancreatitis [426] and in burn patients [427]. In some instances, the diagnosis of infection was further improved by combining HLA-DR expression and IL-10 levels [423, 428]. Surveys were often required to precisely detect the occurrence of sepsis, and it is rather a persistent low level of HLA-DR which is associated with the development of nosocomial infections and sepsis [429–432].

3.1.2 TLR4

Toll-like receptor-4 (TLR4) belongs to the family of transmembrane receptors consisting of an extracellular leucine-rich repeat domain that interacts with PAMPs and DAMPs and an intracellular Toll/IL-1 receptor (TIR) domain, which is required for signaling. TLR4, associated with another molecule, MD2, is the main receptor for LPS that triggers the activation of nuclear factor-κB and production of pro-inflammatory cytokines [433]. Monocytes and neutrophils constitutively express TLR4, which undergo strong upregulation following LPS stimulation [434]. It has been shown that TLR4 expression increases in both monocytes and neutrophils from sepsis patients [435–437]. However, there was no significant correlation between the APACHE II score and the expression of this receptor [438]. In septic neonates, TLR4 showed no significant changes compared to healthy subjects [439]. Moreover, it has been reported that TLR expression is differentially regulated during sepsis between men and women [440]. The different modulation and baseline expression of TLR4 depending on gender and probably on age (observation reported in mice, [441]) in sepsis patients could explain controversial data obtained in other studies where no difference in terms of cell-surface TLR4 expression on monocytes and on neutrophils between sepsis patients and ICU subjects without sepsis were found [442, 443]. In trauma patients, TLR4 expression was even found decreased as compared to healthy controls, while TLR2 expression was unchanged [444]. Finally, it is worth mentioning that during sepsis, TLR4 expression is also modulated in other leukocytes such as NK cells, which express

TLR4 mainly intracellularly. An enhanced intracellular TLR4 expression in both sepsis and SIRS patients compared to healthy controls was reported [445]. Interestingly, an increase in the percentage of NK cells positive for TLR4 surface expression was mainly observed in SIRS patients.

3.1.3 CD14

CD14 is a glycosylphosphatidylinositol (GPI)-anchored protein that functions as a co-receptor for several ligands of TLRs, including LPS. Indeed, CD14 is a direct partner of TLR4. CD14 is mainly expressed on monocytes and at a lesser degree on neutrophils. A very significant ($p < 0.001$) decreased cell-surface CD14 expression was reported in sepsis patients as compared with healthy controls [446]. Interestingly, a study including 142 critically ill patients with community-acquired pneumonia noticed that decreased expression of CD14 on monocytes could predict 28-day mortality [447]. Indeed, the level of CD14 downregulation positively correlated with sepsis severity (APACHE II) and mortality [448]. Membrane CD14 has been poorly studied to differentiate sepsis from SIRS patients.

3.1.4 CD25

CD25 is a type I transmembrane protein, corresponding to the α-chain of the IL-2 receptor. CD25 is expressed in conventional T cells following activation. Indeed, CD25 defines a subset of CD4+ T cells (regulatory T cells, Treg) with suppressor activity through contact-mediated direct inhibition of other cells of the immune system and through secretion of soluble CD25, IL-4, IL-10, and TGF-β. Increased percent of circulating CD4+CD25+ Treg has been reported during septic shock, although the absolute cell count remained similar to healthy control, as a reflection of the decrease of the other T-lymphocyte subsets through apoptosis [449–451]. The proportion of Treg increased 3 days after the onset of shock. Even though there was an inverse correlation between severity scores (SAPS II, SOFA, or arterial lactate level) and Treg proportion, a similar pattern of Treg kinetics was found in infected and non-infected patients, and the time course was similar between survivors and non-survivors [452]. However, in another study of a small cohort of patients, the authors concluded that an increased proportion of CD4+CD25+Foxp3+ Treg could significantly discriminate between sepsis and SIRS patients [240]. Later on, it was confirmed that the percentage of CD4+CD25+Foxp3+ Treg was significantly higher during the early stage of sepsis as compared to SIRS patients and control group ($p = 0.003$) [453].

3.1.5 CD40

CD40 is a member of the TNF receptor family, broadly expressed, including on monocytes and neutrophils, and is the receptor of the CD40 ligand. GM-CSF, IL-3, or IFN-γ triggers upregulation of this receptor in primary human monocytes. Increased CD40 expression has been reported in monocytes from sepsis subjects compared to healthy volunteers. Interestingly, a correlation between the peak of CD40 expression on monocytes of patients with severe sepsis and mortality ($p = 0.05$) was observed [454]. An increased expression of CD40 on circulating monocytes in septic patients, within the first 24 h after admission in ICU, was further confirmed [455]. Higher levels of CD40 were detected in patients with circulatory failure/septic shock and normalization of the receptor expression was observed in later time points when the severity of illness decreased indicating a tight correlation between modulation of CD40 expression and APACHE II and SAPS II scores.

3.1.6 CD48

The CD48 is a GPI protein, belonging to the CD2 family. CD48 is a pan leukocyte cell-surface antigen, and its expression is modulated by bacterial and viral products. Only one study addressed the value of CD48 as biomarker of interest in infectious disease. Interestingly, the authors reported CD48 upregulation in monocytes and neutrophils from infected patients compared to healthy controls, whereas no differences were found on lymphocytes [456]. CD48 expression rather reflects the disease activity of infectious diseases, especially of viral infections.

3.1.7 CD64

CD64 is probably one of the most extensively studied cell-surface marker in sepsis. CD64, the high-affinity IgG receptor FcγRI, is constitutively expressed on monocytes and to a very low extent on resting neutrophils. But increased CD64 cell-surface expression reflects the neutrophil activation status. It has been shown that the plasma collected from septic patients triggers upregulation of neutrophil CD64 expression in vitro [457]. Several groups reported that neutrophil CD64 serves as a highly sensitive and specific marker for systemic infection and sepsis in neonates and children. CD64 expression was found significantly upregulated in sepsis children compared to SIRS infants, with a sensitivity of 70 % and a specificity of 62 % [458]. Another study confirmed the data, reporting that the CD64 index achieved diagnostic accuracy within the first 24 h of suspected sepsis in children (AUC = 0.88) and 24 h later in both neonates (AUC = 0.96) and children (AUC = 0.98) [459]. A recent prospective study of CD64 in a large cohort of critically ill neonates strongly supports the use of this marker in this setting. In a population of 684 neonates with a low prevalence of sepsis (5 %), increased CD64 expression showed a sensitivity of 75 % and a specificity of 77 %, using an optimal cutoff [460].

In adults, during early sepsis, neutrophil CD64 expression was shown to correlate with disease severity and with mortality within 28 days (OR = 1.3, $p = 0.01$ by logistic regression analysis) [197]. In a small group of critically ill adults, CD64 could discriminate sepsis from SIRS patients with a sensitivity comparable to PCT but with higher specificity [461].

These data were confirmed by a larger study, enrolling a higher number of patients ($n = 300$), [462]. However, one recent study reported a lower CD64 sensitivity than previously observed (63 % versus 84 %) [463]. The reason for such a discrepancy could rely on the fact that the later study only included patients with a documented infection, therefore increasing the cutoff of the test. Indeed, the high performance of CD64 as diagnostic tool has been confirmed in a very recent large prospective study. CD64 could detect sepsis over 500 ICU patients with a sensitivity of 89 % and specificity of 87 % [464].

3.1.8 CD69

CD69 is a C-type II lectin receptor expressed as a homodimer. Upon antigenic stimulation, both T lymphocytes and NK cells quickly express CD69. Therefore, CD69 has been mostly considered as one of the earliest activation markers. T cell subsets are found regularly activated during sepsis. Indeed, CD4+ and CD8+ lymphocytes isolated from sepsis patients displayed up to threefold increase in the expression of CD69 compared with controls ($p < 0.05$) [465, 466]. An increased level of CD69 expression was also detected on NK cells from sepsis patients. However, no significant difference was found between SIRS and sepsis patients indicating that CD69 is not able to discriminate an infection from a sterile insult [445]. Following a cohort of 52 septic shock patients during the first 28 days, a significant increase in the counts and percentages of CD69+ NK cells at ICU admission and on day 3 was found in non-survivors indicating that NK cells from non-survivors were very early activated [467].

3.1.9 CD80

CD80 (B7-1) is part of a large family of co-stimulatory molecules critical for T cell activation. CD80 is mainly expressed on APCs. However, freshly isolated human monocytes do not express CD80 at both mRNA and protein level. Indeed, CD80 is induced and increased rapidly in vitro culture in response to cytokines or LPS stimulation. CD80 surface expression levels were found to be upregulated in septic patients at admission in ICU compared to healthy subjects [455]. Interestingly, the authors reported a positive correlation between the level of expression of the receptor and the severity of the disease ($p = 0.002$), suggesting that CD80 could predict development of septic shock. So far, too little is known about the capacity of CD80 to differentiate sepsis from non-infectious SIRS.

3.1.10 TREM-1

TREM-1 belongs to the immunoglobulin superfamily and is expressed on human neutrophils and monocytes. TREM-1 is strongly upregulated when PMNs are exposed to microbial products and play a critical role in sepsis [468]. Contradictory clinical data regarding TREM-1 expression on monocytes have been reported. Among 25 septic shock patients, 15 patients with shock of non-infectious origin, and 7 healthy volunteers, TREM-1 expression was significantly higher in septic shock patients than in non-septic patients, and there was no difference in monocytic TREM-1 expression between non-septic patients and healthy volunteers [469]. In contrast, another study found that TREM-1 expression on monocytes was equally increased in both SIRS ($n = 58$) and sepsis ($n = 14$) patients [470]. Furthermore, it was lately shown that the monocyte expression of TREM-1 was even decreased in patients with severe sepsis or septic shock as compared to patients with sepsis [471]. This study suggested that TREM-1 modulation upon transition from sepsis to shock was dependent on the type of infection and on the causative pathogen. These data could explain the discrepancies observed in the different clinical studies.

3.1.11 CX3CR1

The seven transmembrane-spanning G-protein-coupled receptor, CX3CR1, is the specific receptor for the CX3CL1 chemokine fractalkine. CX3CR1 is expressed on monocytes, NK cells, Th1, and cytotoxic lymphocytes. In vitro, fractalkine triggers CD11b upregulation in monocytes, which represents a key event in monocytes chemotaxis and in their infiltration into the tissues. A microarray study, aimed to identify a signature of genes whose peripheral blood mRNA expression could efficiently discriminate survivor from non-survivor septic shock patients, led to the identification of CX3CR1 as a candidate [472]. Later on, the same group reported that sepsis patients displayed a decrease in CX3CR1 surface expression in comparison with healthy controls, and when the patients were stratified according to mortality, non-survivor patients showed a significantly lower expression during the course of the disease [473]. However, so far, no further studies have addressed the performance of CX3CR1 on monocytes in discriminating SIRS versus sepsis patients.

3.1.12 PD-1

The inhibitory receptor programmed death-1 (PD-1) belongs to the CD28 family of molecules and is a negative regulator of activated T cells. It has been shown that PD-1 is inducible and expressed on B and T lymphocytes, NKT cells, and monocytes upon activation and plays a critical role in the pathophysiology of sepsis, since PD-1 knock-out mice are profoundly resistant to CLP-induced mortality [474]. Increased PD-1 expression on CD4+ lymphocytes and on monocytes was detected in sepsis patients compared with healthy subjects. Interestingly, in the same cohort, patients who developed a secondary nosocomial infection showed higher monocyte PD-1 expression in comparison with patients who did not develop a subsequent infectious episode [475]. A study including 19 septic patients also confirmed increased PD-1 expression on CD4+ lymphocytes in comparison with healthy controls [476]. An increased PD-1 expression on monocytes, granulocytes, and lymphocytes in critically ill surgical patient showed a positive correlation with the severity of the illness [477]. Still, further studies are needed to elucidate the capacity of PD-1 in differentiating between an infection and a sterile insult.

3.1.13 BTLA

B- and T-lymphocyte attenuator (BTLA) is a co-inhibitory receptor belonging to the CD28 family of the immunoglobulin superfamily. BTLA is not expressed in naive T cells, but it is induced during activation and remains expressed on T helper type 1 exclusively. In mice, BTLA is upregulated on CD4+ T cells and B cells following peritonitis [478]. In sepsis patients, an increased percentage of BTLA expressing CD4+ T cells was observed compared with SIRS patients and healthy subjects. Moreover, increased level of BTLA correlated positively with poor outcome and longer hospital length stay [478]. Further investigations are necessary to confirm the value of BTLA as candidate biomarker.

3.2 mRNA (Transcriptomic)

A large number of investigators have studied the global messenger RNA (mRNA) expression in circulating leukocyte sepsis patients. One bias in these studies is that they have mainly been performed with whole blood samples, and accordingly, the transcriptomic analysis is greatly influenced by the relative presence of each leukocyte subset. In most cases, once activated leukocytes leave the bloodstream, circulating leukocytes are not the main source of all mediators/biomarkers found in plasma. This is illustrated for example by the fact that mRNAs coding for IL-6 and TNF are not found in blood leukocytes of sepsis [191] or trauma patients [479], while these cytokines are found in the plasma. Few teams have been able to propose a transcriptomic signature that could help to decipher between patients. For example, 35 genes could help to discriminate between viral and bacterial infection and 30 other genes could allow to distinguish between *E. coli* and *S. aureus* infection [480]. But another study failed to confirm any difference between Gram-negative and Gram-positive sepsis [481]. Among the 459 genes differentially expressed in sepsis and SIRS, 65 were downregulated, whereas 395 were upregulated in sepsis as compared to SIRS. However, most of the fold changes were below two, except for TLR5, IL-18R1, "TRAF-interacting protein with a forkhead-associated domain" (TIFA), "B-cell lymphoma 2" (Bcl-2), and "suppressor of cytokine signaling 3" (SOCS3) [482]. Other studies have been reporting gene signature able to discriminate between SIRS and sepsis. One report identified 138 genes that displayed this property with an 80–91 % accuracy [483]. Another study identified 79 SIRS-specific genes, 42 sepsis-specific genes, and 941 septic shock-specific genes in children below 10 years of age [484]. A set of 50 signature genes correctly identified sepsis among critically ill patients with an 88–91 % accuracy [485]. With only a combination of three genes (*cd3d*, *il1b*, and *tnf*), it was possible with a 90 % specificity and 85 % sensitivity to predict postoperative sepsis [486]. More recently, a panel of 7 genes, three being upregulated (*tlr5*, *cd59*, *clusterin*) and 4 being downregulated (*il7r*, *fibrinogen-like 2*, *major histocompatibility complex class II dp alpha1*, *and carboxypeptidase vitellogenic like*) described the magnitude of immune alterations and was superior to CRP or PCT in discriminating non-infectious patients from sepsis patients [487]. On another hand, transcriptomic analysis can also be employed as prognosis tool in patients with septic shock and the use of 28 genes discriminated non-survivors from survivors with a sensitivity of 100 % and a specificity of 86 % [472]. Surprisingly, none of these studies converged to identify similar key genes. This may reflect the great heterogeneity of the patients (age, gender, underlying disease, time of analysis after onset of sepsis, genetic background, site of infection, nature of the bacteria, etc.). This difficulty is illustrated in a study that addressed the early transcriptomic response of children with septic shock: a huge discrepancy of gene expression was found between neonates, toddlers, infants, and school-age children [488]. However, a systematic review of transcriptomic analysis ended with the important conclusion that the distinction of separating sepsis into pro-inflammatory and anti-inflammatory phases was not supported by gene expression data [489].

3.3 miRNAs

MicroRNAs (miRNAs) are noncoding, single-stranded RNAs (21–24 nucleotides in length) that control gene expression post-transcriptionally by inhibiting the translation of mRNA or by degradation of the mRNA itself [490]. MicroRNAs regulate important processes such as cell proliferation, adhesion, apoptosis, and angiogenesis [491].

A large body of research has highlighted the role of miRNAs in the endotoxin tolerance phenomenon and pointed out the value of circulating microRNAs as diagnostic markers in sepsis. The production of miR-146a was reported for the first time in 2000 in LPS-stimulated human monocytes, and LPS-triggered upregulation of miR-146a was shown to be NF-κB dependent [492]. It was then shown to be associated with capacity of the cells to produce TNF and to play a critical role in endotoxin tolerance in a human monocytic cell line [493]. Furthermore, miR-146a expression was found upregulated in circulating monocytes purified from septic patients [494]. Surprisingly, serum miR-146a levels, measured by quantitative PCR in 50 sepsis patients, 30 SIRS patients and 20 healthy controls, were found lower in septic patients compared with SIRS patients and healthy donors. The predictive value of serum miR-146a levels was better than that of IL-6 (AUC = 0.804 and 0.785, respectively) yielding with a given cutoff to a 100 % specificity and a 63 % sensitivity [495]. These data were later confirmed with a smaller cohort of patients [496]. miR-15a and miR-16 were also reported to serve as diagnostic markers for sepsis. In a group of 166 sepsis patients and 32 SIRS patients, levels of both miRNAs in sepsis and SIRS patients were found significantly higher than in healthy controls. However, miR-15a could be used to distinguish sepsis patients from SIRS patients. Indeed, the area under the ROC curve for miR-15a was 0.858, which was much higher than the curves for CRP (AUC=0.57) and PCT (AUC=0.61) [497]. MicroRNA can also be prognosis markers. A combination of four microRNA markers in the serum (miR-15a, miR-16, miR-193*, and miR-483-5p) could predict sepsis mortality at 28 days in a cohort of 214 sepsis patients with a sensitivity of 88.5 % and a specificity of 90.4 % [498]. Downregulation of miR-150 has been described in leukocytes of human volunteers upon treatment with LPS [499]. Interestingly, circulating miR-150 levels were found reduced in a small cohort of 17 sepsis patients as compared to healthy people [500]. In this study, the authors found that the expression levels of miR-150 correlated with those of main established markers of inflammation, such as TNF, IL-10, and IL-18, and with the SOFA score. Data about miR-150 as a valuable biomarker for the diagnosis of sepsis are quite controversial.

On one hand, a study confirmed that miR-150 was downregulated during sepsis compared with SIRS and healthy subjects and yielding to an AUC of 0.83 [501]. On the other hand, another one failed to confirm those results, finding that median miR-150 serum levels were not significantly different in septic patients compared to non-septic critically ill patients or healthy controls, in a large cohort of patients ($n=223$) [502]. In this study, the authors showed that miRNA150 was rather a prognosis marker, low concentration being associated with a lower prognosis of survival among the critically ill patients. Recently, high miR-133a levels were also found to be predictive of unfavorable prognosis in critically ill patients. Serum miR-133a levels were measured in 223 critically ill patients (138 with sepsis and 85 without sepsis) and found significantly elevated at ICU admission when compared with healthy controls ($n=76$). Even though miR-133a levels could not discriminate between sepsis and SIRS patients, correlation analyses revealed significant association of miR-133a with disease severity, classical markers of inflammation, bacterial infection, and organ failure [503]. The study of miRNA seems a promising approach, although contradictory results illustrate that like for other plasma markers, many parameters including experimental ones still need to be taken into consideration. The genome-wide sequencing of cellular microRNAs rather than circulating microRNA should also provide new insights in the identification of diagnostic sepsis candidates.

4 Combinations

None of the individual markers has the highest expected value to ascertain with 100 % sensitivity and 100 % specificity the occurrence of sepsis in a patient with SIRS. Accordingly, although some may be of some help and can be proposed to be used in routine, an obvious suggestion was to make a combination of few markers to

Table 5
Combinations of few biomarkers proposed to increase accuracy in identifying sepsis patients among noninfectious SIRS patients

IL-6 + CRP	[133]
IL-6 + PMN count	[134]
IL-6 + IL-10	[146]
IL-10 + HLA DR expression	[422, 427]
IL-8 + GM-CSF	[177]
IL-8 + sCD25 + CRP	[207]
CRP + sICAM-1 + sE-selectin	[36]
CRP + temperature	[18]
PCT + mid-regional-proadrenomedullin	[505]
PCT + sTREM-1 + CD64	[462]
PCT + C3a	[312]
PSP + (sCD25 or PCT)	[66]
suPAR, sTREM-1, MIF, CRP, and PCT	[28]
IL-1Ra + protein C + gelatinase-associated lipocalin	[353]

define a better diagnosis tool (Table 5). Thanks to the setup of new technologies that allow the measurement of a large number of markers within a very small volume of samples, the simultaneous measurement of few markers can now be easily proposed. Clinical scores are not good predictors of the incidence of nosocomial infection in SIRS patients [504], and even when associated with some biomarkers, they remain better prognosis indicator than diagnosis marker. This was shown in sepsis patients for whom levels of MCP-1 slightly improved the accuracy of APACHE II score (AUC = 0.89 instead of 0.85) [109]. In a neonatal and pediatric ICU, the combined use of IL-8 and GM-CSF levels to diagnose infection improved the sensitivity of each individual markers (57 % each) to 67 % but did not change the specificity (93 %) [177]. Another combination was proposed for the diagnosis of neonatal sepsis associating CRP, IL-8, and sCD25, with a specificity of 85 % and a specificity of 97 % [207]. In adult, a combination of C3a and PCT led to an AUC = 0.93 [312] for the diagnosis of sepsis. In addition, suPAR, sTREM-1, MIF, CRP, and PCT ended to a AUC = 0.88 for the detection of bacterial infection in patients with SIRS [28]. Another combination, including gelatinase-associated lipocalin, IL-1Ra, and protein C, ended to an AUC = 0.80 for the diagnosis of sepsis in emergency department patients [353]. An impressive AUC = 0.998 was reported when PCT and mid-regional pro-adrenomedullin were combined for the early diagnosis of sepsis [505]. An interesting scoring system was proposed associating two plasma markers, sTREM-1 and PCT, and a cell-surface marker, CD64, on neutrophils [462]. A "bioscore" was developed, attributing one point per biomarker with a value above the optimal cutoff point. With a AUC = 0.95, it was better than any individual marker. At admission, when the bioscore reached 3, 100 % of the patients had sepsis and 93 % of patients with a "bioscore" = 2 had sepsis.

5 Conclusions

Thousands of reports have been published on the use of biomarkers to define sepsis patients. However, the quality of the trials designed to define these biomarkers was not always sufficient to warrant solid and definite conclusions [506]. Among the most frequently bias are the choice of inappropriate statistical tests, the sample size (in fact, most studies have gathered less than 100 patients), and the sampling storage (for how long and at which temperature have the samples been kept before measurement, have they been thawed before). Rarely, the influence of variables such as age, gender, and medication and the presence of various underlying diseases have been addressed to ensure the quality of the biomarkers. Most importantly, the group of SIRS patients is extremely heterogeneous and the questions might be different: Do we wish to discriminate sepsis patients in an emergency ward receiving a large variety of patients or to determine the occurrence of nosocomial infection among a homogenous group of patients after a cardiopulmonary bypass surgery or after transplantation, or the occurrence of sepsis in an ICU among SIRS patients with different type of disorders? While there is hope that an appropriate combination will emerge from the ongoing studies, will it lead to a routine test? Companies are supporting numerous researches in that direction, but usually they prefer their own devices, which would be mandatory for the measurement of the defined biomarkers. While both cell-surface markers and plasma markers could lead to interesting combination as the one defined by S. Gibot et al. [462], what could be the feasibility in a routine lab? Then, efforts and supports are still needed before a trustable test could be used at the bedside.

References

1. Bone RC, Grodzin CJ, Balk RA (1997) Sepsis: a new hypothesis for pathogenesis of the disease process. Chest 121:235–243
2. Adib-Conquy M, Cavaillon JM (2009) Compensatory anti-inflammatory response syndrome. Thromb Haemost 101:36–47
3. Cavaillon JM, Muñoz C, Fitting C, Misset B, Carlet J (1992) Circulating cytokines: the tip of the iceberg ? Circ Shock 38:145–152
4. Pierrakos C, Vincent JL (2010) Sepsis biomarkers: a review. Crit Care 14:R15
5. Kumar A, Roberts D, Wood KE, Light B, Parrillo JE et al (2006) Duration of hypotension before initiation of effective antimicrobial therapy is the critical determinant of survival in human septic shock. Crit Care Med 34:1589–1596
6. Nobre V, Harbarth S, Graf JD, Rohner P, Pugin J (2008) Use of procalcitonin to shorten antibiotic treatment duration in septic patients: a randomized trial. Am J Respir Crit Care Med 177:498–505

7. Vincent JL, Beumier M (2013) Diagnostic and prognostic markers in sepsis. Expert Rev Anti Infect Ther 11:265–275
8. Reinhart K, Wiegand-Lohnert C, Grimminger F, Kaul M, Withington S et al (1996) Assessment of the safety and efficacy of the monoclonal anti-tumor necrosis factor antibody-fragment, MAK 195F, in patients with sepsis and septic shock: a multicenter, randomized, placebo-controlled, dose-ranging study. Crit Care Med 24:733–742
9. Tillet W, Francis T (1930) Serological reactions in pneumonia with a non-protein somatic fraction of pneumococcus. J Exp Med 52:561–571
10. McCarty M (1947) The occurrence during acute infection of a protein not normally present in the blood. IV Crystallization of the C-reactive protein. J Exp Med 85:491–498
11. Osmand AP, Friedenson B, Gewurz H, Painter RH, Hofmann T et al (1977) Characterization of C-reactive protein and the complement subcomponent C1t as homologous proteins displaying cyclic pentameric symmetry (pentraxins). Proc Natl Acad Sci U S A 74:739–743
12. Meynaar IA, Droog W, Batstra M, Vreede R, Herbrink P (2011) In critically Ill patients, serum procalcitonin is more useful in differentiating between sepsis and SIRS than CRP, Il-6, or LBP. Crit Care Res Pract 2011:594645
13. Fitting C, Parlato M, Adib-Conquy M, Memain N, Philippart F et al (2012) DNAemia detection by multiplex PCR and biomarkers for infection in systemic inflammatory response syndrome patients. PLoS One 7:e38916
14. Su L, Han B, Liu C, Liang L, Jiang Z et al (2012) Value of soluble TREM-1, procalcitonin, and C-reactive protein serum levels as biomarkers for detecting bacteremia among sepsis patients with new fever in intensive care units: a prospective cohort study. BMC Infect Dis 12:157
15. Sierra R, Rello J, Bailen MA, Benitez E, Gordillo A et al (2004) C-reactive protein used as an early indicator of infection in patients with systemic inflammatory response syndrome. Intensive Care Med 30:2038–2045
16. Meisner M, Tschaikowsky K, Palmaers T, Schmidt J (1999) Comparison of procalcitonin (PCT) and C-reactive protein (CRP) plasma concentrations at different SOFA scores during the course of sepsis and MODS. Crit Care 3:45–50
17. Ohlin A, Bjorkqvist M, Montgomery SM, Schollin J (2010) Clinical signs and CRP values associated with blood culture results in neonates evaluated for suspected sepsis. Acta Paediatr 99:1635–1640
18. Povoa P, Coelho L, Almeida E, Fernandes A, Mealha R et al (2005) C-reactive protein as a marker of infection in critically ill patients. Clin Microbiol Infect 11:101–108
19. Tsalik EL, Jaggers LB, Glickman SW, Langley RJ, van Velkinburgh JC et al (2012) Discriminative value of inflammatory biomarkers for suspected sepsis. J Emerg Med 43:97–106
20. Tschaikowsky K, Hedwig-Geissing M, Schmidt J, Braun GG (2011) Lipopolysaccharide-binding protein for monitoring of postoperative sepsis: complemental to C-reactive protein or redundant? PLoS One 6:e23615
21. Tschaikowsky K, Hedwig-Geissing M, Braun GG, Radespiel-Troeger M (2011) Predictive value of procalcitonin, interleukin-6, and C-reactive protein for survival in postoperative patients with severe sepsis. J Crit Care 26:54–64
22. Schmit X, Vincent JL (2008) The time course of blood C-reactive protein concentrations in relation to the response to initial antimicrobial therapy in patients with sepsis. Infection 36:213–219
23. Boraey N, Sheneef A, Mohammad M, Yousef L (2012) Procalcitonin and C- reactive protein as diagnostic markers of neonatal sepsis. Aust J Basic Appl Sci 6:108–114
24. Simon L, Gauvin F, Amre DK, Saint-Louis P, Lacroix J (2004) Serum procalcitonin and C-reactive protein levels as markers of bacterial infection: a systematic review and meta-analysis. Clin Infect Dis 39:206–217
25. Yu CW, Juan LI, Wu MH, Shen CJ, Wu JY et al (2013) Systematic review and meta-analysis of the diagnostic accuracy of procalcitonin, C-reactive protein and white blood cell count for suspected acute appendicitis. Br J Surg 100:322–329
26. Yu CW, Juan LI, Hsu SC, Chen CK, Wu CW et al (2013) Role of procalcitonin in the diagnosis of infective endocarditis: a meta-analysis. Am J Emerg Med 31:935–941
27. Lyu YX, Yu XC, Zhu MY (2013) Comparison of the diagnostic value of procalcitonin and C-reactive protein after hematopoietic stem cell transplantation: a systematic review and meta-analysis. Transpl Infect Dis 15:290–299
28. Kofoed K, Andersen O, Kronborg G, Tvede M, Petersen J et al (2007) Use of plasma C-reactive protein, procalcitonin, neutrophils, macrophage migration inhibitory factor, soluble urokinase-type plasminogen activator receptor, and soluble triggering receptor expressed on myeloid cells-1 in combination

to diagnose infections: a prospective study. Crit Care 11:R38
29. Chan T, Gu F (2011) Early diagnosis of sepsis using serum biomarkers. Expert Rev Mol Diagn 11:487–496
30. Uhlar CM, Whitehead AS (1999) Serum amyloid A, the major vertebrate acute-phase reactant. Eur J Biochem 265:501–523
31. Cicarelli DD, Vieira JE, Bensenor FE (2008) Comparison of C-reactive protein and serum amyloid a protein in septic shock patients. Mediators Inflamm 2008:631414
32. Casl MT, Rogina B, Glojnaric-Spasic I, Minigo H, Planinc-Peraica A et al (1994) The differential diagnostic capacity of serum amyloid A protein between infectious and non-infectious febrile episodes of neutropenic patients with acute leukemia. Leuk Res 18:665–670
33. Arnon S, Litmanovitz I, Regev RH, Bauer S, Shainkin-Kestenbaum R et al (2007) Serum amyloid A: an early and accurate marker of neonatal early-onset sepsis. J Perinatol 27:297–302
34. Enguix A, Rey C, Concha A, Medina A, Coto D et al (2001) Comparison of procalcitonin with C-reactive protein and serum amyloid for the early diagnosis of bacterial sepsis in critically ill neonates and children. Intensive Care Med 27:211–215
35. Ucar B, Yildiz B, Aksit MA, Yarar C, Colak O et al (2008) Serum amyloid A, procalcitonin, tumor necrosis factor-alpha, and interleukin-1beta levels in neonatal late-onset sepsis. Mediators Inflamm 2008:737141
36. Edgar JD, Gabriel V, Gallimore JR, McMillan SA, Grant J (2010) A prospective study of the sensitivity, specificity and diagnostic performance of soluble intercellular adhesion molecule 1, highly sensitive C-reactive protein, soluble E-selectin and serum amyloid A in the diagnosis of neonatal infection. BMC Pediatr 10:22
37. Lannergard A, Viberg A, Cars O, Karlsson MO, Sandstrom M et al (2009) The time course of body temperature, serum amyloid A protein, C-reactive protein and interleukin-6 in patients with bacterial infection during the initial 3 days of antibiotic therapy. Scand J Infect Dis 41:663–671
38. Smith K, Bigham M (2011) Biomarkers in pediatric sepsis. Open Inflamm J 4:24–30
39. Prucha M, Herold I, Zazula R, Dubska L, Dostal M et al (2003) Significance of lipopolysaccharide-binding protein (an acute phase protein) in monitoring critically ill patients. Crit Care 7:R154–R159
40. Sakr Y, Burgett U, Nacul FE, Reinhart K, Brunkhorst F (2008) Lipopolysaccharide binding protein in a surgical intensive care unit: a marker of sepsis? Crit Care Med 36:2014–2022
41. Gaini S, Koldkjaer OG, Pedersen C, Pedersen SS (2006) Procalcitonin, lipopolysaccharide-binding protein, interleukin-6 and C-reactive protein in community-acquired infections and sepsis: a prospective study. Crit Care 10:R53
42. Watkin RW, Harper LV, Vernallis AB, Lang S, Lambert PA et al (2007) Pro-inflammatory cytokines IL6, TNF-alpha, IL1beta, procalcitonin, lipopolysaccharide binding protein and C-reactive protein in infective endocarditis. J Infect 55:220–225
43. Albillos A, de-la-Hera A, Alvarez-Mon M (2004) Serum lipopolysaccharide-binding protein prediction of severe bacterial infection in cirrhotic patients with ascites. Lancet 363:1608–1610
44. Porcel JM, Vives M, Cao G, Bielsa S, Ruiz-Gonzalez A et al (2009) Biomarkers of infection for the differential diagnosis of pleural effusions. Eur Respir J 34:1383–1389
45. Muller B, Peri G, Doni A, Torri V, Landmann R et al (2001) Circulating levels of the long pentraxin PTX3 correlate with severity of infection in critically ill patients. Crit Care Med 29:1404–1407
46. Vänskä M, Koivula I, Hamalainen S, Pulkki K, Nousiainen T et al (2011) High pentraxin 3 level predicts septic shock and bacteremia at the onset of febrile neutropenia after intensive chemotherapy of hematologic patients. Haematologica 96:1385–1389
47. Uusitalo-Seppala R, Huttunen R, Aittoniemi J, Koskinen P, Leino A et al (2013) Pentraxin 3 (PTX3) is associated with severe sepsis and fatal disease in emergency room patients with suspected infection: a prospective cohort study. PLoS One 8:e53661
48. Mauri T, Bellani G, Patroniti N, Coppadoro A, Peri G et al (2010) Persisting high levels of plasma pentraxin 3 over the first days after severe sepsis and septic shock onset are associated with mortality. Intensive Care Med 36:621–629
49. Huttunen R, Hurme M, Aittoniemi J, Huhtala H, Vuento R et al (2011) High plasma level of long pentraxin 3 (PTX3) is associated with fatal disease in bacteremic patients: a prospective cohort study. PLoS One 6:e17653
50. Okorie ON, Dellinger P (2011) Lactate: biomarker and potential therapeutic target. Crit Care Clin 27:299–326
51. Douzinas EE, Tsidemiadou PD, Pitaridis MT, Andrianakis I, Bobota-Chloraki A et al (1997) The regional production of cytokines and lactate in sepsis-related multiple organ failure. Am J Respir Crit Care Med 155:53–59

52. Hack CE, Nuijens JH, Strack van Schijndel RJ, Abbink JJ, Eerenberg AJ et al (1990) A model for the interplay of inflammatory mediators in sepsis–a study in 48 patients. Intensive Care Med 16(Suppl 3):S187–S191
53. Gogos CA, Lekkou A, Papageorgiou O, Siagris D, Skoutelis A et al (2003) Clinical prognostic markers in patients with severe sepsis: a prospective analysis of 139 consecutive cases. J Infect 47:300–306
54. Trzeciak S, Dellinger RP, Chansky ME, Arnold RC, Schorr C et al (2007) Serum lactate as a predictor of mortality in patients with infection. Intensive Care Med 33:970–977
55. Nguyen HB, Rivers EP, Knoblich BP, Jacobsen G, Muzzin A et al (2004) Early lactate clearance is associated with improved outcome in severe sepsis and septic shock. Crit Care Med 32:1637–1642
56. Marty P, Roquilly A, Vallee F, Luzi A, Ferre F et al (2013) Lactate clearance for death prediction in severe sepsis or septic shock patients during the first 24 hours in intensive care unit: an observational study. Ann Intensive Care 3:3
57. Jeng JC, Jablonski K, Bridgeman A, Jordan MH (2002) Serum lactate, not base deficit, rapidly predicts survival after major burns. Burns 28:161–166
58. Manikis P, Jankowski S, Zhang H, Kahn RJ, Vincent JL (1995) Correlation of serial blood lactate levels to organ failure and mortality after trauma. Am J Emerg Med 13:619–622
59. McNelis J, Marini CP, Jurkiewicz A, Szomstein S, Simms HH et al (2001) Prolonged lactate clearance is associated with increased mortality in the surgical intensive care unit. Am J Surg 182:481–485
60. del Portal DA, Shofer F, Mikkelsen ME, Dorsey PJ Jr, Gaieski DF et al (2010) Emergency department lactate is associated with mortality in older adults admitted with and without infections. Acad Emerg Med 17:260–268
61. Berg S, Brodin B, Hesselvik F, Laurent TC, Maller R (1988) Elevated levels of plasma hyaluronan in septicaemia. Scand J Clin Lab Invest 48:727–732
62. Sallisalmi M, Tenhunen J, Kultti A, Tammi M, Pettilä V (2013) Plasma hyaluronan and hemorheology in patients with septic shock: a clinical and experimental study. Microcirc, Clin Hemorheol
63. Yagmur E, Koch A, Haumann M, Kramann R, Trautwein C et al (2012) Hyaluronan serum concentrations are elevated in critically ill patients and associated with disease severity. Clin Biochem 45:82–87
64. Keel M, Harter L, Reding T, Sun LK, Hersberger M et al (2009) Pancreatic stone protein is highly increased during posttraumatic sepsis and activates neutrophil granulocytes. Crit Care Med 37:1642–1648
65. Que YA, Delodder F, Guessous I, Graf R, Bain M et al (2012) Pancreatic stone protein as an early biomarker predicting mortality in a prospective cohort of patients with sepsis requiring ICU management. Crit Care 16:R114
66. Llewelyn MJ, Berger M, Gregory M, Ramaiah R, Taylor AL et al (2013) Sepsis biomarkers in unselected patients on admission to intensive or high-dependency care. Crit Care 17:R60
67. Gukasjan R, Raptis DA, Schulz HU, Halangk W, Graf R (2013) Pancreatic stone protein predicts outcome in patients with peritonitis in the ICU. Crit Care Med 41:1027–1036
68. Delogu G, Lo Bosco L, Marandola M, Famularo G, Lenti L et al (1997) Heat shock protein (HSP70) expression in septic patients. J Crit Care 12:188–192
69. Sonna LA, Hawkins L, Lissauer ME, Maldeis P, Towns M et al (2010) Core temperature correlates with expression of selected stress and immunomodulatory genes in febrile patients with sepsis and noninfectious SIRS. Cell Stress Chaperones 15:55–66
70. Gupta A, Cooper ZA, Tulapurkar ME, Potla R, Maity T et al (2013) Toll-like receptor agonists and febrile range hyperthermia synergize to induce heat shock protein 70 expression and extracellular release. J Biol Chem 288:2756–2766
71. Waterer GW, ElBahlawan L, Quasney MW, Zhang Q, Kessler LA et al (2003) Heat shock protein 70-2 + 1267 AA homozygotes have an increased risk of septic shock in adults with community-acquired pneumonia. Crit Care Med 31:1367–1372
72. Pittet J-F, Lee H, Morabito D, Howard MB, Welch WJ et al (2002) Serum levels of Hsp 72 measured early after trauma correlate with survival. J Trauma 52:611–617
73. Wheeler DS, Fisher LE Jr, Catravas JD, Jacobs BR, Carcillo JA et al (2005) Extracellular hsp70 levels in children with septic shock. Pediatr Crit Care Med 6:308–311
74. Gelain DP, De Bittencourt Pasquali MA, Comim MC, Grunwald MS, Ritter C et al (2011) Serum heat shock protein 70 levels, oxidant status, and mortality in sepsis. Shock 35:466–470
75. Adib-Conquy M, Cavaillon JM (2007) Stress molecules in sepsis and systemic inflammatory response syndrome. FEBS Lett 581: 3723–3733
76. Wheeler DS, Lahni P, Odoms K, Jacobs BR, Carcillo JA et al (2007) Extracellular heat shock protein 60 (Hsp60) levels in children with septic shock. Inflamm Res 56:216–219

77. Rhodes A, Wort SJ, Thomas H, Collinson P, Bennett ED (2006) Plasma DNA concentration as a predictor of mortality and sepsis in critically ill patients. Crit Care 10:R60
78. Saukkonen K, Lakkisto P, Pettila V, Varpula M, Karlsson S et al (2008) Cell-free plasma DNA as a predictor of outcome in severe sepsis and septic shock. Clin Chem 54:1000–1007
79. Margraf S, Logters T, Reipen J, Altrichter J, Scholz M et al (2008) Neutrophil-derived circulating free DNA (cf-DNA/NETs): a potential prognostic marker for posttraumatic development of inflammatory second hit and sepsis. Shock 30:352–358
80. Moreira VG, Prieto B, Rodriguez JS, Alvarez FV (2010) Usefulness of cell-free plasma DNA, procalcitonin and C-reactive protein as markers of infection in febrile patients. Ann Clin Biochem 47:253–258
81. Huttunen R, Kuparinen T, Jylhava J, Aittoniemi J, Vuento R et al (2011) Fatal outcome in bacteremia is characterized by high plasma cell free DNA concentration and apoptotic DNA fragmentation: a prospective cohort study. PLoS One 6:e21700
82. Dwivedi DJ, Toltl LJ, Swystun LL, Pogue J, Liaw KL et al (2012) Prognostic utility and characterization of cell-free DNA in patients with severe sepsis. Crit Care 16:R151
83. Kung CT, Hsiao SY, Tsai TC, Su CM, Chang WN et al (2012) Plasma nuclear and mitochondrial DNA levels as predictors of outcome in severe sepsis patients in the emergency room. J Transl Med 10:130
84. Puskarich MA, Shapiro NI, Trzeciak S, Kline JA, Jones AE (2012) Plasma levels of mitochondrial DNA in patients presenting to the emergency department with sepsis. Shock 38:337–340
85. Yamanouchi S, Kudo D, Yamada M, Miyagawa N, Furukawa H et al (2013) Plasma mitochondrial DNA levels in patients with trauma and severe sepsis: time course and the association with clinical status. J Crit Care 28(6): 1027–31
86. Scaffidi P, Misteli T, Bianchi ME (2002) Release of chromatin protein HMGB1 by necrotic cells triggers inflammation. Nature 418:191–195
87. Wang H, Bloom O, Zhang M, Vishnubhakat JM, Ombrellino M et al (1999) HMG-1 as a late mediator of endotoxin lethality in mice. Science 285:248–251
88. Angus DC, Yang L, Kong L, Kellum JA, Delude RL et al (2007) Circulating high-mobility group box 1 (HMGB1) concentrations are elevated in both uncomplicated pneumonia and pneumonia with severe sepsis. Crit Care Med 35:1061–1067
89. Sunden-Cullberg J, Norrby-Teglund A, Rouhiainen A, Rauvala H, Herman G et al (2005) Persistent elevation of high mobility group box-1 protein (HMGB1) in patients with severe sepsis and septic shock. Crit Care Med 33:564–573
90. van Zoelen MA, Laterre PF, van Veen SQ, van Till JW, Wittebole X et al (2007) Systemic and local high mobility group box 1 concentrations during severe infection. Crit Care Med 35:2799–2804
91. Gaini S, Pedersen SS, Koldkaer OG, Pedersen C, Moestrup SK et al (2008) New immunological serum markers in bacteraemia: anti-inflammatory soluble CD163, but not proinflammatory high mobility group-box 1 protein, is related to prognosis. Clin Exp Immunol 151:423–431
92. Foell D, Wittkowski H, Vogl T, Roth J (2007) S100 proteins expressed in phagocytes: a novel group of damage-associated molecular pattern molecules. J Leukoc Biol 81:28–37
93. van Zoelen MA, Vogl T, Foell D, Van Veen SQ, van Till JW et al (2009) Expression and role of myeloid-related protein-14 in clinical and experimental sepsis. Am J Respir Crit Care Med 180:1098–1106
94. Terrin G, Passariello A, Manguso F, Salvia G, Rapacciuolo L et al (2011) Serum calprotectin: an antimicrobial peptide as a new marker for the diagnosis of sepsis in very low birth weight newborns. Clin Dev Immunol 2011:291085
95. Fontaine M, Pachot A, Larue A, Mougin B, Landelle C et al (2009) Delayed increased S100A9 mRNA predicts hospital-acquired infection after septic shock. Crit Care 13:P56
96. ten Oever J, Giamarellos-Bourboulis EJ, van de Veerdonk FL, Stelma FF, Simon A et al (2013) Circulating galectin-3 in infections and non-infectious inflammatory diseases. Eur J Clin Microbiol Infect Dis 32:1605–1610
97. Hotchkiss RS, Nicholson DW (2006) Apoptosis and caspases regulate death and inflammation in sepsis. Nat Rev Immunol 6:813–822
98. Mariano F, Cantaluppi V, Stella M, Romanazzi GM, Assenzio B et al (2008) Circulating plasma factors induce tubular and glomerular alterations in septic burns patients. Crit Care 12:R42
99. Ertel W, Keel M, Infanger M, Ungethum U, Steckholzer U et al (1998) Circulating mediators in serum of injured patients with septic complications inhibit neutrophil apoptosis through up-regulation of protein-tyrosine phosphorylation. J Trauma 44:767–775, discussion 775-766

100. Paunel-Gorgulu A, Flohe S, Scholz M, Windolf J, Logters T (2011) Increased serum soluble Fas after major trauma is associated with delayed neutrophil apoptosis and development of sepsis. Crit Care 15:R20
101. De Freitas I, Fernandez-Somoza M, Essenfeld-Sekler E, Cardier JE (2004) Serum levels of the apoptosis-associated molecules, tumor necrosis factor-alpha/tumor necrosis factor type-I receptor and Fas/FasL, in sepsis. Chest 125:2238–2246
102. Doughty L, Clark RS, Kaplan SS, Sasser H, Carcillo J (2002) sFas and sFas ligand and pediatric sepsis-induced multiple organ failure syndrome. Pediatr Res 52:922–927
103. Huttunen R, Syrjanen J, Vuento R, Laine J, Hurme M et al (2012) Apoptosis markers soluble Fas (sFas), Fas Ligand (FasL) and sFas/FasL ratio in patients with bacteremia: a prospective cohort study. J Infect 64:276–281
104. Moore DJ, Greystoke A, Butt F, Wurthner J, Growcott J et al (2012) A pilot study assessing the prognostic value of CK18 and nDNA biomarkers in severe sepsis patients. Clin Drug Investig 32:179–187
105. Roth GA, Krenn C, Brunner M, Moser B, Ploder M et al (2004) Elevated serum levels of epithelial cell apoptosis-specific cytokeratin 18 neoepitope m30 in critically ill patients. Shock 22:218–220
106. Hofer S, Brenner T, Bopp C, Steppan J, Lichtenstern C et al (2009) Cell death serum biomarkers are early predictors for survival in severe septic patients with hepatic dysfunction. Crit Care 13:R93
107. Haeffner-Cavaillon N, Cavaillon J-M, Ciancioni C, Bacle F, Delons S et al (1989) In vivo induction of interleukin-1 during hemodialysis. Kidney Int 35:1212–1218
108. Muñoz C, Misset B, Fitting C, Bleriot JP, Carlet J et al (1991) Dissociation between plasma and monocyte-associated cytokines during sepsis. Eur J Immunol 21:2177–2184
109. Bozza FA, Salluh JI, Japiassu AM, Soares M, Assis EF et al (2007) Cytokine profiles as markers of disease severity in sepsis: a multiplex analysis. Crit Care 11:R49
110. Fischer E, Van Zee KJ, Marano MA, Rock CS, Kenney JS et al (1992) Interleukin-1 receptor antagonist circulates in experimental inflammation and in human disease. Blood 79:2196–2200
111. Rogy MA, Coyle SM, Oldenburg HS, Rock CS, Barie PS et al (1994) Persistently elevated soluble tumor necrosis factor receptor and interleukin-1 receptor antagonist levels in critically ill patients. J Am Coll Surg 178:132–138
112. van Deuren M, van der Ven-Jongekrijg J, Demacker PN, Bartelink AK, van Dalen R et al (1994) Differential expression of proinflammatory cytokines and their inhibitors during the course of meningococcal infections. J Infect Dis 169:157–161
113. Gardlund B, Sjölin J, Nilsson A, Roll M, Wickerts CJ et al (1995) Plasma levels of cytokines in primary septic shock in humans: correlation with disease severity. J Infect Dis 172:296–301
114. Küster H, Weiss M, Willeitner AE, Detlefsen S, Jeremias I et al (1998) Interleukin-1 receptor antagonist and interleukin-6 for early diagnosis of neonatal sepsis 2 days before clinical manifestation. Lancet 352:1271–1277
115. Marie C, Losser MR, Fitting C, Kermarrec N, Payen D et al (1997) Cytokines and soluble cytokines receptors in pleural effusions from septic and nonseptic patients. Am J Respir Crit Care Med 156:1515–1522
116. Adrie C, Adib-Conquy M, Laurent I, Monchi M, Vinsonneau C et al (2002) Successful cardiopulmonary resuscitation after cardiac arrest as a "sepsis like" syndrome. Circulation 106:562–568
117. Samson LM, Allen UD, Creery WD, Diaz-Mitoma F, Singh RN (1997) Elevated IL-1ra levels in pediatric sepsis syndrome. J Pediatr 131:587–591
118. Cruickshank AM, Fraser WD, Burns HJ, Van Damme J, Shenkin A (1990) Response of serum interleukin-6 in patients undergoing elective surgery of varying severity. Clin Sci (Lond) 79:161–165
119. Naskalski JW, Kusnierz-Cabala B, Panek J, Kedra B (2003) Poly-C specific ribonuclease activity correlates with increased concentrations of IL-6, IL-8 and sTNFR55/sTNFR75 in plasma of patients with acute pancreatitis. J Physiol Pharmacol 54:439–448
120. Cavaillon JM, Poignet JL, Fitting C, Delons S (1992) Serum interleukin-6 in long-term hemodialyzed patients. Nephron 60:307–313
121. Hack C, de Groot E, Felt-Bersma R, Nuijens J, Strack Van Schijndel R et al (1989) Increased plasma levels of interleukin-6 in sepsis. Blood 74:1704–1710
122. Pinsky MR, Vincent JL, Deviere J, Alegre M, Kahn RJ et al (1993) Serum cytokine levels in human septic shock. Relation to multiple-system organ failure and mortality. Chest 103:565–575
123. Rodriguez-Gaspar M, Santolaria F, Jarque-Lopez A, Gonzalez-Reimers E, Milena A et al (2001) Prognostic value of cytokines in SIRS general medical patients. Cytokine 15:232–236
124. Calandra T, Gerain J, Heumann D, Baumgartner JD, Glauser MP (1991) High circulating levels of interleukin-6 in patients with septic shock: evolution during sepsis, prognostic value, and interplay with other cytokines. The Swiss-Dutch J5 Immunoglobulin Study Group. Am J Med 91:23–29

125. Cavaillon JM, Adib-Conquy M, Fitting C, Adrie C, Payen D (2003) Cytokine cascade in sepsis. Scand J Infect Dis 35:535–544
126. Ortqvist A, Hedlund J, Wretlind B, Carlstrom A, Kalin M (1995) Diagnostic and prognostic value of interleukin-6 and C-reactive protein in community-acquired pneumonia. Scand J Infect Dis 27:457–462
127. Waage A, Brandtzaeg P, Halstensen A, Kierulf P, Espevik T (1989) The complex pattern of cytokines in serum from patients with meningococcal septic shock. Association between interleukin-6, interleukin-1, and fatal outcome. J Exp Med 169:333–338
128. Casey LC, Balk RA, Bone RC (1993) Plasma cytokine and endotoxin levels correlate with survival in patients with the sepsis syndrome. Ann Intern Med 119:771–778
129. Friedland JS, Porter JC, Daryanani S, Bland JM, Screaton NJ et al (1996) Plasma proinflammatory cytokine concentrations, acute physiology and chronic health evaluation (APACHE) III scores and survival in patients in an intensive care unit. Crit Care Med 24:1775–1781
130. Riche FC, Cholley BP, Panis YH, Laisne MJ, Briard CG et al (2000) Inflammatory cytokine response in patients with septic shock secondary to generalized peritonitis. Crit Care Med 28:433–437
131. Suarez-Santamaria M, Santolaria F, Perez-Ramirez A, Aleman-Valls MR, Martinez-Riera A et al (2010) Prognostic value of inflammatory markers (notably cytokines and procalcitonin), nutritional assessment, and organ function in patients with sepsis. Eur Cytokine Netw 21:19–26
132. Buck C, Bundschu J, Gallati H, Bartmann P, Pohlandt F (1994) Interleukin-6: a sensitive parameter for the early diagnosis of neonatal bacterial infection. Pediatrics 93:54–58
133. Messer J, Eyer D, Donato L, Gallati H, Matis J et al (1996) Evaluation of interleukin-6 and soluble receptors of tumor necrosis factor for early diagnosis of neonatal infection. J Pediatr 129:574–580
134. Strait RT, Kelly KJ, Kurup VP (1999) Tumor necrosis factor-alpha, interleukin-1 beta, and interleukin-6 levels in febrile, young children with and without occult bacteremia. Pediatrics 104:1321–1326
135. Urbonas V, Eidukaite A, Tamuliene I (2012) The diagnostic value of interleukin-6 and interleukin-8 for early prediction of bacteremia and sepsis in children with febrile neutropenia and cancer. J Pediatr Hematol Oncol 34:122–127
136. Groeneveld AB, Bossink AW, van Mierlo GJ, Hack CE (2001) Circulating inflammatory mediators in patients with fever: predicting bloodstream infection. Clin Diagn Lab Immunol 8:1189–1195
137. Sander M, von Heymann C, von Dossow V, Spaethe C, Konertz WF et al (2006) Increased interleukin-6 after cardiac surgery predicts infection. Anesth Analg 102:1623–1629
138. Giannoudis PV, Smith MR, Evans RT, Bellamy MC, Guillou PJ (1998) Serum CRP and IL-6 levels after trauma. Not predictive of septic complications in 31 patients. Acta Orthop Scand 69:184–188
139. Rintala E, Pulkki K, Mertsola J, Nevalainen T, Nikoskelainen J (1995) Endotoxin, interleukin-6 and phospholipase-A2 as markers of sepsis in patients with hematological malignancies. Scand J Infect Dis 27:39–43
140. Jekarl DW, Lee SY, Lee J, Park YJ, Kim Y et al (2013) Procalcitonin as a diagnostic marker and IL-6 as a prognostic marker for sepsis. Diagn Microbiol Infect Dis 75:342–347
141. Marchant A, Devière J, Byl B, De Groote D, Vincent J et al (1994) Interleukin-10 production during septicaemia. Lancet 343:707–708
142. Glynn P, Coakley R, Kilgallen I, Murphy N, O'Neill S (1999) Circulating interleukin 6 and interleukin 10 in community acquired pneumonia. Thorax 54:51–55
143. Lehmann AK, Halstensen A, Sornes S, Rokke O, Waage A (1995) High levbels of interleukin-10 in serum are associated with fatality in meningococcal disease. Infect Immun 63: 2109–2112
144. Gogos CA, Drosou E, Bassaris HP, Skoutelis A (2000) Pro- versus anti-inflammatory cytokine profile in patients with severe sepsis: a marker for prognosis and future therapeutic options. J Infect Dis 181:176–180
145. van Dissel JT, van Langevelde P, Westendorp RGJ, Kwappenberg K, Frolich M (1998) Anti-inflammatory cytokine profile and mortality in febrile patients. Lancet 351:950–953
146. Kellum JA, Kong L, Fink MP, Weissfeld LA, Yealy DM et al (2007) Understanding the inflammatory cytokine response in pneumonia and sepsis: results of the Genetic and Inflammatory Markers of Sepsis (GenIMS) Study. Arch Intern Med 167:1655–1663
147. van Deuren M, van Der Ven-Jongekrijg H, Baterlink AKN, van Dalen R, Sauerwein RW et al (1995) Correlation between proinflammatory cytokines and antiinflammatory mediators and the severity of disease in meningococcal infections. J Infect Dis 172:433–439
148. Gomez-Jimenez J, Martin MC, Sauri R, Segura RM, Esteban F et al (1995) Interleukin-10 and the monocyte/macrophage-induced inflammatory response in septic shock. J Infect Dis 171:472–475

149. Riordan FA, Marzouk O, Thomson AP, Sills JA, Hart CA (1996) Proinflammatory and anti-inflammatory cytokines in meningococcal disease. Arch Dis Child 75:453–454
150. Tamayo E, Fernandez A, Almansa R, Carrasco E, Heredia M et al (2011) Pro- and anti-inflammatory responses are regulated simultaneously from the first moments of septic shock. Eur Cytokine Netw 22:82–87
151. Cavaillon J-M, Adib-Conquy M, Cloëz-Tayarani I, Fitting C (2001) Immunodepression in sepsis and SIRS assessed by ex vivo cytokine production is not a generalized phenomenon : a review. J Endotoxin Res 7:85–93
152. Matera G, Puccio R, Giancotti A, Quirino A, Pulicari MC et al (2013) Impact of interleukin-10, soluble CD25 and interferon-gamma on the prognosis and early diagnosis of bacteremic systemic inflammatory response syndrome: a prospective observational study. Crit Care 17:R64
153. Presterl E, Staudinger T, Pettermann M, Lassnigg A, Burgmann H et al (1997) Cytokine profile and correlation to the APACHE III and MPM II scores in patients with sepsis. Am J Respir Crit Care Med 156:825–832
154. Vedrine C, Caraion C, Lambert C, Genin C (2004) Cytometric bead assay of cytokines in sepsis: a clinical evaluation. Cytometry B Clin Cytom 60:14–22
155. Mera S, Tatulescu D, Cismaru C, Bondor C, Slavcovici A et al (2010) Multiplex cytokine profiling in patients with sepsis. APMIS 119:155–163
156. Emmanuilidis K, Weighardt H, Matevossian E, Heidecke CD, Ulm K et al (2002) Differential regulation of systemic IL-18 and IL-12 release during postoperative sepsis: high serum IL-18 as an early predictive indicator of lethal outcome. Shock 18:301–305
157. Wu HP, Chen CK, Chung K, Tseng JC, Hua CC et al (2009) Serial cytokine levels in patients with severe sepsis. Inflamm Res 58: 385–393
158. Lavoie PM, Huang Q, Jolette E, Whalen M, Nuyt AM et al (2010) Profound lack of interleukin (IL)-12/IL-23p40 in neonates born early in gestation is associated with an increased risk of sepsis. J Infect Dis 202:1754–1763
159. Zeerleder S, Hack CE, Caliezi C, van Mierlo G, Eerenberg-Belmer A et al (2005) Activated cytotoxic T cells and NK cells in severe sepsis and septic shock and their role in multiple organ dysfunction. Clin Immunol 116:158–165
160. Lauw FN, Simpson AJH, Prins JM, Smith MD, Kurimoto M et al (1999) Elevated plasma concentrations of interferon (IFN) and the IFN-inducing cytokines interleukin (IL)18, IL-12, and IL-15 in severe melioidosis. J Infect Dis 180:1878–1885
161. Grobmyer SR, Lin E, Lowry SF, Rivadeneira DE, Potter S et al (2000) Elevation of IL-18 in human sepsis. J Clin Immunol 20:212–215
162. Oberholzer A, Steckholzer U, Kurimoto M, Trentz O, Ertel W (2001) Interleukin-18 plasma levels are increased in patients with sepsis compared to severely injured patients. Shock 16:411–414
163. Zaki MES, Elgendy MY, El-Mashad NB, Farahat ME (2007) IL-18 level correlates with development of sepsis in surgical patients. Immunol Invest 36:403–411
164. Mommsen P, Frink M, Pape HC, van Griensven M, Probst C et al (2009) Elevated systemic IL-18 and neopterin levels are associated with posttraumatic complications among patients with multiple injuries: a prospective cohort study. Injury 40:528–534
165. Waage A, Halstensen A, Espevik T (1987) Association between tumor necrosis factor in serum and fatal outcome in patients with meningococcal disease. Lancet i: 355–357
166. Girardin E, Grau G, Dayer J, Roux-Lombard P, Lambert P (1988) Tumor necrosis factor and interleukin-1 in the serum of children with severe infectious purpura. N Engl J Med 319:397–400
167. de Groote MA, Martin MA, Densen P, Pfaller MA, Wenzel RP (1989) Plasma tumor necrosis factor levels in patients with presumed sepsis. Results in those treated with antilipid A antibody vs placebo [see comments]. JAMA 262:249–251
168. Offner F, Philippé J, Vogelaers D, Colardyn F, Baele G et al (1990) Serum tumor necrosis factor levels in patients with infectious diseases and septic shock. J Lab Clin Med 116:100–105
169. Calandra T, Baumgartner JD, Grau GE, Wu MM, Lambert PH et al (1990) Prognostic values of tumor necrosis factor/cachectin, interleukin-1, interferon-a, and interferon-g in the serum of patients with septic shock. J Infect Dis 161:982–987
170. Calandra T, Echtenacher B, Le Roy D, Pugin J, Metz CN et al (2000) Protection from septic shock by neutralization of macrophage migration inhibitory factor. Nat Med 6:164–170
171. Emonts M, Sweep FC, Grebenchtchikov N, Geurts-Moespot A, Knaup M et al (2007) Association between high levels of blood macrophage migration inhibitory factor, inappropriate adrenal response, and early

death in patients with severe sepsis. Clin Infect Dis 44:1321–1328
172. Maxime V, Fitting C, Annane D, Cavaillon J-M (2005) Corticoids normalize leukocyte production of macrophage migration inhibitory factor in septic shock. J Infect Dis 191: 138–144
173. Leaver SK, MacCallum NS, Pingle V, Hacking MB, Quinlan GJ et al (2010) Increased plasma thioredoxin levels in patients with sepsis: positive association with macrophage migration inhibitory factor. Intensive Care Med 36:336–341
174. Grieb G, Simons D, Piatkowski A, Bernhagen J, Steffens G et al (2010) Macrophage migration inhibitory factor-A potential diagnostic tool in severe burn injuries? Burns 36:335–342
175. Gessler P, Kirchmann N, Kientsch-Engel R, Haas N, Lasch P et al (1993) Serum concentrations of granulocyte colony-stimulating factor in healthy term and preterm neonates and in those with various diseases including bacterial infections. Blood 82:3177–3182
176. Pauksen K, Elfman L, Ulfgren AK, Venge P (1994) Serum levels of granulocyte-colony stimulating factor (G-CSF) in bacterial and viral infections, and in atypical pneumonia. Br J Haematol 88:256–260
177. Fischer JE, Benn A, Harbarth S, Nadal D, Fanconi S (2002) Diagnostic accuracy of G-CSF, IL-8, and IL-1ra in critically ill children with suspected infection. Intensive Care Med 28:1324–1331
178. Tanaka H, Ishikawa K, Nishino M, Shimazu T, Yoshioka T (1996) Changes in granulocyte colony-stimulating factor concentration in patients with trauma and sepsis. J Trauma 40:718–725, discussion 725-716
179. Kragsbjerg P, Jones I, Vikerfors T, Holmberg H (1995) Diagnostic value of blood cytokine concentrations in acute pneumonia. Thorax 50:1253–1257
180. Presneill JJ, Waring PM, Layton JE, Maher DW, Cebon J et al (2000) Plasma granulocyte colony-stimulating factor and granulocyte-macrophage colony-stimulating factor levels in critical illness including sepsis and septic shock: relation to disease severity, multiple organ dysfunction, and mortality. Crit Care Med 28:2344–2354
181. Torre D, Tambini R, Manfredi M, Mangani V, Livi P et al (2003) Circulating levels of granulocyte macrophage colony-stimulating factor in patients with the systemic inflammatory response syndrome. J Infect 47:296–299
182. Waring P, Wycherley K, Cary D, Nicola N, Metcalf D (1992) Leukemia inhibitory factor levels are elevated in septic shock and various inflammatory body fluids. J Clin Invest 90:2031–2037
183. Villers D, Dao T, Nguyen JM, Bironneau E, Godard A et al (1995) Increased plasma levels of human interleukin for DA1a cells/leukemia inhibitory factor in sepsis correlate with shock and poor prognosis. J Infect Dis 171:232–236
184. Guillet C, Fourcin M, Chevalier S, Pouplard A, Gascan H (1995) ELISA detection of circulating levels of LIF, OSM and CNTF in septic shock. Ann N Y Acad Sci 762:407–412
185. Collighan N, Giannoudis PV, Kourgeraki O, Perry SL, Guillou PJ et al (2004) Interleukin 13 and inflammatory markers in human sepsis. Br J Surg 91:762–768
186. Wong HR, Cvijanovich NZ, Hall M, Allen GL, Thomas NJ et al (2012) Interleukin-27 is a novel candidate diagnostic biomarker for bacterial infection in critically ill children. Crit Care 16:R213
187. Lee KA, Gong MN (2011) Pre-B-cell colony-enhancing factor and its clinical correlates with acute lung injury and sepsis. Chest 140:382–390
188. Bjerre A, Brusletto B, Hoiby EA, Kierulf P, Brandtzaeg P (2004) Plasma interferon-gamma and interleukin-10 concentrations in systemic meningococcal disease compared with severe systemic Gram-positive septic shock. Crit Care Med 32:433–438
189. van der Flier M, van Leeuwen HJ, van Kessel KP, Kimpen JL, Hoepelman AI et al (2005) Plasma vascular endothelial growth factor in severe sepsis. Shock 23:35–38
190. Socha LA, Gowardman J, Silva D, Correcha M, Petrosky N (2006) Elevation in interleukin 13 levels in patients diagnosed with systemic inflammatory response syndrome. Intensive Care Med 32:244–250
191. Friedland J, Suputtamongkol Y, Remick D, Chaowagul W, Strieter R et al (1992) Prolonged elevation of interleukin-8 and interleukin-6 concentrations in plasma and of leukocyte interleukin-8 m-RNA levels during septicemic and localized *Pseudomonas pseudomallei* infection. Infect Immun 60:2402–2408
192. Hack C, Hart M, Strack van Schijndel R, Eerenberg A, Nuijens J et al (1992) Interleukin-8 in sepsis: relation to shock and inflammatory mediators. Infect Immun 60: 2835–2842
193. Bossink AW, Paemen L, Jansen PM, Hack CE, Thijs LG et al (1995) Plasma levels of the chemokines monocyte chemotactic proteins-1 and -2 are elevated in human sepsis. Blood 86:3841–3847
194. Endo S, Inada K, Ceska M, Takakuwa T, Yamada Y et al (1995) Plasma interleukin 8 and polymorphonuclear leukocyte elastase concentrations in patients with septic shock. J Inflamm 45:136–142

195. Marty C, Misset B, Tamion F, Fitting C, Carlet J et al (1994) Circulating interleukin-8 concentrations in patients with multiple organ failure of septic and nonseptic origin. Crit Care Med 22:673–679
196. Fujishima S, Sasaki J, Shinozawa Y, Takuma K, Kimura H et al (1996) Serum MIP-1 alpha and IL-8 in septic patients. Intensive Care Med 22:1169–1175
197. Livaditi O, Kotanidou A, Psarra A, Dimopoulou I, Sotiropoulou C et al (2006) Neutrophil CD64 expression and serum IL-8: sensitive early markers of severity and outcome in sepsis. Cytokine 36:283–290
198. Wong HR, Cvijanovich N, Wheeler DS, Bigham MT, Monaco M et al (2008) Interleukin-8 as a stratification tool for interventional trials involving pediatric septic shock. Am J Respir Crit Care Med 178:276–282
199. de Bont ES, Vellenga E, Swaanenburg JC, Fidler V, Visser-van Brummen PJ et al (1999) Plasma IL-8 and IL-6 levels can be used to define a group with low risk of septicaemia among cancer patients with fever and neutropenia. Br J Haematol 107:375–380
200. Lin KJ, Lin J, Hanasawa K, Tani T, Kodama M (2000) Interleukin-8 as a predictor of the severity of bacteremia and infectious disease. Shock 14:95–100
201. Tromp YH, Daenen SM, Sluiter WJ, Vellenga E (2009) The predictive value of interleukin-8 (IL-8) in hospitalised patients with fever and chemotherapy-induced neutropenia. Eur J Cancer 45:596–600
202. Santolaya ME, Alvarez AM, Aviles CL, Becker A, King A et al (2008) Predictors of severe sepsis not clinically apparent during the first twenty-four hours of hospitalization in children with cancer, neutropenia, and fever: a prospective, multicenter trial. Pediatr Infect Dis J 27:538–543
203. Kurt AN, Aygun AD, Godekmerdan A, Kurt A, Dogan Y et al (2007) Serum IL-1beta, IL-6, IL-8, and TNF-alpha levels in early diagnosis and management of neonatal sepsis. Mediators Inflamm 2007:31397
204. Marie C, Fitting C, Cheval C, Losser MR, Carlet J et al (1997) Presence of high levels of leukocyte-associated interleukin-8 upon cell activation and in patients with sepsis syndrome. Infect Immun 65:865–871
205. Steinbach G, Bolke E, Schulte am Esch J, Peiper M, Zant R et al (2007) Comparison of whole blood interleukin-8 and plasma interleukin-8 as a predictor for sepsis in postoperative patients. Clin Chim Acta 378:117–121
206. Lvovschi V, Arnaud L, Parizot C, Freund Y, Juillien G et al (2011) Cytokine profiles in sepsis have limited relevance for stratifying patients in the emergency department: a prospective observational study. PLoS One 6:e28870
207. Santana Reyes C, Garcia-Munoz F, Reyes D, Gonzalez G, Dominguez C et al (2003) Role of cytokines (interleukin-1beta, 6, 8, tumour necrosis factor-alpha, and soluble receptor of interleukin-2) and C-reactive protein in the diagnosis of neonatal sepsis. Acta Paediatr 92:221–227
208. Franz AR, Bauer K, Schalk A, Garland SM, Bowman ED et al (2004) Measurement of interleukin 8 in combination with C-reactive protein reduced unnecessary antibiotic therapy in newborn infants: a multicenter, randomized, controlled trial. Pediatrics 114:1–8
209. Ng PC, Li K, Chui KM, Leung TF, Wong RP et al (2007) IP-10 is an early diagnostic marker for identification of late-onset bacterial infection in preterm infants. Pediatr Res 61:93–98
210. Olszyna DP, Prins JM, Dekkers PEP, De Jonge E, Speelman P et al (1999) Sequential measurements of chemokines in urosepsis and experimental endotoxemia. J Clin Immunol 19:399–405
211. Moller AS, Bjerre A, Brusletto B, Joo GB, Brandtzaeg P et al (2005) Chemokine patterns in meningococcal disease. J Infect Dis 191:768–775
212. Vermont CL, Hazelzet JA, de Kleijn ED, van den Dobbelsteen GP, de Groot R (2006) CC and CXC chemokine levels in children with meningococcal sepsis accurately predict mortality and disease severity. Crit Care 10:R33
213. Andaluz-Ojeda D, Bobillo F, Iglesias V, Almansa R, Rico L et al (2012) A combined score of pro- and anti-inflammatory interleukins improves mortality prediction in severe sepsis. Cytokine 57:332–336
214. El-Maghraby SM, Moneer MM, Ismail MM, Shalaby LM, El-Mahallawy HA (2007) The diagnostic value of C-reactive protein, interleukin-8, and monocyte chemotactic protein in risk stratification of febrile neutropenic children with hematologic malignancies. J Pediatr Hematol Oncol 29:131–136
215. Nowak JE, Wheeler DS, Harmon KK, Wong HR (2010) Admission chemokine (C-C motif) ligand 4 levels predict survival in pediatric septic shock. Pediatr Crit Care Med 11:213–216
216. Schall TJ, Jongstra J, Dyer BJ, Jorgensen J, Clayberger C et al (1988) A human T cell-specific molecule is a member of a new gene family. J Immunol 141:1018–1025
217. Carrol ED, Thomson AP, Mobbs KJ, Hart CA (2000) The role of RANTES in meningococcal disease. J Infect Dis 182:363–366
218. Ellis M, al-Ramadi B, Hedstrom U, Alizadeh H, Shammas V et al (2005) Invasive fungal

infections are associated with severe depletion of circulating RANTES. J Med Microbiol 54:1017–1022
219. John CC, Opika-Opoka R, Byarugaba J, Idro R, Boivin MJ (2006) Low levels of RANTES are associated with mortality in children with cerebral malaria. J Infect Dis 194:837–845
220. Shouman B, Badr R (2010) Regulated on activation, normal T cell expressed and secreted and tumor necrosis factor-alpha in septic neonates. J Perinatol 30:192–196
221. Ellis M, al-Ramadi B, Hedstrom U, Frampton C, Alizadeh H et al (2005) Significance of the CC chemokine RANTES in patients with haematological malignancy: results from a prospective observational study. Br J Haematol 128:482–489
222. Ng PC, Li K, Leung TF, Wong RP, Li G et al (2006) Early prediction of sepsis-induced disseminated intravascular coagulation with interleukin-10, interleukin-6, and RANTES in preterm infants. Clin Chem 52:1181–1189
223. Bas S, Gauthier BR, Spenato U, Stingelin S, Gabay C (2004) CD14 is an acute-phase protein. J Immunol 172:4470–4479
224. Endo S, Inada K, Kasai T, Takakuwa T, Nakae H et al (1994) Soluble CD14 (sCD14) levels in patients with multiple organ failure (MOF). Res Commun Chem Pathol Pharmacol 84:17–25
225. Landmann R, Zimmerli W, Sansano S, Link S, Hahn A et al (1995) Increased circulating soluble CD14 is associated with high mortality in gram-negative septic shock. J Infect Dis 171:639–644
226. Burgmann H, Winkler S, Locker GJ, Presterl E, Laczika K et al (1996) Increased serum concentration of soluble CD14 is a prognostic marker in gram-positive sepsis. Clin Immunol Immunopathol 80:307–310
227. Berner R, Furll B, Stelter F, Drose J, Muller HP et al (2002) Elevated levels of lipopolysaccharide-binding protein and soluble CD14 in plasma in neonatal early-onset sepsis. Clin Diagn Lab Immunol 9:440–445
228. Blanco A, Solis G, Arranz E, Coto GD, Ramos A et al (1996) Serum levels of CD14 in neonatal sepsis by gram-positive and gram-negative bacteria. Acta Paediatr 85:728–732
229. Pavcnik-Arnol M, Hojker S, Derganc M (2007) Lipopolysaccharide-binding protein, lipopolysaccharide, and soluble CD14 in sepsis of critically ill neonates and children. Intensive Care Med 33:1025–1032
230. Chalupa P, Beran O, Herwald H, Kasprikova N, Holub M (2011) Evaluation of potential biomarkers for the discrimination of bacterial and viral infections. Infection 39:411–417
231. Carrillo EH, Gordon L, Goode E, Davis E, Polk HC Jr (2001) Early elevation of soluble CD14 may help identify trauma patients at high risk for infection. J Trauma 50:810–816
232. Yaegashi Y, Shirakawa K, Sato N, Suzuki Y, Kojika M et al (2005) Evaluation of a newly identified soluble CD14 subtype as a marker for sepsis. J Infect Chemother 11:234–238
233. Urbonas V, Eidukaite A, Tamuliene I (2013) The predictive value of soluble biomarkers (CD14 subtype, interleukin-2 receptor, human leucocyte antigen-G) and procalcitonin in the detection of bacteremia and sepsis in pediatric oncology patients with chemotherapy-induced febrile neutropenia. Cytokine 62:34–37
234. Pugin J, Stern-Voeffray S, Daubeuf B, Matthay MA, Elson G et al (2004) Soluble MD-2 activity in plasma from patients with severe sepsis and septic shock. Blood 104:4071–4079
235. Tissieres P, Dunn-Siegrist I, Schappi M, Elson G, Comte R et al (2008) Soluble MD-2 is an acute-phase protein and an opsonin for Gram-negative bacteria. Blood 111:2122–2131
236. Brunner M, Krenn C, Roth G, Moser B, Dworschak M et al (2004) Increased levels of soluble ST2 protein and IgG1 production in patients with sepsis and trauma. Intensive Care Med 30:1468–1473
237. Hoogerwerf JJ, Tanck MW, van Zoelen MA, Wittebole X, Laterre PF et al (2010) Soluble ST2 plasma concentrations predict mortality in severe sepsis. Intensive Care Med 36:630–637
238. Takala A, Jousela I, Jansson SE, Olkkola KT, Takkunen O et al (1999) Markers of systemic inflammation predicting organ failure in community-acquired septic shock. Clin Sci (Lond) 97:529–538
239. Fleischhack G, Kambeck I, Cipic D, Hasan C, Bode U (2000) Procalcitonin in paediatric cancer patients: its diagnostic relevance is superior to that of C-reactive protein, interleukin 6, interleukin 8, soluble interleukin 2 receptor and soluble tumour necrosis factor receptor II. Br J Haematol 111:1093–1102
240. Saito K, Wagatsuma T, Toyama H, Ejima Y, Hoshi K et al (2008) Sepsis is characterized by the increases in percentages of circulating CD4+CD25+ regulatory T cells and plasma levels of soluble CD25. Tohoku J Exp Med 216:61–68
241. Moller HJ, Moestrup SK, Weis N, Wejse C, Nielsen H et al (2006) Macrophage serum markers in pneumococcal bacteremia: prediction of survival by soluble CD163. Crit Care Med 34:2561–2566
242. Gaini S, Koldkjaer OG, Pedersen SS, Pedersen C, Moestrup SK et al (2006) Soluble haemoglobin scavenger receptor (sCD163) in patients with suspected community-acquired infections. APMIS 114:103–111

243. Schaer DJ, Schleiffenbaum B, Kurrer M, Imhof A, Bachli E et al (2005) Soluble hemoglobin-haptoglobin scavenger receptor CD163 as a lineage-specific marker in the reactive hemophagocytic syndrome. Eur J Haematol 74:6–10
244. Feng L, Zhou X, Su LX, Feng D, Jia YH et al (2012) Clinical significance of soluble hemoglobin scavenger receptor CD163 (sCD163) in sepsis, a prospective study. PLoS One 7:e38400
245. Evans TJ, Moyes D, Carpenter A, Martin R, Loetscher H et al (1994) Protective effect of 55 but not 75 kD soluble tumor necrosis factor receptor-immunoglobulin G fusion proteins in animal model of gram negative sepsis. J Exp Med 180:2173–2179
246. Girardin E, Roux-Lombard P, Grau GE, Suter P, Gallati H et al (1992) Imbalance between tumour necrosis factor-alpha and soluble TNF receptor concentrations in severe meningococcaemia. Immunology 76:20–23
247. Ertel W, Scholl FA, Gallati H, Bonaccio M, Schildberg FW et al (1994) Increased release of soluble tumor necrosis factor receptors into blood during clinical sepsis. Arch Surg 129:1330–1336, discussion 1336-1337
248. van der Poll T, Jansen J, van Leenen D, von der Mohlen M, Levi M et al (1993) Release of soluble receptors for tumor necrosis factor in clinical sepsis and experimental endotoxemia. J Infect Dis 168:955–960
249. Froon AH, Bemelmans MH, Greve JW, van der Linden CJ, Buurman WA (1994) Increased plasma concentrations of soluble tumor necrosis factor receptors in sepsis syndrome: correlation with plasma creatinine values. Crit Care Med 22:803–809
250. de Pablo R, Monserrat J, Reyes E, Diaz-Martin D, Rodriguez Zapata M et al (2011) Mortality in patients with septic shock correlates with anti-inflammatory but not proinflammatory immunomodulatory molecules. J Intensive Care Med 26:125–132
251. Pilz G, Fraunberger P, Appel R, Kreuzer E, Werdan K et al (1996) Early prediction of outcome in score-identified, postcardiac surgical patients at high risk for sepsis, using soluble tumor necrosis factor receptor-p55 concentrations. Crit Care Med 24:596–600
252. Zhang B, Huang YH, Chen Y, Yang Y, Hao ZL et al (1998) Plasma tumor necrosis factor-alpha, its soluble receptors and interleukin-1beta levels in critically burned patients. Burns 24:599–603
253. el-Barbary M, Khabar KS (2002) Soluble tumor necrosis factor receptor p55 predicts cytokinemia and systemic inflammatory response after cardiopulmonary bypass. Crit Care Med 30:1712–1716
254. Spielmann S, Kerner T, Ahlers O, Keh D, Gerlach M et al (2001) Early detection of increased tumour necrosis factor alpha (TNFalpha) and soluble TNF receptor protein plasma levels after trauma reveals associations with the clinical course. Acta Anaesthesiol Scand 45:364–370
255. Hou YQ, Xu P, Zhang M, Han D, Peng L et al (2012) Serum decoy receptor 3, a potential new biomarker for sepsis. Clin Chim Acta 413:744–748
256. Chen CY, Yang KY, Chen MY, Chen HY, Lin MT et al (2009) Decoy receptor 3 levels in peripheral blood predict outcomes of acute respiratory distress syndrome. Am J Respir Crit Care Med 180:751–760
257. Cowley HC, Heney D, Gearing AJ, Hemingway I, Webster NR (1994) Increased circulating adhesion molecule concentrations in patients with the systemic inflammatory response syndrome: a prospective cohort study. Crit Care Med 22:651–657
258. Boldt J, Wollbruck M, Kuhn D, Linke LC, Hempelmann G (1995) Do plasma levels of circulating soluble adhesion molecules differ between surviving and nonsurviving critically ill patients? Chest 107:787–792
259. Froon AH, Bonten MJ, Gaillard CA, Greve JW, Dentener MA et al (1998) Prediction of clinical severity and outcome of ventilator-associated pneumonia. Comparison of simplified acute physiology score with systemic inflammatory mediators. Am J Respir Crit Care Med 158:1026–1031
260. Hein OV, Misterek K, Tessmann JP, van Dossow V, Krimphove M et al (2005) Time course of endothelial damage in septic shock: prediction of outcome. Crit Care 9:R323–R330
261. Sessler CN, Windsor AC, Schwartz M, Watson L, Fisher BJ et al (1995) Circulating ICAM-1 is increased in septic shock. Am J Respir Crit Care Med 151:1420–1427
262. Shapiro NI, Schuetz P, Yano K, Sorasaki M, Parikh SM et al (2010) The association of endothelial cell signaling, severity of illness, and organ dysfunction in sepsis. Crit Care 14:R182
263. Boldt J, Muller M, Kuhn D, Linke LC, Hempelmann G (1996) Circulating adhesion molecules in the critically ill: a comparison between trauma and sepsis patients. Intensive Care Med 22:122–128
264. Leone M, Boutiere B, Camoin-Jau L, Albanese J, Horschowsky N et al (2002) Systemic endothelial activation is greater in septic than in traumatic-hemorrhagic shock but does not correlate with endothelial activation in skin biopsies. Crit Care Med 30:808–814
265. Cummings CJ, Sessler CN, Beall LD, Fisher BJ, Best AM et al (1997) Soluble E-selectin

levels in sepsis and critical illness. Correlation with infection and hemodynamic dysfunction. Am J Respir Crit Care Med 156:431–437
266. Kayal S, Jais JP, Aguini N, Chaudiere J, Labrousse J (1998) Elevated circulating E-selectin, intercellular adhesion molecule 1, and von Willebrand factor in patients with severe infection. Am J Respir Crit Care Med 157:776–784
267. Schuetz P, Jones AE, Aird WC, Shapiro NI (2011) Endothelial cell activation in emergency department patients with sepsis-related and non-sepsis-related hypotension. Shock 36:104–108
268. de Pablo R, Monserrat J, Reyes E, Diaz D, Rodriguez-Zapata M et al (2013) Circulating sICAM-1 and sE-Selectin as biomarker of infection and prognosis in patients with systemic inflammatory response syndrome. Eur J Intern Med 24:132–138
269. Dollner H, Vatten L, Austgulen R (2001) Early diagnostic markers for neonatal sepsis: comparing C-reactive protein, interleukin-6, soluble tumour necrosis factor receptors and soluble adhesion molecules. J Clin Epidemiol 54:1251–1257
270. Geppert A, Zorn G, Karth GD, Haumer M, Gwechenberger M et al (2000) Soluble selectins and the systemic inflammatory response syndrome after successful cardiopulmonary resuscitation. Crit Care Med 28:2360–2365
271. Fiedler U, Reiss Y, Scharpfenecker M, Grunow V, Koidl S et al (2006) Angiopoietin-2 sensitizes endothelial cells to TNF-alpha and has a crucial role in the induction of inflammation. Nat Med 12:235–239
272. Parikh SM, Mammoto T, Schultz A, Yuan HT, Christiani D et al (2006) Excess circulating angiopoietin-2 may contribute to pulmonary vascular leak in sepsis in humans. PLoS Med 3:e46
273. Orfanos SE, Kotanidou A, Glynos C, Athanasiou C, Tsigkos S et al (2007) Angiopoietin-2 is increased in severe sepsis: correlation with inflammatory mediators. Crit Care Med 35:199–206
274. Siner JM, Bhandari V, Engle KM, Elias JA, Siegel MD (2009) Elevated serum angiopoietin 2 levels are associated with increased mortality in sepsis. Shock 31:348–353
275. David S, Mukherjee A, Ghosh CC, Yano M, Khankin EV et al (2012) Angiopoietin-2 may contribute to multiple organ dysfunction and death in sepsis*. Crit Care Med 40:3034–3041
276. Davis JS, Yeo TW, Piera KA, Woodberry T, Celermajer DS et al (2010) Angiopoietin-2 is increased in sepsis and inversely associated with nitric oxide-dependent microvascular reactivity. Crit Care 14:R89
277. Kumpers P, Lukasz A, David S, Horn R, Hafer C et al (2008) Excess circulating angiopoietin-2 is a strong predictor of mortality in critically ill medical patients. Crit Care 12:R147
278. Giuliano JS Jr, Lahni PM, Harmon K, Wong HR, Doughty LA et al (2007) Admission angiopoietin levels in children with septic shock. Shock 28:650–654
279. Agrawal A, Matthay MA, Kangelaris KN, Stein J, Chu JC et al (2013) Plasma angiopoietin-2 predicts the onset of acute lung injury in critically ill patients. Am J Respir Crit Care Med 187:736–742
280. Ganter MT, Cohen MJ, Brohi K, Chesebro BB, Staudenmayer KL et al (2008) Angiopoietin-2, marker and mediator of endothelial activation with prognostic significance early after trauma? Ann Surg 247:320–326
281. Kranidioti H, Orfanos SE, Vaki I, Kotanidou A, Raftogiannis M et al (2009) Angiopoietin-2 is increased in septic shock: evidence for the existence of a circulating factor stimulating its release from human monocytes. Immunol Lett 125:65–71
282. van der Heijden M, van Nieuw Amerongen GP, van Hinsbergh VW, Groeneveld AB (2010) The interaction of soluble Tie2 with angiopoietins and pulmonary vascular permeability in septic and nonseptic critically ill patients. Shock 33:263–268
283. Jesmin S, Wada T, Gando S, Sultana SS, Zaedi S (2013) The dynamics of angiogenic factors and their soluble receptors in relation to organ dysfunction in disseminated intravascular coagulation associated with sepsis. Inflammation 36:186–196
284. Pickkers P, Sprong T, Eijk L, Hoeven H, Smits P et al (2005) Vascular endothelial growth factor is increased during the first 48 hours of human septic shock and correlates with vascular permeability. Shock 24:508–512
285. Karlsson S, Pettila V, Tenhunen J, Lund V, Hovilehto S et al (2008) Vascular endothelial growth factor in severe sepsis and septic shock. Anesth Analg 106:1820–1826
286. Yang KY, Liu KT, Chen YC, Chen CS, Lee YC et al (2011) Plasma soluble vascular endothelial growth factor receptor-1 levels predict outcomes of pneumonia-related septic shock patients: a prospective observational study. Crit Care 15:R11
287. Hamalainen S, Juutilainen A, Matinlauri I, Kuittinen T, Ruokonen E et al (2009) Serum vascular endothelial growth factor in adult haematological patients with neutropenic fever: a comparison with C-reactive protein. Eur J Haematol 83:251–257
288. Pittet JF, Morel DR, Hemsen A, Gunning K, Lacroix JS et al (1991) Elevated plasma

endothelin-1 concentrations are associated with the severity of illness in patients with sepsis. Ann Surg 213:261–264
289. Takakuwa T, Endo S, Nakae H, Kikichi M, Suzuki T et al (1994) Plasma levels of TNF-alpha, endothelin-1 and thrombomodulin in patients with sepsis. Res Commun Chem Pathol Pharmacol 84:261–269
290. Nakae H, Endo S, Inada K, Yamada Y, Takakuwa T et al (1996) Plasma levels of endothelin-1 and thrombomodulin in burn patients. Burns 22:594–597
291. Tschaikowsky K, Sagner S, Lehnert N, Kaul M, Ritter J (2000) Endothelin in septic patients: effects on cardiovascular and renal function and its relationship to proinflammatory cytokines. Crit Care Med 28:1854–1860
292. Schuetz P, Christ-Crain M, Morgenthaler NG, Struck J, Bergmann A et al (2007) Circulating precursor levels of endothelin-1 and adrenomedullin, two endothelium-derived, counteracting substances, in sepsis. Endothelium 14:345–351
293. Figueras-Aloy J, Gomez L, Rodriguez-Miguelez JM, Jordan Y, Salvia MD et al (2003) Plasma nitrite/nitrate and endothelin-1 concentrations in neonatal sepsis. Acta Paediatr 92:582–587
294. Hirata Y, Mitaka C, Sato K, Nagura T, Tsunoda Y et al (1996) Increased circulating adrenomedullin, a novel vasodilatory peptide, in sepsis. J Clin Endocrinol Metab 81:1449–1453
295. Nishio K, Akai Y, Murao Y, Doi N, Ueda S et al (1997) Increased plasma concentrations of adrenomedullin correlate with relaxation of vascular tone in patients with septic shock. Crit Care Med 25:953–957
296. Ehlenz K, Koch B, Preuss P, Simon B, Koop I et al (1997) High levels of circulating adrenomedullin in severe illness: correlation with C-reactive protein and evidence against the adrenal medulla as site of origin. Exp Clin Endocrinol Diabetes 105:156–162
297. Ueda S, Nishio K, Minamino N, Kubo A, Akai Y et al (1999) Increased plasma levels of adrenomedullin in patients with systemic inflammatory response syndrome. Am J Respir Crit Care Med 160:132–136
298. Chen YX, Li CS (2013) Prognostic value of adrenomedullin in septic patients in the ED. Am J Emerg Med 31:1017–1021
299. Oncel MY, Dilmen U, Erdeve O, Ozdemir R, Calisici E et al (2012) Proadrenomedullin as a prognostic marker in neonatal sepsis. Pediatr Res 72:507–512
300. Al Shuaibi M, Bahu RR, Chaftari AM, Al Wohoush I, Shomali W et al (2013) Pro-adrenomedullin as a novel biomarker for predicting infections and response to antimicrobials in febrile patients with hematologic malignancies. Clin Infect Dis 56:943–950
301. Christ-Crain M, Morgenthaler NG, Struck J, Harbarth S, Bergmann A et al (2005) Mid-regional pro-adrenomedullin as a prognostic marker in sepsis: an observational study. Crit Care 9:R816–R824
302. Guignant C, Voirin N, Venet F, Poitevin F, Malcus C et al (2009) Assessment of pro-vasopressin and pro-adrenomedullin as predictors of 28-day mortality in septic shock patients. Intensive Care Med 35:1859–1867
303. Scherpereel A, Depontieu F, Grigoriu B, Cavestri B, Tsicopoulos A et al (2006) Endocan, a new endothelial marker in human sepsis. Crit Care Med 34:532–537
304. De Freitas CN, Legendre B, Parmentier E, Scherpereel A, Tsicopoulos A et al (2013) Identification of a 14 kDa endocan fragment generated by cathepsin G, a novel circulating biomarker in patients with sepsis. J Pharm Biomed Anal 78–79:45–51
305. Linder A, Christensson B, Herwald H, Bjorck L, Akesson P (2009) Heparin-binding protein: an early marker of circulatory failure in sepsis. Clin Infect Dis 49:1044–1050
306. Borgel D, Clauser S, Bornstain C, Bieche I, Bissery A et al (2006) Elevated growth-arrest-specific protein 6 plasma levels in patients with severe sepsis. Crit Care Med 34:219–222
307. Ekman C, Linder A, Akesson P, Dahlback B (2010) Plasma concentrations of Gas6 (growth arrest specific protein 6) and its soluble tyrosine kinase receptor sAxl in sepsis and systemic inflammatory response syndromes. Crit Care 14:R158
308. Duswald KH, Jochum M, Schramm W, Fritz H (1985) Released granulocytic elastase: an indicator of pathobiochemical alterations in septicemia after abdominal surgery. Surgery 98:892–899
309. Tanaka H, Sugimoto H, Yoshioka T, Sugimoto T (1991) Role of granulocyte elastase in tissue injury in patients with septic shock complicated by multiple-organ failure. Ann Surg 213:81–85
310. Gardinali M, Padalino P, Vesconi S, Calcagno A, Ciappellano S et al (1992) Complement activation and polymorphonuclear neutrophil leukocyte elastase in sepsis. Correlation with severity of disease. Arch Surg 127: 1219–1224
311. Bossink AW, Groeneveld AB, Thijs LG (1999) Prediction of microbial infection and mortality in medical patients with fever: plasma procalcitonin, neutrophilic elastase-alpha1-antitrypsin, and lactoferrin compared with clinical variables. Clin Infect Dis 29:398–407
312. Selberg O, Hecker H, Martin M, Klos A, Bautsch W et al (2000) Discrimination of sepsis and systemic inflammatory response syndrome by determination of circulating plasma concentrations of procalcitonin, protein

complement 3a, and interleukin-6. Crit Care Med 28:2793–2798
313. Basu RK, Standage SW, Cvijanovich NZ, Allen GL, Thomas NJ et al (2011) Identification of candidate serum biomarkers for severe septic shock-associated kidney injury via microarray. Crit Care 15:R273
314. Hoffmann U, Bertsch T, Dvortsak E, Liebetrau C, Lang S et al (2006) Matrix-metalloproteinases and their inhibitors are elevated in severe sepsis: prognostic value of TIMP-1 in severe sepsis. Scand J Infect Dis 38:867–872
315. Lorente L, Martin MM, Labarta L, Diaz C, Sole-Violan J et al (2009) Matrix metalloproteinase-9, -10, and tissue inhibitor of matrix metalloproteinases-1 blood levels as biomarkers of severity and mortality in sepsis. Crit Care 13:R158
316. Gaddnas FP, Sutinen MM, Koskela M, Tervahartiala T, Sorsa T et al (2010) Matrix-metalloproteinase-2, -8 and -9 in serum and skin blister fluid in patients with severe sepsis. Crit Care 14:R49
317. Yazdan-Ashoori P, Liaw P, Toltl L, Webb B, Kilmer G et al (2011) Elevated plasma matrix metalloproteinases and their tissue inhibitors in patients with severe sepsis. J Crit Care 26:556–565
318. Lauhio A, Hastbacka J, Pettila V, Tervahartiala T, Karlsson S et al (2011) Serum MMP-8, -9 and TIMP-1 in sepsis: high serum levels of MMP-8 and TIMP-1 are associated with fatal outcome in a multicentre, prospective cohort study. Hypothetical impact of tetracyclines. Pharmacol Res 64:590–594
319. Tressel SL, Kaneider NC, Kasuda S, Foley C, Koukos G et al (2011) A matrix metalloprotease-PAR1 system regulates vascular integrity, systemic inflammation and death in sepsis. EMBO Mol Med 3:370–384
320. Punyadeera C, Schneider EM, Schaffer D, Hsu HY, Joos TO et al (2010) A biomarker panel to discriminate between systemic inflammatory response syndrome and sepsis and sepsis severity. J Emerg Trauma Shock 3:26–35
321. Vadas P (1984) Elevated plasma phospholipase A2 levels: correlation with the hemodynamic and pulmonary changes in gram-negative septic shock. J Lab Clin Med 104:873–881
322. Rintala EM, Nevalainen TJ (1993) Group II phospholipase A2 in sera of febrile patients with microbiologically or clinically documented infections. Clin Infect Dis 17:864–870
323. Takakuwa T, Endo S, Nakae H, Suzuki T, Inada K et al (1994) Relationships between plasma levels of type-II phospholipase A2, PAF-acetylhydrolase, leukotriene B4, complements, endothelin-1, and thrombomodulin in patients with sepsis. Res Commun Chem Pathol Pharmacol 84:271–281
324. Nyman KM, Uhl W, Forsstrom J, Buchler M, Beger HG et al (1996) Serum phospholipase A2 in patients with multiple organ failure. J Surg Res 60:7–14
325. Sorensen J, Kald B, Tagesson C, Lindahl M (1994) Platelet-activating factor and phospholipase A2 in patients with septic shock and trauma. Intensive Care Med 20:555–561
326. Endo S, Inada K, Nakae H, Takakuwa T, Yamada Y et al (1995) Plasma levels of type II phospholipase A2 and cytokines in patients with sepsis. Res Commun Mol Pathol Pharmacol 90:413–421
327. Yamada Y, Endo S, Kamei Y, Minato T, Yokoyama M et al (1998) Plasma levels of type II phospholipase A2 and nitrite/nitrate in patients with burns. Burns 24:513–517
328. Schrama AJ, de Beaufort AJ, Poorthuis BJ, Berger HM, Walther FJ (2008) Secretory phospholipase A(2) in newborn infants with sepsis. J Perinatol 28:291–296
329. Uusitalo-Seppala R, Peuravuori H, Koskinen P, Vahlberg T, Rintala EM (2012) Role of plasma bactericidal/permeability-increasing protein, group IIA phospholipase A(2), C-reactive protein, and white blood cell count in the early detection of severe sepsis in the emergency department. Scand J Infect Dis 44:697–704
330. Hattori N, Oda S, Sadahiro T, Nakamura M, Abe R et al (2009) YKL-40 identified by proteomic analysis as a biomarker of sepsis. Shock 32:393–400
331. Wittenhagen P, Kronborg G, Weis N, Nielsen H, Obel N et al (2004) The plasma level of soluble urokinase receptor is elevated in patients with Streptococcus pneumoniae bacteraemia and predicts mortality. Clin Microbiol Infect 10:409–415
332. Kornblit B, Hellemann D, Munthe-Fog L, Bonde J, Strom JJ et al (2013) Plasma YKL-40 and CHI3L1 in systemic inflammation and sepsis-experience from two prospective cohorts. Immunobiology 218:1227–1234
333. Accardo-Palumbo A, D'Amelio L, Pileri D, D'Arpa N, Mogavero R et al (2010) Reduction of plasma granzyme A correlates with severity of sepsis in burn patients. Burns 36:811–818
334. Lauw FN, Simpson AJ, Hack CE, Prins JM, Wolbink AM et al (2000) Soluble granzymes are released during human endotoxemia and in patients with severe infection due to gram-negative bacteria. J Infect Dis 182:206–213
335. Cohen J (2002) The immunopathogenesis of sepsis. Nature 420:885–891
336. Wilson RF, Farag A, Mammen EF, Fujii Y (1989) Sepsis and antithrombin III,

prekallikrein, and fibronectin levels in surgical patients. Am Surg 55:450–456
337. Leithauser B, Matthias FR, Nicolai U, Voss R (1996) Hemostatic abnormalities and the severity of illness in patients at the onset of clinically defined sepsis. Possible indication of the degree of endothelial cell activation? Intensive Care Med 22:631–636
338. Sakr Y, Reinhart K, Hagel S, Kientopf M, Brunkhorst F (2007) Antithrombin levels, morbidity, and mortality in a surgical intensive care unit. Anesth Analg 105:715–723
339. Wilson RF, Mammen EF, Tyburski JG, Warsow KM, Kubinec SM (1996) Antithrombin levels related to infections and outcome. J Trauma 40:384–387
340. Iba T, Kidokoro A, Fukunaga M, Sugiyama K, Sawada T et al (2005) Association between the severity of sepsis and the changes in hemostatic molecular markers and vascular endothelial damage markers. Shock 23:25–29
341. Kinasewitz GT, Yan SB, Basson B, Comp P, Russell JA et al (2004) Universal changes in biomarkers of coagulation and inflammation occur in patients with severe sepsis, regardless of causative micro-organism [ISRCTN74215569]. Crit Care 8:R82–R90
342. Pettila V, Pentti J, Pettila M, Takkunen O, Jousela I (2002) Predictive value of antithrombin III and serum C-reactive protein concentration in critically ill patients with suspected sepsis. Crit Care Med 30:271–275
343. Okabayashi K, Wada H, Ohta S, Shiku H, Nobori T et al (2004) Hemostatic markers and the sepsis-related organ failure assessment score in patients with disseminated intravascular coagulation in an intensive care unit. Am J Hematol 76:225–229
344. Reade MC, Yende S, D'Angelo G, Kong L, Kellum JA et al (2009) Differences in immune response may explain lower survival among older men with pneumonia. Crit Care Med 37:1655–1662
345. Lauterbach R, Pawlik D, Radziszewska R, Wozniak J, Rytlewski K (2006) Plasma antithrombin III and protein C levels in early recognition of late-onset sepsis in newborns. Eur J Pediatr 165:585–589
346. Ostrowski SR, Berg RM, Windelov NA, Meyer MA, Plovsing RR et al (2013) Coagulopathy, catecholamines, and biomarkers of endothelial damage in experimental human endotoxemia and in patients with severe sepsis: a prospective study. J Crit Care 28:586–596
347. Fisher CJ Jr, Yan SB (2000) Protein C levels as a prognostic indicator of outcome in sepsis and related diseases. Crit Care Med 28:S49–S56
348. Macias WL, Nelson DR (2004) Severe protein C deficiency predicts early death in severe sepsis. Crit Care Med 32:S223–S228
349. Gutovitz S, Papa L, Jimenez E, Falk J, Wieman L et al (2011) Protein C as an early biomarker to distinguish pneumonia from sepsis. J Crit Care 26(330):e339–312
350. Asakura H, Ontachi Y, Mizutani T, Kato M, Ito T et al (2001) Decreased plasma activity of antithrombin or protein C is not due to consumption coagulopathy in septic patients with disseminated intravascular coagulation. Eur J Haematol 67:170–175
351. Borgel D, Bornstain C, Reitsma PH, Lerolle N, Gandrille S et al (2007) A comparative study of the protein C pathway in septic and nonseptic patients with organ failure. Am J Respir Crit Care Med 176:878–885
352. Shaw AD, Vail GM, Haney DJ, Xie J, Williams MD (2011) Severe protein C deficiency is associated with organ dysfunction in patients with severe sepsis. J Crit Care 26:539–545
353. Shapiro NI, Trzeciak S, Hollander JE, Birkhahn R, Otero R et al (2009) A prospective, multicenter derivation of a biomarker panel to assess risk of organ dysfunction, shock, and death in emergency department patients with suspected sepsis. Crit Care Med 37:96–104
354. Iba T, Yagi Y, Kidokoro A, Fukunaga M, Fukunaga T (1995) Increased plasma levels of soluble thrombomodulin in patients with sepsis and organ failure. Surg Today 25:585–590
355. Boldt J, Wollbruck T, Sonneborn S, Welters A, Hempelmann G (1995) Thrombomodulin in intensive care patients. Intensive Care Med 21:645–650
356. Boldt J, Papsdorf M, Rothe A, Kumle B, Piper S (2000) Changes of the hemostatic network in critically ill patients–is there a difference between sepsis, trauma, and neurosurgery patients? Crit Care Med 28:445–450
357. Ikegami K, Suzuki Y, Yukioka T, Matsuda H, Shimazaki S (1998) Endothelial cell injury, as quantified by the soluble thrombomodulin level, predicts sepsis/multiple organ dysfunction syndrome after blunt trauma. J Trauma 44:789–794, discussion 794-785
358. Lin SM, Wang YM, Lin HC, Lee KY, Huang CD et al (2008) Serum thrombomodulin level relates to the clinical course of disseminated intravascular coagulation, multiorgan dysfunction syndrome, and mortality in patients with sepsis. Crit Care Med 36:683–689
359. Kato T, Sakai T, Kato M, Hagihara M, Hasegawa T et al (2013) Recombinant human soluble thrombomodulin administration improves sepsis-induced disseminated intravascular coagulation and mortality: a retrospective cohort study. Thromb J 11:3
360. Pralong G, Calandra T, Glauser MP, Schellekens J, Verhoef J et al (1989) Plasminogen activator inhibitor 1: a new

prognostic marker in septic shock. Thromb Haemost 61:459–462
361. Brandtzaeg P, Joo GB, Brusletto B, Kierulf P (1990) Plasminogen activator inhibitor 1 and 2, alpha-2-antiplasmin, plasminogen, and endotoxin levels in systemic meningococcal disease. Thromb Res 57:271–278
362. Madoiwa S, Nunomiya S, Ono T, Shintani Y, Ohmori T et al (2006) Plasminogen activator inhibitor 1 promotes a poor prognosis in sepsis-induced disseminated intravascular coagulation. Int J Hematol 84:398–405
363. Gando S, Nakanishi Y, Tedo I (1995) Cytokines and plasminogen activator inhibitor-1 in posttrauma disseminated intravascular coagulation: relationship to multiple organ dysfunction syndrome. Crit Care Med 23:1835–1842
364. Dofferhoff AS, Bom VJ, de Vries-Hospers HG, van Ingen J, Meer J et al (1992) Patterns of cytokines, plasma endotoxin, plasminogen activator inhibitor, and acute-phase proteins during the treatment of severe sepsis in humans. Crit Care Med 20:185–192
365. Kruithof E, Calandra T, Pralong G, Heumann D, Gerain J et al (1993) Evolution of plasminogen-activator inhibitor type-1 in patients with septic shock - correlation with cytokine concentrations. Fibrinolysis 7:117–121
366. Menges T, Hermans PW, Little SG, Langefeld T, Boning O et al (2001) Plasminogen-activator-inhibitor-1 4G/5G promoter polymorphism and prognosis of severely injured patients. Lancet 357:1096–1097
367. Garcia-Segarra G, Espinosa G, Tassies D, Oriola J, Aibar J et al (2007) Increased mortality in septic shock with the 4G/4G genotype of plasminogen activator inhibitor 1 in patients of white descent. Intensive Care Med 33:1354–1362
368. Robbie LA, Dummer S, Booth NA, Adey GD, Bennett B (2000) Plasminogen activator inhibitor 2 and urokinase-type plasminogen activator in plasma and leucocytes in patients with severe sepsis. Br J Haematol 109:342–348
369. Bagge L, Haglund O, Wallin R, Borg T, Modig J (1989) Differences in coagulation and fibrinolysis after traumatic and septic shock in man. Scand J Clin Lab Invest 49:63–72
370. Rubin DB, Wiener-Kronish JP, Murray JF, Green DR, Turner J et al (1990) Elevated von Willebrand factor antigen is an early plasma predictor of acute lung injury in nonpulmonary sepsis syndrome. J Clin Invest 86:474–480
371. Ware LB, Eisner MD, Thompson BT, Parsons PE, Matthay MA (2004) Significance of von Willebrand factor in septic and nonseptic patients with acute lung injury. Am J Respir Crit Care Med 170:766–772
372. Claus RA, Bockmeyer CL, Budde U, Kentouche K, Sossdorf M et al (2009) Variations in the ratio between von Willebrand factor and its cleaving protease during systemic inflammation and association with severity and prognosis of organ failure. Thromb Haemost 101:239–247
373. Christeff N, Benassayag C, Carli-Vielle C, Carli A, Nunez EA (1988) Elevated oestrogen and reduced testosterone levels in the serum of male septic shock patients. J Steroid Biochem 29:435–440
374. Fourrier F, Jallot A, Leclerc L, Jourdain M, Racadot A et al (1994) Sex steroid hormones in circulatory shock, sepsis syndrome, and septic shock. Circ Shock 43:171–178
375. Luppa P, Munker R, Nagel D, Weber M, Engelhardt D (1991) Serum androgens in intensive-care patients: correlations with clinical findings. Clin Endocrinol (Oxf) 34:305–310
376. May AK, Dossett LA, Norris PR, Hansen EN, Dorsett RC et al (2008) Estradiol is associated with mortality in critically ill trauma and surgical patients. Crit Care Med 36:62–68
377. Baue AE, Gunther B, Hartl W, Ackenheil M, Heberer G (1984) Altered hormonal activity in severely ill patients after injury or sepsis. Arch Surg 119:1125–1132
378. Bornstein SR, Licinio J, Tauchnitz R, Engelmann L, Negrao AB et al (1998) Plasma leptin levels are increased in survivors of acute sepsis: associated loss of diurnal rhythm, in cortisol and leptin secretion. J Clin Endocrinol Metab 83:280–283
379. Arnalich F, Lopez J, Codoceo R, Jim nez M, Madero R et al (1999) Relationship of plasma leptin to plasma cytokines and human survival in sepsis and septic shock. J Infect Dis 180:908–911
380. Langouche L, Vander Perre S, Frystyk J, Flyvbjerg A, Hansen TK et al (2009) Adiponectin, retinol-binding protein 4, and leptin in protracted critical illness of pulmonary origin. Crit Care 13:R112
381. Koch A, Weiskirchen R, Zimmermann HW, Sanson E, Trautwein C et al (2010) Relevance of serum leptin and leptin-receptor concentrations in critically ill patients. Mediators Inflamm 2010
382. Hillenbrand A, Knippschild U, Weiss M, Schrezenmeier H, Henne-Bruns D et al (2010) Sepsis induced changes of adipokines and cytokines - septic patients compared to morbidly obese patients. BMC Surg 10:26
383. Yousef AA, Amr YM, Suliman GA (2010) The diagnostic value of serum leptin monitoring and its correlation with tumor necrosis factor-alpha in critically ill patients: a prospective observational study. Crit Care 14:R33
384. Jochberger S, Morgenthaler NG, Mayr VD, Luckner G, Wenzel V et al (2006) Copeptin

and arginine vasopressin concentrations in critically ill patients. J Clin Endocrinol Metab 91:4381–4386

385. Lee JH, Chan YH, Lai OF, Puthucheary J (2013) Vasopressin and copeptin levels in children with sepsis and septic shock. Intensive Care Med 39:747–753
386. Seligman R, Papassotiriou J, Morgenthaler NG, Meisner M, Teixeira PJ (2008) Copeptin, a novel prognostic biomarker in ventilator-associated pneumonia. Crit Care 12:R11
387. Muller B, Morgenthaler N, Stolz D, Schuetz P, Muller C et al (2007) Circulating levels of copeptin, a novel biomarker, in lower respiratory tract infections. Eur J Clin Invest 37:145–152
388. Purhonen AK, Vanska M, Hamalainen S, Pulkki K, Lehtikangas M et al (2012) Plasma copeptin in the assessment of febrile neutropenia. Peptides 36:129–132
389. Morgenthaler NG, Struck J, Christ-Crain M, Bergmann A, Muller B (2005) Pro-atrial natriuretic peptide is a prognostic marker in sepsis, similar to the APACHE II score: an observational study. Crit Care 9:R37–R45
390. Ueda S, Nishio K, Akai Y, Fukushima H, Ueyama T et al (2006) Prognostic value of increased plasma levels of brain natriuretic peptide in patients with septic shock. Shock 26:134–139
391. Witthaut R, Busch C, Fraunberger P, Walli A, Seidel D et al (2003) Plasma atrial natriuretic peptide and brain natriuretic peptide are increased in septic shock: impact of interleukin-6 and sepsis-associated left ventricular dysfunction. Intensive Care Med 29:1696–1702
392. Brueckmann M, Huhle G, Lang S, Haase KK, Bertsch T et al (2005) Prognostic value of plasma N-terminal pro-brain natriuretic peptide in patients with severe sepsis. Circulation 112:527–534
393. Charpentier J, Luyt CE, Fulla Y, Vinsonneau C, Cariou A et al (2004) Brain natriuretic peptide: a marker of myocardial dysfunction and prognosis during severe sepsis. Crit Care Med 32:660–665
394. Varpula M, Pulkki K, Karlsson S, Ruokonen E, Pettila V (2007) Predictive value of N-terminal pro-brain natriuretic peptide in severe sepsis and septic shock. Crit Care Med 35:1277–1283
395. Kerbaul F, Giorgi R, Oddoze C, Collart F, Guidon C et al (2004) High concentrations of N-BNP are related to non-infectious severe SIRS associated with cardiovascular dysfunction occurring after off-pump coronary artery surgery. Br J Anaesth 93:639–644
396. McLean AS, Huang SJ, Hyams S, Poh G, Nalos M et al (2007) Prognostic values of B-type natriuretic peptide in severe sepsis and septic shock. Crit Care Med 35:1019–1026
397. Post F, Weilemann LS, Messow CM, Sinning C, Munzel T (2008) B-type natriuretic peptide as a marker for sepsis-induced myocardial depression in intensive care patients. Crit Care Med 36:3030–3037
398. Pirracchio R, Deye N, Lukaszewicz AC, Mebazaa A, Cholley B et al (2008) Impaired plasma B-type natriuretic peptide clearance in human septic shock. Crit Care Med 36: 2542–2546
399. Kandil E, Burack J, Sawas A, Bibawy H, Schwartzman A et al (2008) B-type natriuretic peptide: a biomarker for the diagnosis and risk stratification of patients with septic shock. Arch Surg 143:242–246, discussion 246
400. Chen Y, Li C (2009) Prognostic significance of brain natriuretic peptide obtained in the ED in patients with SIRS or sepsis. Am J Emerg Med 27:701–706
401. Perman SM, Chang AM, Hollander JE, Gaieski DF, Trzeciak S et al (2011) Relationship between B-type natriuretic peptide and adverse outcome in patients with clinical evidence of sepsis presenting to the emergency department. Acad Emerg Med 18:219–222
402. Hama N, Itoh H, Shirakami G, Suga S, Komatsu Y et al (1994) Detection of C-type natriuretic peptide in human circulation and marked increase of plasma CNP level in septic shock patients. Biochem Biophys Res Commun 198:1177–1182
403. Bahrami S, Pelinka L, Khadem A, Maitzen S, Hawa G et al (2010) Circulating NT-proCNP predicts sepsis in multiple-traumatized patients without traumatic brain injury. Crit Care Med 38:161–166
404. Koch A, Voigt S, Sanson E, Duckers H, Horn A et al (2011) Prognostic value of circulating amino-terminal pro-C-type natriuretic peptide in critically ill patients. Crit Care 15:R45
405. Rubli E, Bussard S, Frei E, Lundsgaard-Hansen P, Pappova E (1983) Plasma fibronectin and associated variables in surgical intensive care patients. Ann Surg 197:310–317
406. Ekindjian OG, Marien M, Wassermann D, Bruxelle J, Cazalet C et al (1984) Plasma fibronectin time course in burned patients: influence of sepsis. J Trauma 24:214–219
407. Blanco A, Guisasola JA, Solis P, Bachiller R, Gonzalez H (1990) Fibronectin in meningococcal sepsis. Correlation with antithrombin III and protein C. Acta Paediatr Scand 79:73–76
408. Ruiz Martin G, Prieto Prieto J, Veiga de Cabo J, Gomez Lus L, Barberan J et al (2004) Plasma fibronectin as a marker of sepsis. Int J Infect Dis 8:236–243

409. Glattard E, Welters ID, Lavaux T, Muller AH, Laux A et al (2010) Endogenous morphine levels are increased in sepsis: a partial implication of neutrophils. PLoS One 5:e8791
410. Sakr Y, Reinhart K, Bloos F, Marx G, Russwurm S et al (2007) Time course and relationship between plasma selenium concentrations, systemic inflammatory response, sepsis, and multiorgan failure. Br J Anaesth 98:775–784
411. Vaschetto R, Nicola S, Olivieri C, Boggio E, Piccolella F et al (2008) Serum levels of osteopontin are increased in SIRS and sepsis. Intensive Care Med 34:2176–2184
412. Wang H, Cheng B, Chen Q, Wu S, Lv C et al (2008) Time course of plasma gelsolin concentrations during severe sepsis in critically ill surgical patients. Crit Care 12:R106
413. Stove S, Welte T, Wagner TO, Kola A, Klos A et al (1996) Circulating complement proteins in patients with sepsis or systemic inflammatory response syndrome. Clin Diagn Lab Immunol 3:175–183
414. Wolk K, Döcke WD, von Baehr V, Volk HD, Sabat R (2000) Impaired antigen presentation by human monocytes during endotoxin tolerance. Blood 96(1):218–223
415. Cavaillon JM, Adib-Conquy M (2006) Bench-to-bedside review: endotoxin tolerance as a model of leukocyte reprogramming in sepsis. Crit Care 10:233
416. Hershman MJ, Cheadle WG, Wellhausen SR, Davidon P, Polk HC (1990) Monocyte HLA-DR antigen expression characterizes clinical outcome in the trauma patients. Br J Surg 77:204–207
417. Kim OY, Monsel A, Bertrand M, Coriat P, Cavaillon JM et al (2010) Differential down-regulation of HLA-DR on monocyte subpopulations during systemic inflammation. Crit Care 14:R61
418. Fumeaux T, Pugin J (2002) Role of interleukin-10 in the intracellular sequestration of human leukocyte antigen-DR in monocytes during septic shock. Am J Respir Crit Care Med 166:1475–1482
419. Le Tulzo Y, Pangault C, Amiot L, Guilloux V, Tribut O et al (2004) Monocyte human leukocyte antigen-DR transcriptional downregulation by cortisol during septic shock. Am J Respir Crit Care Med 169:1144–1151
420. Ditschkowski M, Kreuzfelder E, Rebmann V, Ferencik S, Majetschak M et al (1999) HLA-DR expression and soluble HLA-DR levels in septic patients after trauma. Ann Surg 229:246–254
421. Tschaikowsky K, Hedwig-Geissing M, Schiele A, Bremer F, Schywalsky M et al (2002) Coincidence of pro- and anti-inflammatory responses in the early phase of severe sepsis: longitudinal study of mononuclear histocompatibility leukocyte antigen-DR expression, procalcitonin, C-reactive protein, and changes in T-cell subsets in septic and postoperative patients. Crit Care Med 30:1015–1023
422. Perry SE, Mostafa SM, Wenstone R, Shenkin A, McLaughlin PJ (2003) Is low monocyte HLA-DR expression helpful to predict outcome in severe sepsis? Intensive Care Med 29:1245–1252
423. Muehlstedt SG, Lyte M, Rodriguez JL (2002) Increased IL-10 production and HLA-DR suppression in the lungs of injured patients precede the development of nosocomial pneumonia. Shock 17:443–450
424. Cheadle WG, Hershman MJ, Wellhausen SR, Polk HC Jr (1991) HLA-DR antigen expression on peripheral blood monocytes correlates with surgical infection. Am J Surg 161:639–645
425. van den Berk JM, Oldenburger RH, van den Berg AP, Klompmaker IJ, Mesander G et al (1997) Low HLA-DR expression on monocytes as a prognostic marker for bacterial sepsis after liver transplantation. Transplantation 63:1846–1848
426. Satoh A, Miura T, Satoh K, Masamune A, Yamagiwa T et al (2002) Human leukocyte antigen-DR expression on peripheral monocytes as a predictive marker of sepsis during acute pancreatitis. Pancreas 25:245–250
427. Venet F, Tissot S, Debard AL, Faudot C, Crampe C et al (2007) Decreased monocyte human leukocyte antigen-DR expression after severe burn injury: correlation with severity and secondary septic shock. Crit Care Med 35:1910–1917
428. Strohmeyer JC, Blume C, Meisel C, Doecke WD, Hummel M et al (2003) Standardized immune monitoring for the prediction of infections after cardiopulmonary bypass surgery in risk patients. Cytometry B Clin Cytom 53:54–62
429. Monneret G, Lepape A, Voirin N, Bohe J, Venet F et al (2006) Persisting low monocyte human leukocyte antigen-DR expression predicts mortality in septic shock. Intensive Care Med 32:1175–1183
430. Lukaszewicz AC, Grienay M, Resche-Rigon M, Pirracchio R, Faivre V et al (2009) Monocytic HLA-DR expression in intensive care patients: interest for prognosis and secondary infection prediction. Crit Care Med 37:2746–2752
431. Cheron A, Floccard B, Allaouchiche B, Guignant C, Poitevin F et al (2010) Lack of recovery in monocyte human leukocyte antigen-DR expression is independently associated

with the development of sepsis after major trauma. Crit Care 14:R208

432. Landelle C, Lepape A, Voirin N, Tognet E, Venet F et al (2010) Low monocyte human leukocyte antigen-DR is independently associated with nosocomial infections after septic shock. Intensive Care Med 36:1859–1866
433. Medzhitov R, Preston-Hurlburt P, Janeway CA (1997) A human homologue of the *Drosophila* Toll protein signals activation of adaptative immunity. Nature 388:394–397
434. Wittebole X, Coyle SM, Kumar A, Goshima M, Lowry SF et al (2005) Expression of tumour necrosis factor receptor and Toll-like receptor 2 and 4 on peripheral blood leucocytes of human volunteers after endotoxin challenge: a comparison of flow cytometric light scatter and immunofluorescence gating. Clin Exp Immunol 141:99–106
435. Harter L, Mica L, Stocker R, Trentz O, Keel M (2004) Increased expression of toll-like receptor-2 and -4 on leukocytes from patients with sepsis. Shock 22:403–409
436. Brandl K, Gluck T, Huber C, Salzberger B, Falk W et al (2005) TLR-4 surface display on human monocytes is increased in septic patients. Eur J Med Res 10:319–324
437. Tsujimoto H, Ono S, Majima T, Efron PA, Kinoshita M et al (2006) Differential toll-like receptor expression after ex vivo lipopolysaccharide exposure in patients with sepsis and following surgical stress. Clin Immunol 119:180–187
438. Tsujimoto H, Ono S, Majima T, Kawarabayashi N, Takayama E et al (2005) Neutrophil elastase, MIP-2, and TLR-4 expression during human and experimental sepsis. Shock 23:39–44
439. Viemann D, Dubbel G, Schleifenbaum S, Harms E, Sorg C et al (2005) Expression of toll-like receptors in neonatal sepsis. Pediatr Res 58:654–659
440. Ono S, Tsujimoto H, Hiraki S, Takahata R, Kinoshita M et al (2005) Sex differences in cytokine production and surface antigen expression of peripheral blood mononuclear cells after surgery. Am J Surg 190:439–444
441. Renshaw M, Rockwell J, Engleman C, Gewirtz A, Katz J et al (2002) Cutting edge: impaired Toll-like receptor expression and function in aging. J Immunol 169:4697–4701
442. Armstrong L, Medford AR, Hunter KJ, Uppington KM, Millar AB (2004) Differential expression of Toll-like receptor (TLR)-2 and TLR-4 on monocytes in human sepsis. Clin Exp Immunol 136:312–319
443. Martins PS, Brunialti MK, Martos LS, Machado FR, Assuncao MS et al (2008) Expression of cell surface receptors and oxidative metabolism modulation in the clinical continuum of sepsis. Crit Care 12:R25
444. Adib-Conquy M, Moine P, Asehnoune K, Edouard A, Espevik T et al (2003) Toll-like receptor-mediated tumor necrosis factor and interleukin-10 production differ during systemic inflammation. Am J Respir Crit Care Med 168:158–164
445. Souza-Fonseca-Guimaraes F, Parlato M, Philippart F, Misset B, Cavaillon JM et al (2012) Toll-like receptors expression and interferon-gamma production by NK cells in human sepsis. Crit Care 16:R206
446. Brunialti MK, Martins PS, Barbosa de Carvalho H, Machado FR, Barbosa LM et al (2006) TLR2, TLR4, CD14, CD11B, and CD11C expressions on monocytes surface and cytokine production in patients with sepsis, severe sepsis, and septic shock. Shock 25:351–357
447. Aalto H, Takala A, Kautiainen H, Siitonen S, Repo H (2007) Monocyte CD14 and soluble CD14 in predicting mortality of patients with severe community acquired infection. Scand J Infect Dis 39:596–603
448. Schaaf B, Luitjens K, Goldmann T, van Bremen T, Sayk F et al (2009) Mortality in human sepsis is associated with downregulation of Toll-like receptor 2 and CD14 expression on blood monocytes. Diagn Pathol 4:12
449. Monneret G, Debard AL, Venet F, Bohe J, Hequet O et al (2003) Marked elevation of human circulating CD4+CD25+ regulatory T cells in sepsis-induced immunoparalysis. Crit Care Med 31:2068–2071
450. Venet F, Pachot A, Debard AL, Bohe J, Bienvenu J et al (2004) Increased percentage of CD4+CD25+ regulatory T cells during septic shock is due to the decrease of CD4+CD25- lymphocytes. Crit Care Med 32:2329–2331
451. Venet F, Chung CS, Kherouf H, Geeraert A, Malcus C et al (2009) Increased circulating regulatory T cells (CD4(+)CD25 (+)CD127 (-)) contribute to lymphocyte anergy in septic shock patients. Intensive Care Med 35: 678–686
452. Hein F, Massin F, Cravoisy-Popovic A, Barraud D, Levy B et al (2010) The relationship between CD4+CD25+CD127- regulatory T cells and inflammatory response and outcome during shock states. Crit Care 14:R19
453. Leng FY, Liu JL, Liu ZJ, Yin JY, Qu HP (2013) Increased proportion of CD4(+) CD25(+)Foxp3(+) regulatory T cells during early-stage sepsis in ICU patients. J Microbiol Immunol Infect 46:338–344
454. Sugimoto K, Galle C, Preiser JC, Creteur J, Vincent JL et al (2003) Monocyte CD40 expression in severe sepsis. Shock 19:24–27
455. Nolan A, Weiden M, Kelly A, Hoshino Y, Hoshino S et al (2008) CD40 and CD80/86

act synergistically to regulate inflammation and mortality in polymicrobial sepsis. Am J Respir Crit Care Med 177:301–308
456. Katsuura M, Shimizu Y, Akiba K, Kanazawa C, Mitsui T et al (1998) CD48 expression on leukocytes in infectious diseases: flow cytometric analysis of surface antigen. Acta Paediatr Jpn 40:580–585
457. Lewis SM, Treacher DF, Bergmeier L, Brain SD, Chambers DJ et al (2009) Plasma from patients with sepsis up-regulates the expression of CD49d and CD64 on blood neutrophils. Am J Respir Cell Mol Biol 40:724–732
458. Bhandari V, Wang C, Rinder C, Rinder H (2008) Hematologic profile of sepsis in neonates: neutrophil CD64 as a diagnostic marker. Pediatrics 121:129–134
459. Groselj-Grenc M, Ihan A, Pavcnik-Arnol M, Kopitar AN, Gmeiner-Stopar T et al (2009) Neutrophil and monocyte CD64 indexes, lipopolysaccharide-binding protein, procalcitonin and C-reactive protein in sepsis of critically ill neonates and children. Intensive Care Med 35:1950–1958
460. Streimish I, Bizzarro M, Northrup V, Wang C, Renna S et al (2014) Neutrophil CD64 with Hematologic Criteria for Diagnosis of Neonatal Sepsis. Am J Perinatol 31:21–30
461. Cardelli P, Ferraironi M, Amodeo R, Tabacco F, De Blasi RA et al (2008) Evaluation of neutrophil CD64 expression and procalcitonin as useful markers in early diagnosis of sepsis. Int J Immunopathol Pharmacol 21:43–49
462. Gibot S, Bene MC, Noel R, Massin F, Guy J et al (2012) Combination biomarkers to diagnose sepsis in the critically ill patient. Am J Respir Crit Care Med 186:65–71
463. Gros A, Roussel M, Sauvadet E, Gacouin A, Marque S et al (2012) The sensitivity of neutrophil CD64 expression as a biomarker of bacterial infection is low in critically ill patients. Intensive Care Med 38:445–452
464. Dimoula A, Pradier O, Kassengera Z, Dalcomune D, Turkan H et al (2014) Serial determinations of neutrophil CD64 expression for the diagnosis and monitoring of sepsis in critically ill patients. Clin Infect Dis 58(6):820–9
465. Schwulst SJ, Muenzer JT, Chang KC, Brahmbhatt TS, Coopersmith CM et al (2008) Lymphocyte phenotyping to distinguish septic from nonseptic critical illness. J Am Coll Surg 206:335–342
466. Roger PM, Hyvernat H, Ticchioni M, Kumar G, Dellamonica J et al (2012) The early phase of human sepsis is characterized by a combination of apoptosis and proliferation of T cells. J Crit Care 27:384–393
467. de Pablo R, Monserrat J, Torrijos C, Martin M, Prieto A et al (2012) The predictive role of early activation of natural killer cells in septic shock. Crit Care 16:413
468. Bouchon A, Facchetti F, Weigand MA, Colonna M (2001) TREM-1 amplifies inflammation and is a crucial mediator of septic shock. Nature 410:1103–1107
469. Gibot S, Le Renard PE, Bollaert PE, Kolopp-Sarda MN, Bene MC et al (2005) Surface triggering receptor expressed on myeloid cells 1 expression patterns in septic shock. Intensive Care Med 31:594–597
470. Ferat-Osorio E, Esquivel-Callejas N, Wong-Baeza I, Aduna-Vicente R, Arriaga-Pizano L et al (2008) The increased expression of TREM-1 on monocytes is associated with infectious and noninfectious inflammatory processes. J Surg Res 150:110–117
471. Poukoulidou T, Spyridaki A, Mihailidou I, Kopterides P, Pistiki A et al (2011) TREM-1 expression on neutrophils and monocytes of septic patients: relation to the underlying infection and the implicated pathogen. BMC Infect Dis 11:309
472. Pachot A, Lepape A, Vey S, Bienvenu J, Mougin B et al (2006) Systemic transcriptional analysis in survivor and non-survivor septic shock patients: a preliminary study. Immunol Lett 106:63–71
473. Pachot A, Cazalis MA, Venet F, Turrel F, Faudot C et al (2008) Decreased expression of the fractalkine receptor CX3CR1 on circulating monocytes as new feature of sepsis-induced immunosuppression. J Immunol 180:6421–6429
474. Huang X, Venet F, Wang YL, Lepape A, Yuan Z et al (2009) PD-1 expression by macrophages plays a pathologic role in altering microbial clearance and the innate inflammatory response to sepsis. Proc Natl Acad Sci U S A 106:6303–6308
475. Guignant C, Lepape A, Huang X, Kherouf H, Denis L et al (2011) Programmed death-1 levels correlate with increased mortality, nosocomial infection and immune dysfunctions in septic shock patients. Crit Care 15:R99
476. Zhang Y, Li J, Lou J, Zhou Y, Bo L et al (2011) Upregulation of programmed death-1 on T cells and programmed death ligand-1 on monocytes in septic shock patients. Crit Care 15:R70
477. Monaghan SF, Thakkar RK, Tran ML, Huang X, Cioffi WG et al (2012) Programmed death 1 expression as a marker for immune and physiological dysfunction in the critically ill surgical patient. Shock 38:117–122
478. Shubin NJ, Monaghan SF, Heffernan DS, Chung CS, Ayala A (2013) B and T lymphocyte attenuator expression on CD4+ T-cells associates with sepsis and subsequent infections in ICU patients. Crit Care 17:R276

479. Hauser CJ, Lagoo S, Lagoo A, Hale E, Hardy KJ et al (1995) Human peripheral mononuclear cells do not show proinflammatory patterns of cytokine transcription in early trauma: a preliminary report. Shock 4:247–250
480. Ramilo O, Allman W, Chung W, Mejias A, Ardura M et al (2007) Gene expression patterns in blood leukocytes discriminate patients with acute infections. Blood 109:2066–2077
481. Tang BM, McLean AS, Dawes IW, Huang SJ, Cowley MJ et al (2008) Gene-expression profiling of gram-positive and gram-negative sepsis in critically ill patients. Crit Care Med 36:1125–1128
482. Johnson SB, Lissauer M, Bochicchio GV, Moore R, Cross AS et al (2007) Gene expression profiles differentiate between sterile SIRS and early sepsis. Ann Surg 245:611–621
483. Tang BM, McLean AS, Dawes IW, Huang SJ, Lin RC (2009) Gene-expression profiling of peripheral blood mononuclear cells in sepsis. Crit Care Med 37:882–888
484. Wong HR, Cvijanovich N, Allen GL, Lin R, Anas N et al (2009) Genomic expression profiling across the pediatric systemic inflammatory response syndrome, sepsis, and septic shock spectrum. Crit Care Med 37:1558–1566
485. Tang BM, McLean AS, Dawes IW, Huang SJ, Lin RC (2007) The use of gene-expression profiling to identify candidate genes in human sepsis. Am J Respir Crit Care Med 176:676–684
486. Hinrichs C, Kotsch K, Buchwald S, Habicher M, Saak N et al (2010) Perioperative gene expression analysis for prediction of postoperative sepsis. Clin Chem 56:613–622
487. Bauer M, Giamarellos-Bourboulis E, Kortgen A, Möller E, Felsmann K, et al. (submitted) A transcriptomic biomarker to quantify systemic inflammation of sepsis
488. Wynn JL, Cvijanovich NZ, Allen GL, Thomas NJ, Freishtat RJ et al (2011) The influence of developmental age on the early transcriptomic response of children with septic shock. Mol Med 17:1146–1156
489. Tang BM, Huang SJ, McLean AS (2010) Genome-wide transcription profiling of human sepsis: a systematic review. Crit Care 14:R237
490. Bartel DP (2004) MicroRNAs: genomics, biogenesis, mechanism, and function. Cell 116:281–297
491. Bartel DP (2009) MicroRNAs: target recognition and regulatory functions. Cell 136: 215–233
492. Taganov KD, Boldin MP, Chang KJ, Baltimore D (2006) NF-kappaB-dependent induction of microRNA miR-146, an inhibitor targeted to signaling proteins of innate immune responses. Proc Natl Acad Sci U S A 103:12481–12486
493. Nahid MA, Pauley KM, Satoh M, Chan EK (2009) miR-146a is critical for endotoxin-induced tolerance: implication in innate immunity. J Biol Chem 284:34590–34599
494. Rossato M, Curtale G, Tamassia N, Castellucci M, Mori L et al (2012) IL-10-induced microRNA-187 negatively regulates TNF-alpha, IL-6, and IL-12p40 production in TLR4-stimulated monocytes. Proc Natl Acad Sci U S A 109:E3101–E3110
495. Wang JF, Yu ML, Yu G, Bian JJ, Deng XM et al (2010) Serum miR-146a and miR-223 as potential new biomarkers for sepsis. Biochem Biophys Res Commun 394:184–188
496. Wang L, Wang HC, Chen C, Zeng J, Wang Q et al (2013) Differential expression of plasma miR-146a in sepsis patients compared with non-sepsis-SIRS patients. Exp Ther Med 5:1101–1104
497. Wang H, Zhang P, Chen W, Feng D, Jia Y et al (2012) Serum microRNA signatures identified by Solexa sequencing predict sepsis patients' mortality: a prospective observational study. PLoS One 7:e38885
498. Wang H, Zhang P, Chen W, Feng D, Jia Y et al (2012) Evidence for serum miR-15a and miR-16 levels as biomarkers that distinguish sepsis from systemic inflammatory response syndrome in human subjects. Clin Chem Lab Med 50:1423–1428
499. Schmidt WM, Spiel AO, Jilma B, Wolzt M, Muller M (2009) In vivo profile of the human leukocyte microRNA response to endotoxemia. Biochem Biophys Res Commun 380:437–441
500. Vasilescu C, Rossi S, Shimizu M, Tudor S, Veronese A et al (2009) MicroRNA fingerprints identify miR-150 as a plasma prognostic marker in patients with sepsis. PLoS One 4:e7405
501. Ma Y, Vilanova D, Atalar K, Delfour O, Edgeworth J et al (2013) Genome-wide sequencing of cellular microRNAs identifies a combinatorial expression signature diagnostic of sepsis. PLoS One 8:e75918
502. Roderburg C, Luedde M, Vargas Cardenas D, Vucur M, Scholten D et al (2013) Circulating microRNA-150 serum levels predict survival in patients with critical illness and sepsis. PLoS One 8:e54612
503. Tacke F, Roderburg C, Benz F, Cardenas DV, Luedde M et al (2014) Levels of circulating miR-133a are elevated in sepsis and predict mortality in critically ill patients. Crit Care Med 42(5):1096–104
504. Hurr H, Hawley HB, Czachor JS, Markert RJ, McCarthy MC (1999) APACHE II and ISS scores as predictors of nosocomial infections in trauma patients. Am J Infect Control 27:79–83

505. Angeletti S, Battistoni F, Fioravanti M, Bernardini S, Dicuonzo G (2013) Procalcitonin and mid-regional pro-adrenomedullin test combination in sepsis diagnosis. Clin Chem Lab Med 51:1059–1067
506. Casserly B, Read R, Levy MM (2011) Multimarker panels in sepsis. Crit Care Clin 27:391–405
507. Waage A, Espevik T, Lamvik J (1986) Detection of tumour necrosis factor-like cytotoxicity in serum from patients with septicaemia but not from untreated cancer patients. Scand J Immunol 24(6):739–743
508. Marchant A, Devière J, Byl B, De Groote D, Vincent JL et al (1994) Interleukin-10 production during septicaemia. Lancet 343(8899):707–708
509. Waring PM, Presneill J, Maher DW, Layton JE, Cebon J et al (1995) Differential alterations in plasma colony-stimulating factor concentrations in meningococcaemia. Clin Exp Immunol 102(3):501–506
510. DiPiro JT, Howdieshell TR, Goddard JK, Callaway DB, Hamilton RG et al (1995) Association of interleukin-4 plasma levels with traumatic injury and clinical course. Arch Surg 130(11):1159–1162
511. Zeni F, Vindimian M, Pain P, Gery P, Tardy B et al (1995) Antiinflammatory and proinflammatory cytokines in patients with severe sepsis. J Infect Dis 172(4):1171–1172
512. Marie C, Cavaillon JM, Losser MR (1996) Elevated levels of circulating transforming growth factor-beta 1 in patients with the sepsis syndrome. Ann Intern Med 125(6): 520–521
513. Sriskandan S, Moyes D, Cohen J (1996) Detection of circulating bacterial superantigen and lymphotoxin-alpha in patients with streptococcal toxic-shock syndrome. Lancet 348(9037):1315–1316
514. Bingold TM, Ziesché E, Scheller B, Sadik CD, Franck K et al (2010) Interleukin-22 detected in patients with abdominal sepsis. Shock 34(4):337–340
515. Kasai T, Inada K, Takakuwa T, Yamada Y, Inoue Y et al (1997) Anti-inflammatory cytokine levels in patients with septic shock. Res Commun Mol Pathol Pharmacol 98(1): 34–42

Chapter 16

Host Response Biomarkers in Sepsis: The Role of Procalcitonin

Jean-Louis Vincent, Marc Van Nuffelen, and Christophe Lelubre

Abstract

Procalcitonin is the prohormone of calcitonin and present in minute quantities in health. However, during infection, its levels rise considerably and are correlated with the severity of the infection. Several assays have been developed for measurement of procalcitonin levels; in this article, we will briefly present the PCT-sensitive Kryptor® test (Brahms, Hennigsdorf, Germany), one of the most widely used assays for procalcitonin in recent studies. Many studies have demonstrated the value of procalcitonin levels for diagnosing sepsis and assessing disease severity. Procalcitonin levels have also been successfully used to guide antibiotic administration. However, procalcitonin is not specific for sepsis, and values need to be interpreted in the context of a full clinical examination and the presence of other signs and symptoms of sepsis.

Key words Procalcitonin, Prohormone, Diagnostic and prognostic marker

1 Introduction

Sepsis is the leading cause of death in intensive care unit (ICU) patients. Rapid diagnosis is important because delayed antibiotic therapy is associated with worse outcomes [1]. However, diagnosis is often not straightforward in critically ill patients in whom clinical signs of sepsis may be absent or associated with other pathologies, and microbiological cultures are frequently negative because of recent or ongoing antimicrobial therapy. Biological markers, or biomarkers, a term first introduced in the late 1970s, have been proposed as a means of aiding diagnosis in patients with sepsis. They have also been suggested for use in predicting disease severity and outcome and for monitoring need for and response to therapy. Procalcitonin is one of more than 170 biomarkers that have been investigated for potential use in septic patients [2] and is one of the most widely studied.

Nicasio Mancini (ed.), *Sepsis: Diagnostic Methods and Protocols*, Methods in Molecular Biology, vol. 1237,
DOI 10.1007/978-1-4939-1776-1_16, © Springer Science+Business Media New York 2015

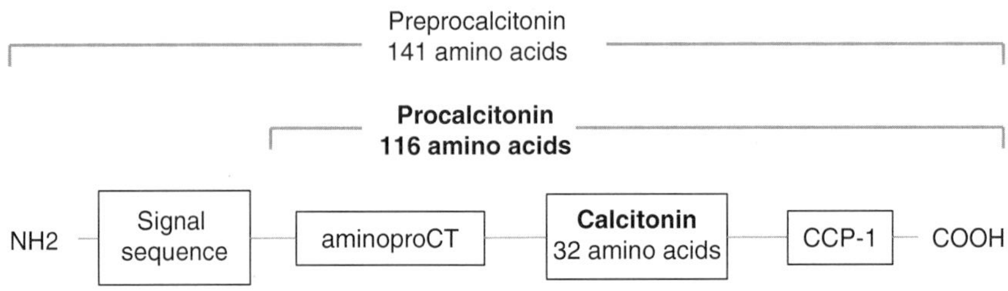

Fig. 1 Simplified schematic of the structure of procalcitonin

1.1 Structure and Release

Procalcitonin is a 116-amino acid peptide precursor or prohormone of calcitonin, a hormone involved in calcium homeostasis. Procalcitonin itself is derived from a preprohormone consisting of 141-amino acid residues, preprocalcitonin [3] (Fig. 1). At the carboxyl terminus of procalcitonin is a 21-amino acid peptide termed the calcitonin carboxy-terminal peptide-1 (CCP-1), and at the amino terminus is a 57-amino acid peptide called aminoprocalcitonin. First identified in the early 1980s in patients with toxic shock syndrome [4, 5], procalcitonin was first proposed for use as a biomarker of sepsis and infection by Assicot et al. in 1993 [6]. In health, procalcitonin is produced mainly in the C-cells of the thyroid gland and is cleaved to calcitonin so that levels in the blood are undetectable at <0.1 ng/mL. However, in infection, regulation of procalcitonin synthesis is altered, mediated by microbial toxins and cytokines. The *CALC1* gene, normally expressed only in the thyroid C-cells, is expressed in other cells, including liver, kidney, and adipose cells [7, 8], in which the released procalcitonin is not spliced to form calcitonin. Serum levels of procalcitonin thus increase rapidly, within about 3–6 h, during endotoxemia and sepsis [9]. Importantly, viral infections do not seem to have the same effect on procalcitonin levels [10], possibly because interferon, a cytokine released in larger amounts in viral infections, attenuates procalcitonin release [11, 12].

Procalcitonin is highly conserved in evolution suggesting that it has an important physiological role, but exactly what that role is remains unclear. Studies have shown that procalcitonin increases the expression of proinflammatory cytokines by leukocytes and reduces neutrophil migration [13], augments sepsis-induced increases in nitric oxide release [14], and increases mortality in septic animals [15]. Moreover, antibodies to procalcitonin have been shown to improve survival in various animal models of sepsis [15–17]. However, other studies have suggested that procalcitonin can neutralize lipopolysaccharide and decrease proinflammatory cytokine release [18, 19].

1.2 Assay

Quantitative and qualitative assays for procalcitonin are available, but many of the more recent studies have used the quantitative PCT-sensitive Kryptor® test (Brahms, Hennigsdorf, Germany).

The Kryptor assay is based on time-resolved amplified cryptate emission (TRACE) technology [20]. Essentially, TRACE technology relies on measurement of the nonradioactive energy transfer from a "donor" molecule to an "acceptor" molecule. In the Brahms' assay, the donor molecule is a europium cryptate-labeled polyclonal sheep antibody that recognizes epitopes in the immature calcitonin region of procalcitonin, and the acceptor molecule is an XL665-labeled monoclonal antibody against the CCP-1 region of procalcitonin. When the sample is excited with a nitrogen laser at 337 nm, the donor emits a long-lived fluorescent signal (used as a reference) in the millisecond range at 620 nm, and the acceptor emits a short-lived signal (the specific signal for procalcitonin) in the nanosecond range at 665 nm. If procalcitonin is present in the sample, it is sandwiched between the two molecules forming an immunocomplex enabling a transfer of energy between the donor and acceptor molecules. When the immunocomplex is formed, the fluorescent signal is intensified, and the resultant signal amplified at 665 nm and prolonged to last for a few microseconds. The specific fluorescence, proportional to the concentration of procalcitonin, is obtained from the spectral and temporal selection [21].

In practical terms, a sample size of about 50 μl is required. Serum, EDTA, or heparin plasma can be used, but the same type should be used for subsequent analyses. If the sample is not used within 24 h, it must be frozen and stored at −20 °C [20]. The measuring equipment must be calibrated with every new reagent kit, and controls should be run ideally every day. The test takes about 19 min [22]. The Brahms' Kryptor measures PCT values in the range 0.02–50 ng/mL. Sensitivity has been measured at 0.06 ng/mL.

1.3 Procalcitonin as a Biomarker in Sepsis

1.3.1 For Diagnosis

The ability to diagnose infection early in intensive care unit patients is critical to enable appropriate therapy to be started and to maximize chances of survival [1]; however, this is not always easy as microbiological cultures are frequently negative, and interpretation of the typical signs of sepsis, e.g., tachycardia, tachypnea, fever, and raised white cell count, can be confusing because they can be present in many other conditions frequently present in ICU patients.

Procalcitonin has been widely studied for use as a possible biomarker for sepsis, initially more in neonates but increasingly also in adult populations, much as troponins are used to diagnose myocardial infarction. Procalcitonin levels are raised in healthy individuals after administration of endotoxin [9, 23] and in patients with sepsis [24, 25]. After endotoxin injection in healthy volunteers, levels of procalcitonin precursors increased within 3 h to reach a peak at 24 h and then decreased slowly, taking up to 2 weeks to return to baseline values [23]. Although generally measured in serum, procalcitonin levels may also be raised in other body fluids during infection, e.g., in the saliva of patients with periodontitis [26], in the exudates from patients with wound infection [27], and in the cerebrospinal fluid of patients with meningitis [28]; however, procalcitonin levels in bronchoalveolar lavage fluid were no different in patients with and without ventilator-associated pneumonia (VAP) [29].

In adult ICU patients, many studies have evaluated the diagnostic value of serum procalcitonin levels for sepsis, and studies specifically using the Kryptor test have reported sensitivity and specificity values ranging from 44 to 100 and 53 to 100, respectively, with cutoff values ranging from 0.25 to 9.7 ng/mL (Table 1). In a recent meta-analysis of 30 studies that investigated the role of procalcitonin, measured using various assays, to differentiate patients with sepsis from those with a noninfectious inflammatory response, Wacker and colleagues reported a mean sensitivity of 77 % (95 % confidence interval 72–81 %) and mean specificity of 79 % (95 % CI 74–87 %), with a median cutoff of 1.1 ng/mL [30]. All the included studies were relatively small, with more than half having less than 100 patients; there was also considerable heterogeneity among the studies [30].

Several studies, but not all [31–33], have suggested that procalcitonin has greater diagnostic value than other biomarkers, including C-reactive protein (CRP) [34–37]. Although procalcitonin is raised in most bacterial and fungal infections, studies have reported significantly higher levels in infections caused by Gram-negative organisms than Gram-positive or fungal infections [38, 39].

Repeated procalcitonin measurements may be of more use than single values, particularly for identifying healthcare-associated infection. In 70 patients with proven or suspected nosocomial infection, Charles et al. [40] reported that the difference between the procalcitonin level on the day of diagnosis and that on the preceding day was predictive of nosocomial infection with a 100 % positive predictive value using a threshold of +0.26 ng/mL. More recently in a small study of 46 ICU patients, an increase >0.20 ng/mL of procalcitonin on the day of diagnosis and any of the 4 preceding days was associated with a positive predictive value for intravascular catheter-related bloodstream infections of >96 % [41].

Similar to other biomarkers, procalcitonin levels may also increase in other noninfectious inflammatory conditions, e.g., after cardiac surgery [42] or cardiac arrest [43], and in patients with medullary cell carcinoma [44]. Levels were also raised in patients receiving antithymocyte globulin prior to allogeneic hematopoietic stem cell transplantation [45]. Although the American College of Critical Care Medicine and the Infectious Diseases Society of America suggest that "serum procalcitonin levels… can be employed as an adjunctive diagnostic tool for discriminating infection as the cause for fever or sepsis presentations (level 2)" in their Guidelines for evaluation of new fever in critically ill adult patients [46], the

Table 1
Some of the recent studies using the Kryptor PCT test (Brahms) to assess the value of procalcitonin levels for the diagnosis of sepsis

First author [ref.]	Year of publication	Patients	Cutoff	Sensitivity (%)	Specificity (%)	Comments
Clec'h [74]	2006	36 medical ICU patients with septic shock versus 40 with SIRS	1.00 ng/mL	80	94	
Clec'h [74]	2006	31 surgical patients with septic shock versus 36 with SIRS	9.70 ng/mL	92	74	
Gaini [32]	2006	106 ICU patients with community-acquired infection or sepsis versus 88 noninfected	1.00 ng/mL	76	53	Procalcitonin worse than CRP and IL-6 for differentiating sepsis from SIRS but better at differentiating severity of sepsis
Kofoed [75]	2007	151 patients with SIRS suspected of having community-acquired infections	0.25 ng/mL	80	58	
Ruiz-Alvarez [76]	2009	103 ICU patients with suspected sepsis	0.32 ng/mL	83	64	Procalcitonin independently associated with infection in multivariateble analysis
Tsangaris [77]	2009	50 ICU patients with duration of stay >10 days and fever >38 °C	1.0 ng/mL	70	91	
Hsu [78]	2011	66 mechanically ventilated ICU patients	2.2 ng/mL	56	100	
Meynaar [79]	2011	76 ICU patients with SIRS or sepsis	2 ng/mL	97	80	

(continued)

Table 1 (continued)

First author [ref.]	Year of publication	Patients	Cutoff	Sensitivity (%)	Specificity (%)	Comments
Bele [80]	2011	119 immunocompromised ICU patients with suspected sepsis	0.5 ng/mL	100	63	Procalcitonin concentrations of >0.5 ng/mL independently predicted bacterial sepsis in multivariable analysis
Hoeboer [81]	2012	101 ICU patients with new fever >38 °C	1.98 ng/mL for septic shock	44	88	
			2.44 ng/mL for blood stream infection	58	85	
Sakran [82]	2012	102 ICU patients with severe trauma	0.82 ng/mL	87	82	Procalcitonin independent predictor of sepsis in multivariable analysis

CRP C-reactive protein, *IL* interleukin, *SIRS* systemic inflammatory response syndrome most recent Surviving Sepsis Campaign Guidelines note that "The utility of procalcitonin levels or other biomarkers (such as CRP) to discriminate the acute inflammatory pattern of sepsis from other causes of generalized inflammation... has not been demonstrated" and "No recommendation can be given for the use of these markers to distinguish between severe infection and other acute inflammatory states" [47].

1.3.2 For Prognosis

Procalcitonin levels increase with the severity of sepsis [33], and several studies have suggested that procalcitonin may be particularly useful in evaluating disease severity and prognosis. Already in 1999, we [33] reported that procalcitonin was a better prognostic marker than CRP. In 234 ICU patients with sepsis, Giamarellos-Bourboulis et al. [48] reported mortality rates of 26 % in those with procalcitonin ≤0.85 ng/mL but 45 % in those with procalcitonin >0.85 ng/mL ($p=0.002$). Bloos et al. reported that procalcitonin levels were significantly higher on admission in patients with VAP who died than in survivors [49]. Trends in concentrations over time are again of more value than single measurements. Karlsson et al. [50] reported that although initial procalcitonin concentrations did not differ between hospital survivors and non-survivors, mortality rates were lower in patients whose procalcitonin concentration decreased by more than 50 % in 72 h than in those in whom levels decreased by less than 50 % (12.2 % vs. 29.8 %, $p=0.007$). In critically ill patients with sepsis, Schuetz et al. [51] reported that change in procalcitonin levels over the first 72 h of sepsis was associated with prognosis: when procalcitonin decreased by at least 80 %, the negative predictive value for ICU mortality was 91 %, and when procalcitonin showed no decrease or increase, the positive predictive value was 36 %. In 289 patients with sepsis, the mortality rates were 12.3 % in patients in whom procalcitonin decreased by more than 30 % or was below 0.25 ng/mL on day 3 compared to day 1 of sepsis diagnosis and 29.9 % in patients in whom procalcitonin on day 3 was either >0.25 ng/mL or had decreased <30 % ($p<0.0001$) [52]. A decrease by more than 30 % between days 1 and 3 was independently associated with a favorable prognosis (OR, 0.408; 95 % CI 0.202–0.822; $p=0.012$). In addition, increase in procalcitonin was associated with inappropriate antimicrobial therapy [52]. In the largest study on this topic, in 472 ICU patients, an increase in procalcitonin level for 1 day was an independent predictor of 90-day all-cause mortality; moreover, the risk of death increased for every additional day procalcitonin levels were increased: hazard ratio for death after 1 day increase, 1.8 (95 % CI 1.4–2.4); after 2 days increase, 2.2 (95 % CI 1.6–3.0); and after 3 days increase, 2.8 (95 % CI 2.0–3.8) [53].

1.3.3 For Antibiotic Guidance

With observational studies demonstrating that procalcitonin levels decrease with recovery from sepsis and appropriate antibiotic therapy [50–52, 54], interest in the possible role of procalcitonin to guide antimicrobial therapy, so-called antibiotic stewardship, increased and several randomized trials have now been conducted in different groups of patients using different protocols, cutoffs, and endpoints. In the first of these studies, published in 2004 by Christ-Crain et al. [55], 243 patients with suspected lower respiratory tract infection were randomized to standard care or procalcitonin-guided treatment, in which use of antibiotics was discouraged if serum procalcitonin was <0.25 μg/L or encouraged if the level was >0.25 μg/L. Procalcitonin guidance substantially reduced antibiotic use, the primary endpoint, but had no effect on mortality rates [55]. Since this first study, studies have shown benefit in terms of reduced antimicrobial use in patients with community-acquired pneumonia [56], with exacerbations of chronic obstructive pulmonary disease (COPD) [57], with lower respiratory tract infections [58, 59], with suspected bacterial infection [60], with VAP [61], with severe acute pancreatitis [62], and with severe sepsis [63–66]. Importantly, none of these studies suggested any detrimental effect on survival of this approach. However, not all studies have reported benefit. In a general population of 509 ICU patients, use of procalcitonin levels to guide initiation of antimicrobial therapy was not associated with a reduction in antibiotic use [67]. And in 1,200 critically ill patients, Jensen et al. reported increased length of ICU stay and mechanical ventilation and risk of renal injury in patients randomized to a procalcitonin-guided antimicrobial escalation protocol [68].

Several meta-analyses of the studies that have used procalcitonin to guide antimicrobial therapy have now been conducted. In 14 trials with a total of 4,221 patients with acute respiratory infection in primary care, the emergency department, and the ICU, Schuetz et al. [69] reported that procalcitonin guidance was associated with reduced antibiotic exposure (adjusted difference in days, −3.47 (95 % CI −3.78 to −3.17)) in all patients, with no adverse effect on treatment failure or mortality. Other meta-analyses of studies in ICU patients have reported similar findings [70, 71]. However, in a meta-analysis of five studies conducted in the adult ICU setting, Heyland and colleagues noted that although procalcitonin-guided strategies were associated with a reduction in antibiotic usage and no overall effect on mortality, the results could not rule out a possible associated 7 % increase in hospital mortality [72]. This meta-analysis also suggested reduced costs, but the authors note that this is difficult to assess as the magnitude of the cost savings will depend on the costs of the antibiotics being used, their prescribed duration, the local cost of procalcitonin testing, and the number of tests required by the protocol [72].

Interestingly, in a recent study comparing procalcitonin- and CRP-based antibiotic protocols in ICU patients with sepsis, there were no differences between the two biomarkers in terms of antibiotic exposure or outcomes [73]. As CRP remains a more widely available and cheaper test, further study is needed to evaluate the potential benefits of procalcitonin over CRP in this setting.

2 Conclusion

There is no doubt that better techniques are needed to diagnose sepsis and strategies to guide antimicrobial prescription could have clear benefit in terms of antimicrobial resistance and economic savings. Procalcitonin has been widely studied in critically ill patients, and procalcitonin levels have been shown to be a useful indicator of the presence and severity of sepsis. Nevertheless, levels can be raised in other noninfectious inflammatory conditions, and serum procalcitonin levels are, as with other biomarkers, more useful to exclude sepsis from the differential diagnosis. The time course of procalcitonin levels is more important than a single value, and levels must be interpreted in the context of clinical examination, other signs of sepsis, and other biomarker levels when available. Procalcitonin levels may be useful for guiding antibiotic therapy, but more study is needed to better define cutoff points and standardize protocols.

References

1. Gaieski DF, Mikkelsen ME, Band RA et al (2010) Impact of time to antibiotics on survival in patients with severe sepsis or septic shock in whom early goal-directed therapy was initiated in the emergency department. Crit Care Med 38:1045–1053
2. Pierrakos C, Vincent JL (2010) Sepsis biomarkers: a review. Crit Care 14:R15
3. Vincent JL (2000) Procalcitonin: THE marker of sepsis? Crit Care Med 28:1226–1228
4. Chesney RW, McCarron DM, Haddad JG et al (1983) Pathogenic mechanisms of the hypocalcemia of the staphylococcal toxic-shock syndrome. J Lab Clin Med 101:576–585
5. Becker KL, Nylen ES, White JC et al (2004) Clinical review 167: procalcitonin and the calcitonin gene family of peptides in inflammation, infection, and sepsis—a journey from calcitonin back to its precursors. J Clin Endocrinol Metab 89:1512–1525
6. Assicot M, Gendrel D, Carsin H et al (1993) High serum procalcitonin concentrations in patients with sepsis and infection. Lancet 341:515–518
7. Muller B, White JC, Nylen ES et al (2001) Ubiquitous expression of the calcitonin-i gene in multiple tissues in response to sepsis. J Clin Endocrinol Metab 86:396–404
8. Linscheid P, Seboek D, Nylen ES et al (2003) In vitro and in vivo calcitonin I gene expression in parenchymal cells: a novel product of human adipose tissue. Endocrinology 144:5578–5584
9. Dandona P, Nix D, Wilson MF et al (1994) Procalcitonin increase after endotoxin injection in normal subjects. J Clin Endocrinol Metab 79:1605–1608
10. Gendrel D, Raymond J, Coste J et al (1999) Comparison of procalcitonin with C-reactive protein, interleukin 6 and interferon-alpha for differentiation of bacterial vs. viral infections. Pediatr Infect Dis J 18:875–881
11. Gilbert DN (2010) Use of plasma procalcitonin levels as an adjunct to clinical microbiology. J Clin Microbiol 48:2325–2329
12. Schuetz P, Albrich WC, Mueller B (2011) Procalcitonin for diagnosis of infection and guide to antibiotic decisions: past, present and future. BMC Med 9:107

13. Liappis AP, Gibbs KW, Nylen ES et al (2011) Exogenous procalcitonin evokes a proinflammatory cytokine response. Inflamm Res 60:203–207
14. Hoffmann G, Czechowski M, Schloesser M, Schobersberger W (2002) Procalcitonin amplifies inducible nitric oxide synthase gene expression and nitric oxide production in vascular smooth muscle cells. Crit Care Med 30: 2091–2095
15. Nylen ES, Whang KT, Snider RH Jr et al (1998) Mortality is increased by procalcitonin and decreased by an antiserum reactive to procalcitonin in experimental sepsis. Crit Care Med 26:1001–1006
16. Martinez JM, Wagner KE, Snider RH et al (2001) Late immunoneutralization of procalcitonin arrests the progression of lethal porcine sepsis. Surg Infect (Larchmt) 2:193–202
17. Wagner KE, Martinez JM, Vath SD et al (2002) Early immunoneutralization of calcitonin precursors attenuates the adverse physiologic response to sepsis in pigs. Crit Care Med 30: 2313–2321
18. Monneret G, Pachot A, Laroche B et al (2000) Procalcitonin and calcitonin gene-related peptide decrease LPS-induced tnf production by human circulating blood cells. Cytokine 12: 762–764
19. Matera G, Quirino A, Giancotti A et al (2012) Procalcitonin neutralizes bacterial LPS and reduces LPS-induced cytokine release in human peripheral blood mononuclear cells. BMC Microbiol 12:68
20. PCT sensitive Kryptor. Available at: http://www.procalcitonin.com/pdf/PCT-SENSITIVE_KRYPTOR/ifu_825.050_en_R10_brahms-kryptor-pct-sensitive.pdf
21. Jin M, Khan AI (2010) Procalcitonin: uses in the clinical laboratory for the diagnosis of sepsis. Lab Med 41:173–177
22. Meisner M (2002) Pathobiochemistry and clinical use of procalcitonin. Clin Chim Acta 323:17–29
23. Preas HL, Nylen ES, Snider RH et al (2001) Effects of anti-inflammatory agents on serum levels of calcitonin precursors during human experimental endotoxemia. J Infect Dis 184: 373–376
24. Harbarth S, Holeckova K, Froidevaux C et al (2001) Diagnostic value of procalcitonin, interleukin-6, and interleukin-8 in critically ill patients admitted with suspected sepsis. Am J Respir Crit Care Med 164:396–402
25. Aikawa N, Fujishima S, Endo S et al (2005) Multicenter prospective study of procalcitonin as an indicator of sepsis. J Infect Chemother 11:152–159
26. Bassim CW, Redman RS, DeNucci DJ et al (2008) Salivary procalcitonin and periodontitis in diabetes. J Dent Res 87:630–634
27. Forsberg JA, Elster EA, Andersen RC et al (2008) Correlation of procalcitonin and cytokine expression with dehiscence of wartime extremity wounds. J Bone Joint Surg Am 90: 580–588
28. Jereb M, Muzlovic I, Hojker S, Strle F (2001) Predictive value of serum and cerebrospinal fluid procalcitonin levels for the diagnosis of bacterial meningitis. Infection 29:209–212
29. Linssen CF, Bekers O, Drent M, Jacobs JA (2008) C-reactive protein and procalcitonin concentrations in bronchoalveolar lavage fluid as a predictor of ventilator-associated pneumonia. Ann Clin Biochem 45:293–298
30. Wacker C, Prkno A, Brunkhorst FM, Schlattmann P (2013) Procalcitonin as a diagnostic marker for sepsis: a systematic review and meta-analysis. Lancet Infect Dis 13: 426–435
31. Daniels JM, Schoorl M, Snijders D et al (2010) Procalcitonin vs. C-reactive protein as predictive markers of response to antibiotic therapy in acute exacerbations of COPD. Chest 138: 1108–1115
32. Gaini S, Koldkjaer OG, Pedersen C, Pedersen SS (2006) Procalcitonin, lipopolysaccharide-binding protein, interleukin-6 and C-reactive protein in community-acquired infections and sepsis: a prospective study. Crit Care 10:R53
33. Ugarte H, Silva E, Mercan D et al (1999) Procalcitonin used as a marker of infection in the intensive care unit. Crit Care Med 27: 498–504
34. Luzzani A, Polati E, Dorizzi R et al (2003) Comparison of procalcitonin and C-reactive protein as markers of sepsis. Crit Care Med 31: 1737–1741
35. Uzzan B, Cohen R, Nicolas P et al (2006) Procalcitonin as a diagnostic test for sepsis in critically ill adults and after surgery or trauma: a systematic review and meta-analysis. Crit Care Med 34:1996–2003
36. Simon L, Gauvin F, Amre DK et al (2004) Serum procalcitonin and C-reactive protein levels as markers of bacterial infection: a systematic review and meta-analysis. Clin Infect Dis 39:206–217
37. Castelli GP, Pognani C, Cita M, Paladini R (2009) Procalcitonin as a prognostic and diagnostic tool for septic complications after major trauma. Crit Care Med 37:1845–1849

38. Brodska H, Malickova K, Adamkova V et al (2013) Significantly higher procalcitonin levels could differentiate Gram-negative sepsis from Gram-positive and fungal sepsis. Clin Exp Med 13:165–170
39. Charles PE, Ladoire S, Aho S et al (2008) Serum procalcitonin elevation in critically ill patients at the onset of bacteremia caused by either Gram negative or Gram positive bacteria. BMC Infect Dis 8:38
40. Charles PE, Kus E, Aho S et al (2009) Serum procalcitonin for the early recognition of nosocomial infection in the critically ill patients: a preliminary report. BMC Infect Dis 9:49
41. Theodorou VP, Papaioannou VE, Tripsianis GA et al (2012) Procalcitonin and procalcitonin kinetics for diagnosis and prognosis of intravascular catheter-related bloodstream infections in selected critically ill patients: a prospective observational study. BMC Infect Dis 12:247
42. Sponholz C, Sakr Y, Reinhart K, Brunkhorst F (2006) Diagnostic value and prognostic implications of serum procalcitonin after cardiac surgery: a systematic review of the literature. Crit Care 10:R145
43. Schuetz P, Affolter B, Hunziker S et al (2010) Serum procalcitonin, C-reactive protein and white blood cell levels following hypothermia after cardiac arrest: a retrospective cohort study. Eur J Clin Invest 40:376–381
44. Giovanella L, Verburg FA, Imperiali M et al (2013) Comparison of serum calcitonin and procalcitonin in detecting medullary thyroid carcinoma among patients with thyroid nodules. Clin Chem Lab Med 51:1477–1481
45. Brodska H, Drabek T, Malickova K et al (2009) Marked increase of procalcitonin after the administration of anti-thymocyte globulin in patients before hematopoietic stem cell transplantation does not indicate sepsis: a prospective study. Crit Care 13:R37
46. O'Grady NP, Barie PS, Bartlett JG et al (2008) Guidelines for evaluation of new fever in critically ill adult patients: 2008 update from the American College of Critical Care Medicine and the Infectious Diseases Society of America. Crit Care Med 36:1330–1349
47. Dellinger RP, Levy MM, Rhodes A et al (2013) Surviving sepsis campaign: international guidelines for management of severe sepsis and septic shock—2012. Crit Care Med 41: 580–637
48. Giamarellos-Bourboulis EJ, Tsangaris I, Kanni T et al (2011) Procalcitonin as an early indicator of outcome in sepsis: a prospective observational study. J Hosp Infect 77:58–63
49. Bloos F, Marshall JC, Dellinger RP et al (2011) Multinational, observational study of procalcitonin in ICU patients with pneumonia requiring mechanical ventilation: a multicenter observational study. Crit Care 15:R88
50. Karlsson S, Heikkinen M, Pettila V et al (2010) Predictive value of procalcitonin decrease in patients with severe sepsis: a prospective observational study. Crit Care 14:R205
51. Schuetz P, Maurer P, Punjabi V et al (2013) Procalcitonin decrease over 72 hours in US critical care units predicts fatal outcome in sepsis patients. Crit Care 17:R115
52. Georgopoulou AP, Savva A, Giamarellos-Bourboulis EJ et al (2011) Early changes of procalcitonin may advise about prognosis and appropriateness of antimicrobial therapy in sepsis. J Crit Care 26:331–337
53. Jensen JU, Heslet L, Jensen TH et al (2006) Procalcitonin increase in early identification of critically ill patients at high risk of mortality. Crit Care Med 34:2596–2602
54. Charles PE, Tinel C, Barbar S et al (2009) Procalcitonin kinetics within the first days of sepsis: relationship with the appropriateness of antibiotic therapy and the outcome. Crit Care 13:R38
55. Christ-Crain M, Jaccard-Stolz D, Bingisser R et al (2004) Effect of procalcitonin-guided treatment on antibiotic use and outcome in lower respiratory tract infections: cluster-randomised, single-blinded intervention trial. Lancet 363:600–607
56. Christ-Crain M, Stolz D, Bingisser R et al (2006) Procalcitonin guidance of antibiotic therapy in community-acquired pneumonia: a randomized trial. Am J Respir Crit Care Med 174:84–93
57. Stolz D, Christ-Crain M, Bingisser R et al (2007) Antibiotic treatment of exacerbations of COPD: a randomized, controlled trial comparing procalcitonin-guidance with standard therapy. Chest 131:9–19
58. Schuetz P, Christ-Crain M, Thomann R et al (2009) Effect of procalcitonin-based guidelines vs. standard guidelines on antibiotic use in lower respiratory tract infections: the ProHOSP randomized controlled trial. JAMA 302: 1059–1066
59. Kristoffersen KB, Sogaard OS, Wejse C et al (2009) Antibiotic treatment interruption of suspected lower respiratory tract infections based on a single procalcitonin measurement at hospital admission: a randomized trial. Clin Microbiol Infect 15:481–487
60. Bouadma L, Luyt CE, Tubach F et al (2010) Use of procalcitonin to reduce patients' exposure to antibiotics in intensive care units (PRORATA trial): a multicentre randomised controlled trial. Lancet 375:463–474

61. Stolz D, Smyrnios N, Eggimann P et al (2009) Procalcitonin for reduced antibiotic exposure in ventilator-associated pneumonia: a randomised study. Eur Respir J 34:1364–1375
62. Qu R, Ji Y, Ling Y et al (2012) Procalcitonin is a good tool to guide duration of antibiotic therapy in patients with severe acute pancreatitis. A randomized prospective single-center controlled trial. Saudi Med J 33:382–387
63. Nobre V, Harbarth S, Graf JD et al (2008) Use of procalcitonin to shorten antibiotic treatment duration in septic patients: a randomized trial. Am J Respir Crit Care Med 177: 498–505
64. Schroeder S, Hochreiter M, Koehler T et al (2009) Procalcitonin (PCT)-guided algorithm reduces length of antibiotic treatment in surgical intensive care patients with severe sepsis: results of a prospective randomized study. Langenbecks Arch Surg 394:221–226
65. Hochreiter M, Kohler T, Schweiger AM et al (2009) Procalcitonin to guide duration of antibiotic therapy in intensive care patients: a randomized prospective controlled trial. Crit Care 13:R83
66. Stocker M, Fontana M, El HS et al (2010) Use of procalcitonin-guided decision-making to shorten antibiotic therapy in suspected neonatal early-onset sepsis: prospective randomized intervention trial. Neonatology 97:165–174
67. Layios N, Lambermont B, Canivet JL et al (2012) Procalcitonin usefulness for the initiation of antibiotic treatment in intensive care unit patients. Crit Care Med 40:2304–2309
68. Jensen JU, Hein L, Lundgren B et al (2011) Procalcitonin-guided interventions against infections to increase early appropriate antibiotics and improve survival in the intensive care unit: a randomized trial. Crit Care Med 39: 2048–2058
69. Schuetz P, Briel M, Christ-Crain M et al (2012) Procalcitonin to guide initiation and duration of antibiotic treatment in acute respiratory infections: an individual patient data meta-analysis. Clin Infect Dis 55:651–662
70. Kopterides P, Siempos II, Tsangaris I et al (2010) Procalcitonin-guided algorithms of antibiotic therapy in the intensive care unit: a systematic review and meta-analysis of randomized controlled trials. Crit Care Med 38: 2229–2241
71. Matthaiou DK, Ntani G, Kontogiorgi M et al (2012) An ESICM systematic review and meta-analysis of procalcitonin-guided antibiotic therapy algorithms in adult critically ill patients. Intensive Care Med 38:940–949
72. Heyland DK, Johnson AP, Reynolds SC, Muscedere J (2011) Procalcitonin for reduced antibiotic exposure in the critical care setting: a systematic review and an economic evaluation. Crit Care Med 39:1792–1799
73. Oliveira CF, Botoni FA, Oliveira CR et al (2013) Procalcitonin versus C-reactive protein for guiding antibiotic therapy in sepsis: a randomized trial. Crit Care Med 41:2336–2343
74. Clec'h C, Fosse JP, Karoubi P et al (2006) Differential diagnostic value of procalcitonin in surgical and medical patients with septic shock. Crit Care Med 34:102–107
75. Kofoed K, Andersen O, Kronborg G et al (2007) Use of plasma C-reactive protein, procalcitonin, neutrophils, macrophage migration inhibitory factor, soluble urokinase-type plasminogen activator receptor, and soluble triggering receptor expressed on myeloid cells-1 in combination to diagnose infections: a prospective study. Crit Care 11:R38
76. Ruiz-Alvarez MJ, Garcia-Valdecasas S, De Pablo R et al (2009) Diagnostic efficacy and prognostic value of serum procalcitonin concentration in patients with suspected sepsis. J Intensive Care Med 24:63–71
77. Tsangaris I, Plachouras D, Kavatha D et al (2009) Diagnostic and prognostic value of procalcitonin among febrile critically ill patients with prolonged ICU stay. BMC Infect Dis 9:213
78. Hsu KH, Chan MC, Wang JM et al (2011) Comparison of Fcgamma receptor expression on neutrophils with procalcitonin for the diagnosis of sepsis in critically ill patients. Respirology 16:152–160
79. Meynaar IA, Droog W, Batstra M et al (2011) In critically ill patients, serum procalcitonin is more useful in differentiating between sepsis and SIRS than CRP, Il-6, or LBP. Crit Care Res Pract 2011:594645
80. Bele N, Darmon M, Coquet I et al (2011) Diagnostic accuracy of procalcitonin in critically ill immunocompromised patients. BMC Infect Dis 11:224
81. Hoeboer SH, Alberts E, van den Hul I et al (2012) Old and new biomarkers for predicting high and low risk microbial infection in critically ill patients with new onset fever: a case for procalcitonin. J Infect 64:484–493
82. Sakran JV, Michetti CP, Sheridan MJ, et al (2012) The utility of procalcitonin in critically ill trauma patients. J Trauma Acute Care Surg 73:413–418

Chapter 17

Host Response Biomarkers in Sepsis: Overview on sTREM-1 Detection

Jérémie Lemarié, Damien Barraud, and Sébastien Gibot

Abstract

The diagnosis of sepsis, and especially its differentiation from sterile inflammation, may be challenging. TREM-1, the triggering receptor expressed on myeloid cells-1, is an amplifier of the innate immune response. Its soluble form acts as a decoy for the natural TREM-1 ligand and dampens its activation. In this chapter, we review the numerous studies that have evaluated the usefulness of sTREM-1 concentration determination for the diagnosis and the prognosis evaluation of sepsis or localized infection. Nowadays, sandwich ELISA kits are available and the assay is described.

Key words Sepsis, Diagnostic, Biomarkers, sTREM-1, SIRS, Sandwich ELISA

1 Introduction

Sepsis is a common cause of morbidity and mortality, especially in the intensive care units. Clinical and laboratory signs of systemic inflammation, including changes in body temperature, tachycardia, or leukocytosis, are neither sensitive nor specific enough for the diagnosis of sepsis and can often be misleading. Major trauma, burns, pancreatitis, acute autoimmune disorders, and many other conditions may elicit clinical signs of a systemic inflammatory response syndrome (SIRS) in the absence of microbial infection. There is no gold standard for diagnosing sepsis because culture results may be negative especially in cases of antibiotic pretreatment, inadequate sampling, or other preanalytical difficulties. Indeed, nearly half of infected patients remain without a clear microbial documentation. Moreover, results of microbiological studies are not immediately available. Clinicians feel uncomfortable about the diagnosis and may administer unneeded antibiotics awaiting laboratory results. However, the empirical use of broad-spectrum antibiotics in patients without infection is potentially harmful, facilitating colonization and superinfection with multiresistant bacteria. Thus, there is an unsatisfied need for laboratory

Nicasio Mancini (ed.), *Sepsis: Diagnostic Methods and Protocols*, Methods in Molecular Biology, vol. 1237, DOI 10.1007/978-1-4939-1776-1_17, © Springer Science+Business Media New York 2015

tools allowing distinguishing between SIRS and sepsis. Among the markers of sepsis currently in use, procalcitonin (PCT) has been suggested to be the most promising one. However, several investigators have questioned the diagnostic and prognostic accuracy of PCT measurements, reporting inconsistent and variable results depending on the severity of illness and infection in the studied population [1].

A biomarker has been defined as "a characteristic that is objectively measured and evaluated as an indicator of normal biological processes, pathogenic processes, or pharmacologic responses to a therapeutic intervention [2]." The International Sepsis Forum Colloquium on Biomarkers of Sepsis characterized the roles that may be served by any given biomarker as followed: screening, diagnosis, risk stratification, monitoring, and surrogate end point [3].

In this review we will focus on the diagnostic accuracy of a recently discovered biomarker, the soluble form of TREM-1 (sTREM-1).

1.1 The Triggering Receptor Expressed on Myeloid Cells-1

Recently, a new family of receptors expressed on myeloid cells, distantly related to NKp44, has been described: the triggering receptor expressed on myeloid cell (TREM) family. The TREMs' isoforms share low-sequence homology with each other or with other immunoglobulin superfamily members and are characterized by having only one immunoglobulin-like domain. Five *trem* genes have been identified, with four encoding putative functional type I transmembrane glycoproteins. The *trem* genes are clustered on human chromosome 6 (and mouse chromosome 17). All TREMs associate with the adaptor DNA activating protein 12 (DAP12, also called KARAP) for signaling [4]. Engagement of TREMs triggers a signaling pathway involving ZAP70 (ζ-chain-associated protein 70) and SYK (spleen tyrosine kinase) and an ensuing recruitment and tyrosine phosphorylation of adaptor molecules such as GRB2 (growth factor receptor-binding protein 2) and the activation of PI3K (phosphatidylinositol 3-kinase), PLC-γ (phospholipase C-γ), ERK-1 and ERK-2 (extracellular signal-regulated kinase), p38 MAPK (p38 mitogen-associated protein kinase), Akt serine/threonine kinase, STAT5 (signal transducer and activator of transcription 5), and CARD9-MALT1-BCL10 complex formation [5, 6]. The activation of these pathways ultimately leads to a mobilization of intracellular calcium, a rearrangement of the actin cytoskeleton, and the activation of transcriptional factors such as NFκB. This finally results in the production of metalloproteases [7], proinflammatory cytokines, and chemokines, including monocyte chemoattractant proteins 1 and 3 (MCP-1, MCP-3), macrophage inflammatory protein 1α (MIP1-α), interleukin 1β (IL-1β), IL-6, IL-8, and TNFα, along with rapid neutrophil degranulation and oxidative burst, with a parallel negative regulation of the anti-inflammatory IL-10 [8, 9].

Among the TREM family, TREM-1 has been identified on both human and murine polymorphonuclear cells and mature monocytes. Its expression by these effector cells is dramatically increased in skin, biological fluids, and tissues infected by Gram-positive or Gram-negative bacteria as well as by fungi. By contrast, TREM-1 is not upregulated in samples from patients with noninfectious inflammatory disorders such as psoriasis, ulcerative colitis, or vasculitis caused by immune complexes [10]. The activation of TREM-1 in the presence of toll-like receptor 2 (TLR2) or TLR4 ligands amplifies the production of proinflammatory cytokines [tumor necrosis factor alpha (TNFα), IL-1β, granulocyte-macrophage colony-stimulating factor], together with the inhibition of IL-10 release. In addition, the activation of these TLRs upregulates TREM-1 expression [9]. Thus, TREM-1 and TLRs appear to cooperate in producing an inflammatory response. The role of TREM-1 as an amplifier of the inflammatory response has been confirmed in a mouse model of septic shock in which blocking signaling through TREM-1 partially protected animals from death [10, 11]. Both in vitro and in vivo, synthetic peptides mimicking short highly interspecies-conserved domains of TREM-1 attenuated the cytokine production of human monocytes and protected septic animals from hyperresponsiveness and death. These peptides were efficient not only in preventing but also in down-modulating the deleterious effects of proinflammatory cytokines [11, 12].

Besides its membrane-bound form, a soluble form of TREM-1 is liberated by cleavage of its extracellular domain [13]. Soluble TREM-1 acts as a decoy receptor, sequestering TREM-1 ligand, which may exist in soluble form in the serum of septic patients [14], and dampening TREM-1 activation [10, 15]. To counteract excessive inflammatory reaction, several mechanisms exist, one of which involving another TREM member. Hamerman et al. suggested that one or more DAP12-associated receptors could negatively regulate TLR signaling [16]. One of these receptors could be TREM-2: when expressed on monocytes/macrophages, its activation downregulates TLR signaling through DAP12 [17]. These data suggest that immune cells are able to integrate the sum of different signals through sensor receptors, like TREM-1 and TREM-2, in order to induce a balanced inflammatory response.

TREM-1 is also implicated in the platelet/neutrophil dialogue. Indeed, a TREM-1 ligand is constitutively expressed on platelets and megakaryocytes [15]. Although the TREM-1 ligand (expressed on platelets) interaction with the TREM-1 receptor (expressed on neutrophils) is not responsible for platelet/neutrophil complex formation, it mediates platelet-induced neutrophil activation.

2 Materials

Initial publications used immunoblot techniques [10, 18]. But since 2005, most of published studies use the convenient commercially available sandwich enzyme-linked immunosorbent assay kits (sandwich ELISA), and one should follow the instructions provided by the manufacturer. Briefly, 96-well plates that have a high affinity for proteins are coated with a specific anti-TREM-1 antibody, the so-called capture antibody. After an incubation period of calibrators, controls, and samples into each well to permit a complete binding of TREM-1 contained into each sample, a second biotinylated anti-TREM-1 antibody is added, the so-called detection antibody. Streptavidin-HRP is then added in order to complex with biotinylated antibody. Color reaction is initiated by the addition of chromogenic HRP substrate and stopped by H_2SO_4 solution. Quantification is achieved by a spectrophotometer and appropriate quantification software by comparison between intensity of samples and calibrators.

Materials and solutions required are listed below:

1. Capture antibody: goat anti-TREM-1.
2. Detection antibody: biotinylated goat anti-TREM-1.
3. Standard of recombinant TREM-1 (known concentration) for calibration.
4. Streptavidin-HRP.
5. Phosphate buffered saline (PBS).
6. ELISA wash buffer: 0.05 % Tween® 20 in PBS.
7. ELISA diluent solution: 20 % bovine serum albumin in PBS.
8. Substrate solution.
9. Stop solution (H_2SO_4).
10. Normal goat serum.

3 Methods

3.1 Plate Preparation

1. Dilute the capture antibody in PBS and coat a 96-well microplate with 100 μL (2 μg/mL) per well. Seal the plate and incubate overnight at room temperature.
2. Aspirate and wash each well three times with wash buffer.

3.2 Assay Procedure

1. Add 100 μL of sample or standards (for calibration curve) in diluent solution per well. The diluent solution is used to avoid nonspecific binding of other proteins to unoccupied spaces on the surface of the plate. In our experience, we use protein blockers (high-quality bovine serum albumin). Incubate for 2 h at room temperature.

2. Aspirate/wash three times (as in Subheading 3.1, **step 2**).
3. Add 100 μL of the detection antibody, diluted in normal goat serum. Incubate for 2 h at room temperature.
4. Aspirate/wash three times (as in Subheading 3.1, **step 2**).
5. Add 100 μL of streptavidin-HRP solution diluted in diluent solution to each well. Incubate in the dark for 20 min at room temperature.
6. Aspirate/wash three times (as in Subheading 3.1, **step 2**).
7. Add 100 μL of substrate solution per well. Incubate in the dark for 20 min at room temperature.
8. Add 50 μL of stop solution per well.
9. Determine the optical density of each well by the use of a spectrophotometer reading at the appropriate wavelength for the color produced.

4 sTREM-1 as a Diagnostic Biomarker of Infection

Considering the modest reliability of traditional biomarkers, such as C-reactive protein (CRP) and PCT, and the a priori specific involvement of TREM-1 during infectious processes, the usefulness of sTREM-1 in diagnosing sepsis has been the focus of several studies during the last decade.

4.1 sTREM-1 and the Diagnosis of Systemic Sepsis

Since the initial publication by Gibot and colleagues in 2004 [19], many studies were performed aiming at distinguishing between sepsis and SIRS in various populations of patients. In the field of critically ill patients, Gibot and colleagues determined that plasma concentrations of CRP, PCT, and sTREM-1 were higher in infected patients than in those with noninfectious SIRS, in a cohort of 76 patients admitted to an adult ICU with a suspicion of infection. sTREM-1 performed better than other markers in diagnosing infection, with sensitivity, specificity, positive predictive value, and negative predictive value at 96 %, 89 %, 94 %, and 93 %, respectively. The same encouraging results were reported by Wang and colleagues in a cohort of 56 ICU patients (32 septic patients and 24 SIRS patients): the area under the ROC curve of sTREM-1 was 0.935, much larger than that of PCT or CRP [20]. Su and colleagues also reached the same conclusion in 144 ICU patients in which 84 were septic [21]. These encouraging results were not confirmed in two subsequent studies by Latour-Perez and colleagues [22] and Barati and colleagues [23], involving a total of 246 critically ill patients, in which the sensitivity ranged from 49 to 70 % and the specificity from 60 to 79 %. In these studies, sTREM-1 was inferior to CRP and PCT. In the emergency room, the measurement of sTREM-1 concentrations alone also proved disappointing with an area under the receiver operating characteristic (ROC) curve at 0.61. Interestingly, the combined determination of sTREM-1 and three or six other markers' levels performed far better than each marker taken alone [24]. A similar discordance was found in pediatric patients. In 44 neonates, Chen and colleagues found that sTREM-1 was superior to CRP or immature to total neutrophil ratio in diagnosing severe bacterial infections [25], whereas Sarafidis and colleagues determined that in a neonatal ICU setting, sTREM-1 performed lower than interleukin 6 (IL-6) [26].

Most of these studies were listed in a recent review and meta-analysis by Wu and colleagues published in late 2012 [27]. Its conclusion was that plasma sTREM-1 had a moderate diagnostic performance in differentiating sepsis from SIRS and was not sufficient for sepsis diagnosis in systemic inflammatory patients, especially when pretest probability of SIRS is high.

Interestingly, a recent work from Su and colleagues focused on the diagnostic value of sTREM-1 for differentiating sepsis from SIRS in 104 ICU patients [28]. The specificity of this study was that sTREM-1 was sampled from urine and not from plasma; urine sTREM-1 was found to share a higher diagnostic value than serum CRP or PCT and to provide an early warning of possible secondary acute kidney injury.

Therefore, the measurement of plasma sTREM-1 concentrations does not seem to hold its initial promises in diagnosing systemic infections. Indeed, it now seems that many inflammatory conditions may be responsible for an elevation of plasma sTREM-1 concentrations (see below). Nevertheless, the determination of plasma sTREM-1 concentrations in combination with other markers may be promising. Indeed, in a recent study, Gibot and colleagues reported the construction of a bioscore combining plasma sTREM-1 concentration, PCT concentration, and the expression of the high-affinity immunoglobulin-Fc fragment receptor I CD64 on neutrophils in 300 consecutive ICU patients. This bioscore was then externally validated in another cohort [29].

4.2 sTREM-1 and Localized Infections

Since 2004, with the publication by Gibot and colleagues in the setting of pneumonia [18], many studies have dealt with the local measurement of sTREM-1 concentrations during a variety of localized infections.

Pleuropulmonary infections constitute the core of research on TREM-1 diagnostic performance. The first study was published in 2004 by Richeldi and colleagues [30]. It showed that the expression of TREM-1 at the surface of alveolar neutrophils and macrophages determined by flow cytometry was increased during bacterial pneumonia as compared with levels found in patients with noninfectious interstitial lung diseases. Gibot and colleagues investigated alveolar sTREM-1 as a marker of infectious pneumonia in

148 consecutive patients under mechanical ventilation [18]. In this study, alveolar sTREM-1 concentrations were highly predictive of lung infection and performed better than any other clinical or biological finding in both community-acquired pneumonia (CAP) and ventilator-associated pneumonia (VAP), with a diagnostic odds ratio of 41.5. Several other studies then confirmed these preliminary results. Hue and colleagues found that sTREM-1 concentrations were useful during bacterial or fungal pneumonias, while sTREM-1 concentrations remained low in case of viral infection [31]. The study by Determann and colleagues focusing on VAP added kinetics data; alveolar sTREM-1 concentrations increased a few days before the clinical diagnosis of VAP, and the investigators concluded that the combination of more than 200 pg/mL of sTREM-1 with an increase of more than 100 pg/mL as compared with the value obtain 2 days earlier was highly predictive of the diagnosis of VAP [32]. El Solh and colleagues showed that alveolar sTREM-1 allowed for the discrimination between aspiration pneumonia and pneumonitis [33]. Recently, Ramirez and colleagues found that alveolar sTREM-1 concentration had the capacity to discriminate between a pulmonary and an extrapulmonary infection in the context of acute respiratory failure; a cutoff point of 900 pg/mL had a sensitivity of 81 % and a specificity of 80 % for the diagnosis of pneumonia [34].

However, several other studies, although confirming the elevation of alveolar sTREM-1 concentrations during lung infections, reported a lower discriminative value of measurement of sTREM-1 concentrations. During VAP, alveolar sTREM-1 performed lower than the usual clinical pulmonary infection score, and clearly, plasma concentrations were not informative [35]. In a population of 23 patients clinically suspected of having VAP, Horonenko and colleagues reported a very low specificity of sTREM-1 concentration from BAL samples, with much lower informative value than sTREM-1 concentrations from exhaled ventilator condensates [36]. In the same way, Oudhuis and colleagues found a significant difference between sTREM-1 concentrations from an alveolar sample but an area under the ROC curve at 0.58 only [37]. Finally, some studies revealed no predictive value for the diagnosis of VAP from alveolar samples [38, 39].

There is much less controversy over the diagnosis of pleural effusions. Indeed, 7 different studies, pooled into a recent meta-analysis by Summah and colleagues [40], including a total of 733 patients, demonstrated the role of pleural sTREM-1 in discriminating between infectious (due to empyema, and parapneumonia) and noninfectious pleural effusions (due to congestive heart failure and cancer) with sensitivity of 78 % and specificity of 84 %, positive likelihood ratio of 6.0, and negative likelihood ratio of 0.22.

Identifying the bacterial cause of meningitis can also be challenging, especially when patients have already received antibiotics.

Three different studies showed that the increase in sTREM-1 concentrations in the cerebrospinal fluid was able to discriminate between infectious and viral meningitis, with cutoff values ranging from 20 to 25 pg/mL. Of note, sTREM-1 concentrations were similar during pneumococcal and meningococcal infections, and concentration was normal in a culture-proven tuberculous meningitis [41–43].

Only one study investigated sTREM-1 concentrations in urine for the diagnosis of lower urinary tract infections, and the results were inconclusive [44].

The usefulness of local concentrations of sTREM-1 for the diagnosis of intra-abdominal infections has also been investigated. Determann and colleagues showed in a cohort of 83 patients operated for secondary peritonitis that the peritoneal concentration of sTREM-1 progressively decreased in patients with good outcome but remained persistently elevated and even increased in cases of patients with residual sepsis and tertiary peritonitis [45]. Lu and colleagues recently investigated the diagnosis value of sTREM-1 concentration in peripancreatic necrotic tissue to differentiate between infected necrosis and sterile necrosis in 30 patients with suspected secondary infection of necrotic tissue [46]. They reported an interesting AUC at 0.972 and sensitivity and specificity of 94.4 % and 91.7 %, respectively, for a cutoff value of 285.6 pg/mL. In a mixed population of patients with acute respiratory distress and acute or chronic abdominal diseases, Ramirez and colleagues reported the same encouraging results with sTREM-1 concentrations measured from echography-guided fine-needle aspiration of peritoneal fluid but with a much higher cutoff value of 900 pg/mL [34].

Finally, elevated sTREM-1 concentrations were also reported from gingival crevicular fluid from patients with periodontitis [47–49].

Nearly all the above-discussed studies, most of which were included in the positive meta-analysis from Jiyong and colleagues [50], suggest that the determination of sTREM-1 concentrations at the site of the presumed infection may be useful in clinical practice, but obviously, more research is necessary before implementing this assay into practical diagnosis algorithms. These encouraging data should now be translated into interventional studies, with the demonstration that measurement of sTREM-1 concentrations can safely guide and reduce the use of antibiotics by analogy to what is suggested for PCT [51].

5 sTREM-1 as a Prognostic Marker of Infection

Beyond the use of sTREM-1 as a diagnostic biomarker, the determination of its concentration may also be helpful to prognosticate the outcome of a septic patient. Gibot and colleagues sequentially measured plasma sTREM-1 concentrations and monocytic TREM-1 expression in 63 consecutive septic patients. The baseline (at admission) value of monocytic TREM-1 expression was unable to discriminate between survivors and nonsurvivors. By contrast, the baseline plasma sTREM-1 concentration was higher in survivors and was found to be an independent factor associated with good outcome. The patterns of evolution were also different according to the outcome, with a progressive decrease in sTREM-1 concentrations in survivors, whereas concentrations remained high in nonsurvivors. Two different studies from Giamarellos-Bourboulis and colleagues [52] and Wu and colleagues [53] confirmed the prognostic value of sTREM-1 in VAP. In this last study, the absence of decrease in sTREM-1 concentrations in BAL fluid was also associated with worse outcome. Tejera and colleagues investigated serum levels of sTREM-1 in a cohort of 226 patients with CAP and reported significantly lower values in survivors than in nonsurvivors [54]. In a mixed population of 52 septic patients, half of them suffering from lower respiratory tract infection, Zhang and colleagues demonstrated that serum sTREM-1 concentrations reflected the severity of sepsis more accurately than those of CRP and PCT and were more sensitive for dynamic evaluations of sepsis prognosis [55]. A recent study from Su and colleagues shared the same conclusions [56]. In the setting of chemotherapy-associated febrile neutropenia, a retrospective study from Kwofie and colleagues reported that sTREM-1 levels were potentially useful to predict the clinical course of these patients [57].

Nevertheless, two studies led to a different conclusion. Studying patients with CAP who were admitted to an emergency room, Muller and colleagues did not find any relationship between plasma sTREM-1 concentrations and severity or outcomes [58]. In a surgical setting, Bopp and colleagues found that plasma sTREM-1 was useless to predict outcome in SIRS, sepsis, or severe sepsis [59].

6 Limitations for the Diagnosis of Sepsis

Objective analysis of the published literature on this subject is tricky because of a huge heterogeneity between studies: many did not take into account the Bayes theory, the case mix is highly variable (e.g., immunodepression, previous antibiotics, neonates), the selected cutoff ranges from picograms to nanograms per mL, and so on. Most importantly, the techniques used to measure the sTREM-1 concentrations are not always comparable with large variations both during the preanalytical (such as the technique of sampling and conservation) and the analytical periods. Some commercial kits have been withdrawn from the market during the fall of 2008 due to unreliable results [60]. Nevertheless, most of the recent studies currently use reliable commercial kits.

A recent growing body of evidence suggests that sTREM-1 concentrations could increase in biological fluids even in the absence of infection. Indeed, TREM-1 expression depends on the activation of several TLRs or NOD-like receptors, and it has become clear that many danger-associated molecular patterns (or alarmins, such as high-mobility group box nuclear protein, heat shock proteins, and free cyclic AMP) that activate these receptors may be produced during aseptic inflammatory conditions such as hemorrhagic shock, ischemia-reperfusion, or inflammatory intestinal diseases.

In the surgical patient, the diagnostic usefulness of sTREM-1 has been explored in various pathological conditions. During the postoperative period of cardiac surgery under extracorporeal assistance, Adib-Conquy and colleagues showed an early increase in plasma sTREM-1 concentrations, although at a lower level than those encountered during severe sepsis. These concentrations correlated neither with the length of aortic clamping nor with the length of extracorporeal circulation [61]. In the same study, the authors were able to demonstrate that up to 60 % of a cohort of 54 patients resuscitated from a cardiac arrest presented with elevated plasma sTREM-1 concentrations. Such an elevation was especially present among patients who had multiorgan failure. Recently, in a cohort of 45 adults with multiple trauma and lung contusion, Bingold and colleagues reported an increase in sTREM-1 levels in culture-negative BAL fluid [62]. Of note, the levels of sTREM-1 in the BAL correlated well with both the severity of radiological pulmonary tissue damage and functional impairment of gas exchange.

Physiologically, TREM-1 is not expressed by the macrophages infiltrating the lamina propria of the digestive tract. This phenomenon could be explained by the presence of IL-10 and transforming growth factor-β that refrain TREM-1 expression and oppose to an excessive immune activation in response to intestinal flora [63]. The development of chronic inflammatory bowel disease (IBD) may be the result of aberrations of the innate intestinal immune response to endogenous intestinal flora. Indeed, Schenk and colleagues demonstrated that TREM-1 was overexpressed on the surface of intestinal macrophages in patients with IBD [64]. This upregulation was responsible for a huge production of proinflammatory cytokines and correlated to the disease severity. These data seemed to be confirmed by two studies in which plasma sTREM-1 concentrations were shown to correlate with disease activity [65–67], but not in Crohn's disease [68, 69].

The TREM-1 involvement in gastric ulcer has been pointed out by Koussoulas and colleagues. This group found that sTREM-1 concentrations were elevated in the gastric juice of patients with peptic ulcer, independently of the presence of *Helicobacter pylori* infection, and that this increase correlated with the histologic score [70]. These data suggested that TREM-1 could be implicated in the pathogenesis of gastric ulcer. These authors also reported the diagnostic value of serum sTREM-1 concentrations as a surrogate end point of healing in patients with peptic ulcer disease [71].

TREM-1 has also been investigated in chronic obstructive pulmonary disease (COPD) with inconclusive data since sTREM-1 levels were not different between stable COPD patients and ones with acute exacerbation [72, 73].

Finally, and despite the initial thought that TREM-1 was not involved in vasculitis, several recent studies reported on the role of TREM-1 in pure aseptic inflammatory disorders such as vasculitis and autoimmune diseases [74–76].

7 Therapeutic Manipulation of the TREM-1 Pathway

A relevant biomarker should provide diagnostic, or prognostic, information and should be of physiologic relevance. The therapeutic modulation of the TREM-1 pathway has been the subject of many experimental studies.

Since the synthesis of TREM-1 antagonist peptide, numerous studies aimed to assess the potentially beneficial effects of the modulation of the inflammatory response during various pathological conditions. Most of them are related to sepsis or LPS challenge in rodents. For example, in a rat model of *Pseudomonas aeruginosa-induced* pneumonia as well as in melioidosis, TREM-1 antagonist administration was associated with hemodynamic improvement, as well as with the dampening of the tissue and systemic inflammatory responses and a decrease in coagulation activation. In fine, antagonist administration improved survival [77, 78]. Same encouraging results were reported during experimental hemorrhagic shock, ischemia-reperfusion, or severe acute pancreatitis [79–81].

Finally, data recently obtained by the authors' laboratory (unpublished data, 2013) confirm that inflammatory modulation by TREM-1 inhibition improves myocardial function through a limitation of ventricular remodeling after experimental myocardial infarction in mice and rats.

References

1. Tang BM, Eslick GD, Craig JC, McLean AS (2007) Accuracy of procalcitonin for sepsis diagnosis in critically ill patients: systematic review and meta-analysis. Lancet Infect Dis 7:210–217
2. Atkinson AJ, Colburn WA, DeGruttola VG et al (2001) Biomarkers and surrogate endpoints: preferred definitions and conceptual framework. Clin Pharmacol Ther 69:89–95
3. Marshall JC, Reinhart K (2009) Biomarkers of sepsis. Crit Care Med 37:2290–2298
4. Ford JW, McVicar DW (2009) TREM and TREM-like receptors in inflammation and disease. Curr Opin Immunol 21:38–46
5. Hara H, Ishihara C, Takeuchi A et al (2007) The adaptor protein CARD9 is essential for the activation of myeloid cells through ITAM-associated and Toll-like receptors. Nat Immunol 8:619–629

6. Tessarz AS, Cerwenka A (2008) The TREM-1/DAP12 pathway. Immunol Lett 116:111–116
7. Dower K, Ellis DK, Saraf K et al (2008) Innate immune responses to TREM-1 activation: overlap, divergence, and positive and negative cross-talk with bacterial lipopolysaccharide. J Immunol 180:3520–3534
8. Bouchon A, Dietrich J, Colonna M (2000) Cutting edge: inflammatory responses can be triggered by TREM-1, a novel receptor expressed on neutrophils and monocytes. J Immunol 164:4991–4995
9. Bleharski JR, Kiessler V, Buonsanti C et al (2003) A role for triggering receptor expressed on myeloid cells-1 in host defense during the early-induced and adaptive phases of the immune response. J Immunol 170:3812
10. Bouchon A, Facchetti F, Weigand MA, Colonna M (2001) TREM-1 amplifies inflammation and is a crucial mediator of septic shock. Nature 410:1103–1107
11. Gibot S, Kolopp-Sarda M-N, Béné M-C et al (2004) A soluble form of the triggering receptor expressed on myeloid cells-1 modulates the inflammatory response in murine sepsis. J Exp Med 200:1419–1426
12. Lanier LL, Bakker AB (2000) The ITAM-bearing transmembrane adaptor DAP12 in lymphoid and myeloid cell function. Immunol Today 21:611–614
13. Gómez-Piña V, Soares-Schanoski A, Rodríguez-Rojas A et al (2007) Metalloproteinases shed TREM-1 ectodomain from lipopolysaccharide-stimulated human monocytes. J Immunol 179:4065–4073
14. Wong-Baeza I, Gonz'alez-Rold'an N, Ferat-Osorio E et al (2006) Triggering receptor expressed on myeloid cells (TREM-1) is regulated post-transcriptionally and its ligand is present in the sera of some septic patients. Clin Exp Immunol 145:448–455
15. Haselmayer P, Grosse-Hovest L, von Landenberg P et al (2007) TREM-1 ligand expression on platelets enhances neutrophil activation. Blood 110:1029–1035
16. Hamerman JA, Tchao NK, Lowell CA, Lanier LL (2005) Enhanced Toll-like receptor responses in the absence of signaling adaptor DAP12. Nat Immunol 6:579–586
17. Hamerman JA, Jarjoura JR, Humphrey MB et al (2006) Cutting edge: inhibition of TLR and FcR responses in macrophages by triggering receptor expressed on myeloid cells (TREM)-2 and DAP12. J Immunol 177:2051–2055
18. Gibot S, Cravoisy A, Levy B et al (2004) Soluble triggering receptor expressed on myeloid cells and the diagnosis of pneumonia. N Engl J Med 350:451–458
19. Gibot S, Kolopp-Sarda MN, Béné MC et al (2004) Plasma level of a triggering receptor expressed on myeloid cells-1: its diagnostic accuracy in patients with suspected sepsis. Ann Intern Med 141:9–15
20. Wang H, Chen B (2011) Diagnostic role of soluble triggering receptor expressed on myeloid cell-1 in patients with sepsis. World J Emerg Med 2:190–194
21. Su L, Han B, Liu C et al (2012) Value of soluble TREM-1, procalcitonin, and C-reactive protein serum levels as biomarkers for detecting bacteremia among sepsis patients with new fever in intensive care units: a prospective cohort study. BMC Infect Dis 12:157
22. Latour-Pérez J, Alcalá-López A, García-García MÁ et al (2010) Diagnostic accuracy of sTREM-1 to identify infection in critically ill patients with systemic inflammatory response syndrome. Clin Biochem 43:720–724
23. Barati M, Bashar FR, Shahrami R et al (2010) Soluble triggering receptor expressed on myeloid cells 1 and the diagnosis of sepsis. J Crit Care 25:362.e1–362.e6
24. Kofoed K, Andersen O, Kronborg G et al (2007) Use of plasma C-reactive protein, procalcitonin, neutrophils, macrophage migration inhibitory factor, soluble urokinase-type plasminogen activator receptor, and soluble triggering receptor expressed on myeloid cells-1 in combination to diagnose infections: a prospective study. Crit Care 11:R38
25. Chen H, Hung C, Tseng H, Yang R (2008) Soluble form of triggering receptor expressed on myeloid cells-1 (sTREM-1) as a diagnostic marker of serious bacterial infection in febrile infants less than three months of age. Jpn J Infect Dis 61:31–35
26. Sarafidis K, Soubasi-Griva V, Piretzi K et al (2010) Diagnostic utility of elevated serum soluble triggering receptor expressed on myeloid cells (sTREM)-1 in infected neonates. Intensive Care Med 36:864–868
27. Wu Y, Wang F, Fan X et al (2012) Accuracy of plasma sTREM-1 for sepsis diagnosis in systemic inflammatory patients: a systematic review and meta-analysis. Crit Care 16:R229
28. Su L, Feng L, Zhang J et al (2011) Diagnostic value of urine sTREM-1 for sepsis and relevant acute kidney injuries: a prospective study. Crit Care 15:R250
29. Gibot S, Béné MC, Noel R et al (2012) Combination biomarkers to diagnose sepsis in the critically Ill patient. Am J Respir Crit Care Med 186:65–71

30. Richeldi L, Mariani M, Losi M et al (2004) Triggering receptor expressed on myeloid cells: role in the diagnosis of lung infections. Eur Respir J 24:247–250
31. Huh J, Lim C-M, Koh Y et al (2008) Diagnostic utility of the soluble triggering receptor expressed on myeloid cells-1 in bronchoalveolar lavage fluid from patients with bilateral lung infiltrates. Crit Care 12:R6
32. Determann RM, Millo JL, Gibot S et al (2005) Serial changes in soluble triggering receptor expressed on myeloid cells in the lung during development of ventilator-associated pneumonia. Intensive Care Med 31:1495–1500
33. El Solh AA, Akinnusi ME, Peter M et al (2008) Triggering receptors expressed on myeloid cells in pulmonary aspiration syndromes. Intensive Care Med 34:1012–1019
34. Ramirez P, Kot P, Marti V et al (2011) Diagnostic implications of soluble triggering receptor expressed on myeloid cells-1 in patients with acute respiratory distress syndrome and abdominal diseases: a preliminary observational study. Crit Care 15:1–8
35. Phua J, Koay ESC, Zhang D et al (2006) Soluble triggering receptor expressed on myeloid cells-1 in acute respiratory infections. Eur Respir J 28:695–702
36. Horonenko G, Hoyt JC, Robbins RA et al (2007) Soluble triggering receptor expressed on myeloid cell-1 is increased in patients with ventilator-associated pneumonia: a preliminary report. Chest 132:58–63
37. Oudhuis GJ, Beuving J, Bergmans D et al (2009) Soluble triggering receptor expressed on myeloid cells-1 in bronchoalveolar lavage fluid is not predictive for ventilator-associated pneumonia. Intensive Care Med 35: 1265–1270
38. Anand NJ, Zuick S, Klesney-Tait J, Kollef MH (2009) Diagnostic implications of soluble triggering receptor expressed on myeloid cells-1 in bal fluid of patients with pulmonary infiltrates in the icu. Chest 135:641–647
39. Palazzo SJ, Simpson TA, Simmons JM, Schnapp LM (2012) Soluble triggering receptor expressed on myeloid cells-1 (sTREM-1) as a diagnostic marker of ventilator-associated pneumonia. Respir Care 57:2052–2058
40. Summah H, Tao L-L, Zhu Y-G et al (2011) Pleural fluid soluble triggering receptor expressed on myeloid cells-1 as a marker of bacterial infection: a meta-analysis. BMC Infect Dis 11:280
41. Determann RM, Weisfelt M, de Gans J et al (2006) Soluble triggering receptor expressed on myeloid cells 1: a biomarker for bacterial meningitis. Intensive Care Med 32: 1243–1247
42. Bishara J, Hadari N, Shalita-Chesner M et al (2007) Soluble triggering receptor expressed on myeloid cells-1 for distinguishing bacterial from aseptic meningitis in adults. Eur J Clin Microbiol Infect Dis 26:647–650
43. Guanghui G, Qiaoya J, Zhenhua S (2011) Value of soluble triggering receptor expressed on myeloid cells-1 in the diagnosis of bacterial meningitis. Int J Lab Med 9:12
44. Determann RM, Schultz MJ, Geerlings SE (2007) Soluble triggering receptor expressed on myeloid cells-1 is not a sufficient biological marker for infection of the urinary tract. J Infect 54:e249
45. Determann RM, Olivier van Till JW, van Ruler O et al (2009) sTREM-1 is a potential useful biomarker for exclusion of ongoing infection in patients with secondary peritonitis. Cytokine 46:36–42
46. Lu Z, Liu Y, Dong Y et al (2011) Soluble triggering receptor expressed on myeloid cells in severe acute pancreatitis: a biological marker of infected necrosis. Intensive Care Med 38: 69–75
47. Belibasakis GN, Öztürk V-Ö, Emingil G, Bostanci N (2014) Soluble triggering receptor expressed on myeloid cells 1 (sTREM-1) in gingival crevicular fluid: association with clinical and microbiologic parameters. J Periodontol 85:204–210
48. Bisson C, Massin F, Lefevre PA et al (2012) Increased gingival crevicular fluid levels of soluble triggering receptor expressed on myeloid cells (sTREM)-1 in severe periodontitis. J Clin Periodontol 39:1141–1148
49. Bostanci N, Öztürk VÖ, Emingil G, Belibasakis GN (2013) Elevated oral and systemic levels of soluble triggering receptor expressed on myeloid cells-1 (sTREM-1) in periodontitis. J Dent Res 92:161–165
50. Jiyong J, Tiancha H, Wei C, Huahao S (2008) Diagnostic value of the soluble triggering receptor expressed on myeloid cells-1 in bacterial infection: a meta-analysis. Intensive Care Med 35:587–595
51. Bouadma L, Luyt CE, Tubach F et al (2010) Use of procalcitonin to reduce patients' exposure to antibiotics in intensive care units (PRORATA trial): a multicentre randomised controlled trial. Lancet 375:463–474
52. Giamarellos-Bourboulis EJ, Zakynthinos S, Baziaka F et al (2006) Soluble triggering receptor expressed on myeloid cells 1 as an anti-inflammatory mediator in sepsis. Intensive Care Med 32:237–243

53. Wu C-L, Lu Y-T, Kung Y-C et al (2011) Prognostic value of dynamic soluble triggering receptor expressed on myeloid cells in bronchoalveolar lavage fluid of patients with ventilator-associated pneumonia. Respirology 16:487–494
54. Tejera A, Santolaria F, Diez M-L et al (2007) Prognosis of community acquired pneumonia (CAP): value of triggering receptor expressed on myeloid cells-1 (TREM-1) and other mediators of the inflammatory response. Cytokine 38:117–123
55. Zhang J, She D, Feng D et al (2011) Dynamic changes of serum soluble triggering receptor expressed on myeloid cells-1 (sTREM-1) reflect sepsis severity and can predict prognosis: a prospective study. BMC Infect Dis 11:53
56. Su L, Liu C, Li C et al (2012) Dynamic changes in serum soluble triggering receptor expressed on myeloid cells-1 (sTREM-1) and its gene polymorphisms are associated with sepsis prognosis. Inflammation 35:1833–1843
57. Kwofie L, Rapoport BL, Fickl H et al (2012) Evaluation of circulating soluble triggering receptor expressed on myeloid cells-1 (sTREM-1) to predict risk profile, response to antimicrobial therapy, and development of complications in patients with chemotherapy-associated febrile neutropenia: a pilot study. Ann Hematol 91:605–611
58. Müller B, Gencay MM, Gibot S et al (2007) Circulating levels of soluble triggering receptor expressed on myeloid cells (sTREM)-1 in community-acquired pneumonia. Crit Care Med 35:990–991
59. Bopp C, Hofer S, Bouchon A et al (2009) Soluble TREM-1 is not suitable for distinguishing between systemic inflammatory response syndrome and sepsis survivors and nonsurvivors in the early stage of acute inflammation. Eur J Anaesthesiol 26:504–507
60. Gibot S (2009) Soluble triggering receptor expressed on myeloid cells 1 and the diagnosis of ventilator-associated pneumonia. Chest 136:320
61. Adib-Conquy M, Monchi M, Goulenok C et al (2007) Increased plasma levels of soluble triggering receptor expressed on myeloid cells 1 and procalcitonin after cardiac surgery and cardiac arrest without infection. Shock 28: 406–410
62. Bingold TM, Pullmann B, Sartorius S et al (2011) Soluble triggering receptor on myeloid cells-1 is expressed in the course of non-infectious inflammation after traumatic lung contusion: a prospective cohort study. Crit Care 15:1–7
63. Schenk M, Bouchon A, Birrer S et al (2005) Macrophages expressing triggering receptor expressed on myeloid cells-1 are underrepresented in the human intestine. J Immunol 174:517–524
64. Schenk M, Bouchon A, Seibold F, Mueller C (2007) TREM-1–expressing intestinal macrophages crucially amplify chronic inflammation in experimental colitis and inflammatory bowel diseases. J Clin Invest 117:3097–3106
65. Park JJ, Cheon JH, Kim BY et al (2009) Correlation of serum-soluble triggering receptor expressed on myeloid cells-1 with clinical disease activity in inflammatory bowel disease. Dig Dis Sci 54:1525–1531
66. Jung YS, Park JJ, Kim SW et al (2012) Correlation between soluble triggering receptor expressed on myeloid cells-1 (sTREM-1) expression and endoscopic activity in inflammatory bowel diseases. Dig Liver Dis 44:897–903
67. Tzivras M, Koussoulas V, Giamarellos-Bourboulis EJ et al (2006) Role of soluble triggering receptor expressed on myeloid cells in inflammatory bowel disease. World J Gastroenterol 12:3416
68. Billioud V, Gibot S, Massin F et al (2012) Plasma soluble triggering receptor expressed on myeloid cells-1 in Crohn's disease. Dig Liver Dis 44:466–470
69. Saurer L, Rihs S, Birrer M et al (2012) Elevated levels of serum-soluble triggering receptor expressed on myeloid cells-1 in patients with IBD do not correlate with intestinal TREM-1 mRNA expression and endoscopic disease activity. J Crohns Colitis 6:913–923
70. Koussoulas V, Tzivras M, Giamarellos-Bourboulis EJ et al (2007) Can soluble triggering receptor expressed on myeloid cells (sTREM-1) be considered an anti-inflammatory mediator in the pathogenesis of peptic ulcer disease? Dig Dis Sci 52: 2166–2169
71. Koussoulas V, Giamarellos-Bourboulis EJ, Barbatzas C, Pimentel M (2011) Serum sTREM-1 as a surrogate marker of treatment outcome in patients with peptic ulcer disease. Dig Dis Sci 56:3590–3595
72. Radsak MP, Taube C, Haselmayer P et al (2007) Soluble triggering receptor expressed on myeloid cells 1 is released in patients with stable chronic obstructive pulmonary disease. Clin Dev Immunol 2007:52040
73. Rohde G, Radsak MP, Borg I et al (2012) Levels of soluble triggering receptor expressed on myeloid cells 1 in infectious exacerbations of chronic obstructive pulmonary disease. Respiration 83:133–139

74. Collins CE, La DT, Yang H-T et al (2009) Elevated synovial expression of triggering receptor expressed on myeloid cells 1 in patients with septic arthritis or rheumatoid arthritis. Ann Rheum Dis 68:1768–1774
75. Daikeler T, Regenass S, Tyndall A et al (2008) Increased serum levels of soluble triggering receptor expressed on myeloid cells-1 in anti-neutrophil cytoplasmic antibody-associated vasculitis. Ann Rheum Dis 67:723–724
76. Molad Y, Pokroy-Shapira E, Kaptzan T et al (2013) Serum soluble triggering receptor on myeloid cells-1 (sTREM-1) is elevated in systemic lupus erythematosus but does not distinguish between lupus alone and concurrent infection. Inflammation 1–6:1519–1524
77. Gibot S, Alauzet C, Massin F et al (2006) Modulation of the triggering receptor expressed on myeloid cells–1 pathway during pneumonia in rats. J Infect Dis 194:975–983
78. Wiersinga WJ, van't Veer C, Wieland CW et al (2007) Expression profile and function of triggering receptor expressed on myeloid cells-1 during melioidosis. J Infect Dis 196:1707–1716
79. Kamei K, Yasuda T, Ueda T et al (2010) Role of triggering receptor expressed on myeloid cells-1 in experimental severe acute pancreatitis. J Hepatobiliary Pancreat Sci 17:305–312
80. Gibot S, Massin F, Alauzet C et al (2009) Effects of the TREM 1 pathway modulation during hemorrhagic shock in rats. Shock 32: 633–637
81. Gibot S, Massin F, Alauzet C et al (2008) Effects of the TREM-1 pathway modulation during mesenteric ischemia-reperfusion in rats. Crit Care Med 36:504–510

Chapter 18

Host Response Biomarker in Sepsis: suPAR Detection

Evangelos J. Giamarellos-Bourboulis and Marianna Georgitsi

Abstract

Recent studies of our group have shown that suPAR may complement APACHE II score for risk assessment in sepsis. suPAR may be measured in serum of patients by an enzyme immunosorbent assay developed by Virogates (suPARnostic™). Production of suPAR from circulating neutrophils and monocytes may be assessed after isolation of neutrophils and monocytes and ex vivo culture. This is followed by measurement of suPAR in culture supernatants.

Key words suPAR, Sepsis, Neutrophils, Severity, Immunoassay

1 Introduction

Severe sepsis and septic shock are among the leading causes of morbidity and mortality in the world. It is estimated that almost 1.5 million people develop severe sepsis annually in North America and another 1.5 million people in Europe; 35–50 % of them die [1]. Cornerstone of efficient patient management is early recognition of patients and early start of therapy. Everyday clinical practice suggests that this is often difficult because initial symptoms are often subtle. It is suggested that biomarkers should be used to improve the changes for early diagnosis and risk assessment. More than 170 protein molecules have been studied as biomarkers of sepsis. However, none seems to be the ideal marker for diagnosis and prognosis of patients at risk of severe sepsis complications [2].

suPAR is the soluble counterpart of the urokinase plasminogen activator receptor (uPAR) that is expressed on myeloid cells, namely, neutrophils and monocytes. uPAR is participating in a variety of inflammatory conditions and in the process of coagulation and fibrinolysis [3]. Two studies of our group in large cohorts of patients have shown that suPAR may be a major tool of risk assessment that can efficiently complement clinical scores like APACHE II. Acute physiology and chronic health evaluation (APACHE) II score is a clinical score implemented in everyday

Nicasio Mancini (ed.), *Sepsis: Diagnostic Methods and Protocols*, Methods in Molecular Biology, vol. 1237,
DOI 10.1007/978-1-4939-1776-1_18, © Springer Science+Business Media New York 2015

clinical practice for the assessment of the severity of critically ill patients. One major hurdle of APACHE II is that it cannot differentiate severity between patients who abruptly deteriorate and who score equal to patients with chronic health problems. suPAR may help overcome this difficulty. More precisely, analysis of a large cohort of 1,914 Greek patients using both suPAR and APACHE II managed to reclassify patients into four strata of severity: those with APACHE II <17 and suPAR <12 ng/ml and mortality 5.5 %, those with APACHE II <17 and suPAR ≥ 12 ng/ml and mortality 17.4 %, those with APACHE II ≥ 17 and suPAR <12 ng/ml and mortality 37.2 %, and those with APACHE II ≥ 17 and suPAR ≥ 12 ng/ml and mortality 51.2 %. These findings were fully confirmed in an independent cohort of 196 patients from Sweden [4].

Using a homogeneous cohort of 180 patients with sepsis developing in the field of ventilator-associated pneumonia (VAP), production of suPAR was measured in supernatants coming from cultured monocytes and neutrophils isolated from peripheral blood. Production from neutrophils of patients was significantly greater than from healthy controls; this was further enhanced after stimulation with lipopolysaccharide (LPS) in the culture medium [5].

The present chapter is aiming to present the methods for measurement of suPAR production by circulating myeloid cells and of circulating suPAR in serum of patients with sepsis.

2 Materials

All materials should be brought into room temperature (18–25 °C) 1 h before use. For the preparation of cultures for the measuring suPAR production, the following material are required:

1. Clean, sterile, and pyrogen-free plastic tubes.
2. Sterile and pyrogen-free plastic tubes of 10 ml volume coated with EDTA (ethyldiaminetetracetic acid).
3. Sterile and pyrogen-free Falcon tubes of 15 and 50 ml volume.
4. Ammonium chloride.
5. Ficoll-Hypaque ready-made solution.
6. Fetal bovine serum.
7. RPMI1640 supplemented with glutamine 2 mM ready to use.
8. Powders of gentamicin and penicillin G.
9. Phosphate buffered saline (PBS) adjusted to pH 7.2.
10. Flasks of 75 cm^2.
11. Trypsin 0.2 %/EDTA 0.02 % ready-to-use solutions for cell cultures.

12. 24-well plates of 1 ml final volume.
13. Lyophilized lipopolysaccharide (LPS) of *Escherichia coli* O55:B5.

All procedures should be done using distilled or deionized water. Wash buffer can be used according to the instructions of the manufacturer of the diagnostic kit or by using phosphate buffered saline (PBS) and bovine serum albumin (BSA). The suPARnostic™ kit should be used (Virogates, Lygby, Denmark). This contains the following ready-made reagents:

1. 96-well microplates of 300 μl final volume per well covered with suPAR capture antibody. The plate is quoted with eight rows signed A to H and with 12 columns numbered 1–12.
2. Ready-made wash buffer provided in the kit.
3. Plastic plate covers.
4. Five standards of recombinant suPAR.
5. One curve control.
6. Peroxidase conjugate of human anti-suPAR detection antibody.
7. Plastic tubes for preparation of peroxidase conjugate solution.
8. 3′, 3′, 5′, 5′tetramethylbenzyl (TMB) substrate.
9. 0.45 M sulfuric acid to be used as stop solution.

3 Methods

3.1 Blood Sampling and Processing

Ten to 20 ml microliters of whole blood is sampled after venipuncture of one forearm vein under sterile conditions. Five milliliters is immediately poured into one pyrogen-free tube, and the remaining is poured into one tube coated with EDTA. The first tube is left for 30 min at room temperature. The second tube should be processed within 30 min. Processing of the first tube involves centrifugation at 800 × *g* in room temperature for 10 min. Serum is aliquoted into 0.5–1.0 ml volume tubes; aliquots are kept refrigerated at −70 °C. The second tube is processed as follows:

1. Pour 10 ml of whole blood into one Falcon tube.
2. Add 10 ml of PBS (pH 7.2).
3. Using a syringe and a needle, apply vertically and slowly at the bottom of the tube 4 ml of Ficoll-Hypaque. Ficoll remains at the bottom and blood is layered above.
4. Centrifuge for 20 min at 1,700 × *g* at room temperature.
5. Then three layers are formed: an upper containing plasma, a middle containing peripheral blood mononuclear cells (PBMCs), and a lower containing neutrophils.
6. Slowly discard the upper layer.

3.2 suPAR Stimulation of Monocytes

1. Collect the middle layer into one 15 ml Falcon tube and add ice-cold PBS (pH 7.2) to a final volume of 10 ml. Wash two times at 1,700 × *g* and at 4 °C.
2. Prepare a culture medium dilution by RMPI 1640 with antibiotics and FBS so that the final concentrations of FBS will be 10 %, of penicillin G 100,000 U/ml and of gentamicin 100 μg/ml.
3. Dilute PBMCs with 20 ml of medium dilution and pour them gently into a 75 cm^3 flask. Allow them to incubate for 1 h at 37 °C at 5 % CO_2 (*see* **Note 1**).
4. Discard the medium and add 4 ml of trypsin/EDTA solution. Incubate for 10 min at 37 °C at 5 % CO_2 (*see* **Notes 2** and **3**).
5. Add 1 ml of FBS, aspirate into a 15 ml Falcon tube, and centrifuge at 1,700 × *g* for 10 min at room temperature.
6. Dilute the cell pellet with 1 ml of prepared medium and count monocytes in a Neubauer chamber. Then add another 1 ml of medium.
7. Use the prepared medium dilution to dilute LPS to 20 ng/ml.
8. Pour 0.5 ml of medium-diluted monocytes into the four wells of a 24-well plate (*see* **Note 4**).
9. Add 0.5 ml of 20 ng/ml LPS solution in two of the wells.
10. Incubate for 24 h at 37 °C at 5 % CO_2. Then centrifuge the plate and aliquot the supernatants. Keep refrigerated at −70 °C.

3.3 suPAR Stimulation of Neutrophils

1. Collect the lower layer into a 15 ml Falcon tube and add 1.0 mM ammonium chloride solution to a final volume of 10 ml. Invert gently the tube and leave at room temperature for 5 min. Then centrifuge the tube at 1,700 × *g* at room temperature. Wash the pellet containing the neutrophils three times with ice-cold PBS (pH 7.2) at the washing conditions mentioned above.
2. Dilute the cell pellet with 1 ml of prepared medium and count neutrophils in a Neubauer chamber (*see* **Note 3**). Then add another 1 ml of medium.
3. Pour 0.5 ml of medium-diluted neutrophils into the four wells of a 24-well plate (*see* **Note 4**).
4. Add 0.5 ml of 20 ng/ml LPS solution in two of the wells.
5. Incubate for 24 h at 37 °C at 5 % CO_2. Then centrifuge the plate and aliquot the supernatants. Keep refrigerated at −70 °C.

3.4 Preparation of Wash Buffer

Dilute the stock solution one plus nine parts (1:10) with water.

3.5 Enzyme Immunosorbent Assay (According to the Instructions of the Manufacturer with Slight Modifications)

1. All samples should be added in duplicate.
2. Map the plate so that positions A1 and A2 are blanks: B1/B2, C1/C2, D1/D2. E1/E2 and F1/F2 are standards; G1/G2 are curve controls; and H1/H2 are positive controls; and the remaining are unknown samples.
3. Use a clean 96-well plate of 200 μl volume per well to dilute unknown samples 1:5 using water.
4. Add 25 μl of water into positions A1 and A2.
5. Add 25 μl of standard 20.7 ng/ml at positions B1/B2, of standard 10 ng/ml at position C1/C2, of standard 5 ng/ml at position D1/D2, of standard 2.5 ng/ml at position E1/E2, and of standard 1.1 ng/ml at position F1/F2.
6. Add 25 μl of curve controls at positions G1 and G2.
7. Add 25 μl of two samples of known concentrations undiluted at positions H1 and H2, respectively (*see* **Note 5**).
8. Add 25 μl of undiluted unknown samples at columns 3, 5, 7, 9, and 11 and of respective diluted unknown samples at columns 4, 6, 8, 10, and 12.
9. Dilute 65 μl into of peroxidase conjugate detection antibody into 13 ml of water using the plastic tubes provided in the suPARnostic™ kit. Add in all wells 225 μl of solution into each well.
10. Use a rotator to gently mix the plate and seal with a plastic cover.
11. Incubate for 1 h at room temperature.
12. Unseal the plate and aspirate the content of wells with the aspiration function of an automated washer.
13. Wash and decant five times.
14. Tap the plate and add 100 μl of TMB solution into each well.
15. Cover the plate with a plastic cover and incubate for 20 min at room temperature in the dark. At the end of the incubation, the content of wells should bring a bluish color.
16. Add 100 μl of sulfuring acid solution. This works to stop the reaction and changes the color of the content of wells into yellow.
17. Read absorbance of wells at 450 nm with a microplate reader against wells A1 and A2. The reader should be provided with the mapping of the plate and with the concentrations of the standard so that the concentration of each well should be provided as the output (*see* **Notes 6–8**).

4 Notes

1. Make sure that monocytes have adhered to the plastic bottom of the flask. To this end before discarding the medium from the flask, microscope the flask with an inverted microscope, and during microscopy move gently the flask to ascertain that monocytes are firmly adherent and they do not move (Subheading 3.2).
2. Use a scraper to remove monocytes from the flask at the end of the incubation period with trypsin/EDTA and before addition of FBS (Subheading 3.2).
3. Isolated neutrophils and monocytes are purified at more than 99 %, as assessed after staining with anti-CD15 and anti-CD14, respectively, and analysis through flow cytometer (Subheadings 3.2 and 3.3).
4. Note the number of cells added per well. Then suPAR is adjusted per 10,000 monocytes or 100,000 neutrophils (Subheadings 3.2 and 3.3).
5. The positive controls are selected from the biobank of previously aliquoted serum. Selected controls have concentration close to 15 ng/ml and 4 ng/ml. These controls should have run at least four times in the past (Subheading 3.5).
6. Report the mean of undiluted and diluted samples (Subheading 3.5).
7. The intraday variation of the assay is actually less than 1 % (Subheading 3.5).
8. Although aliquots should not be frozen and re-thawed, practice suggests that this procedure rarely affects results (Subheading 3.5).

References

1. Martin GS, Mannino DM, Eaton S, Moss M (2003) The epidemiology of sepsis in the United States from 1979 through 2000. N Engl J Med 348:1546–1554
2. Pierrakos C, Vincent JL (2010) Sepsis biomarkers: a review. Crit Care 14:R15
3. Backes Y, van der Sluijs KF, Mackie DP, Tacke F, Koch A, Tenhunen JJ et al (2012) Usefulness of suPAR as a biological marker in patients with systemic inflammation or infection: a systemic review. Intensive Care Med 38: 1418–1428
4. Giamarellos-Bourboulis EJ, Norrby-Teglund A, Mylona V, Savva A, Tsangaris I, Dimopoulou I et al (2012) Risk assessment in sepsis: a new prognostication score by APACHE II score and serum soluble urokinase plasminogen activator receptor. Crit Care 16:R149
5. Savva A, Raftogiannis M, Baziaka F, Routsi C, Antonopoulou A, Koutoukas P et al (2011) Soluble urokinase plasminogen activator receptor (suPAR) for assessment of disease severity in ventilator-associated pneumonia and sepsis. J Infect 63:344–350

Chapter 19

Clinical Diagnosis of Sepsis and the Combined Use of Biomarkers and Culture- and Non-Culture-Based Assays

Frank Bloos

Abstract

Sepsis is among the most common causes of death in hospitalized patients, and early recognition followed by immediate initiation of therapy is an important concept to improve survival in these patients. According to the definition of sepsis, diagnosis of sepsis requires the recognition of the systemic inflammatory response syndrome (SIRS) caused by infection as well as recognition of possible infection-related organ dysfunctions for diagnosis of severe sepsis or septic shock. Both SIRS and organ dysfunctions may occur frequently in hospitalized patients for various reasons. However, the fast recognition of acute infection as a cause of SIRS and newly developed organ dysfunction may be a demanding task since culture-based results of microbiological samples will be available only days after onset of symptoms. Biomarkers and PCR-based pathogen detection may help the physician in differentiating SIRS from sepsis. Procalcitonin (PCT) is the best investigated biomarker for this purpose. Furthermore, the current data support the usage of PCT for guidance of antimicrobial therapy. C-reactive protein (CRP) may be used to monitor the course of infection but has only limited discriminative capabilities. Interleukin-6 is widely used for its fast response to the infectious stimulus, but conclusive data for the application of this biomarker are missing. None of the available biomarkers can by itself reliably differentiate SIRS from sepsis but can aid and shorten the decision process. PCR-based pathogen detection can theoretically shorten the recognition of the underlying pathogen to about 8 h. However, this technique is expensive and requires additional staff in the laboratory; controlled prospective studies are missing. Although current studies suggest that PCR-based pathogen detection may be useful to shorten time to adequate antimicrobial therapy and diagnose invasive *Candida* infections, no general recommendations about the application of PCR for the diagnosis of sepsis can be given.

Key words Sepsis, Diagnosis, Biomarker, Cytokines, Procalcitonin, PCR

Sepsis is among the most common causes of death in hospitalized patients, and its incidence is likely to increase substantially as the population ages [1]. Hospital mortality of patients with sepsis ranges from 28.3 to 41.1 % in North America and Europe [2]. The population-based incidence for severe sepsis has been estimated in several European countries to be 66–78 per 100,000 inhabitants [3–5]. Severe sepsis often remains unrecognized outside of intensive care services [6]. This may be partly due to missing documentation of a new onset of organ dysfunction but may also be explained

Nicasio Mancini (ed.), *Sepsis: Diagnostic Methods and Protocols*, Methods in Molecular Biology, vol. 1237, DOI 10.1007/978-1-4939-1776-1_19, by the complex diagnostic procedures necessary to initiate adequate therapy of these patients. However, diagnosis of sepsis may be challenging even in the intensive care unit since the onset of sepsis may be misinterpreted as not being associated with a new infection. Culture-based pathogen detection cannot guide the physician in the first decision whether and how antimicrobial therapy needs to be initiated since results will be available after several days only. Although recent data might suggest that culture results should be taken into account before starting anti-infectious treatment in uncomplicated infection [7], current guidelines recommend to start antimicrobials within one hour after diagnosis of severe sepsis or septic shock [8]. Several new techniques such as biomarkers and molecular methods like PCR have been developed to improve and fasten the diagnostic process. However, there are only few clinical studies available which investigate the impact of these techniques on the clinical course of the patient. This chapter attempts to illustrate how diagnostic tools could affect the diagnostic workup in sepsis.

1 A Simple Case of Sepsis

A 52-year-old confused female patient presents in the emergency department with fever and shortness of breath. She has a breathing rate of 32 breaths/min, a heart rate of 130 beats/min, and a blood pressure of 130/65 mmHg. The pulse oximeter shows an arterial oxygen saturation of 85% when breathing room air, and the temperature is 38.5° C. Lung auscultation reveals crackles over the lower right lobe. White blood cell count is 15 Gpts/l.

1.1 What Is Sepsis?

Sepsis was defined by Roger Bone as an invasion of microorganisms or their toxins into the bloodstream together with the host response to this invasion [9]. In 1992, the *American Society of Chest Physicians* (ACCP) and the *Society of Critical Care Medicine* (SCCM) developed diagnostic criteria [10] which were supposed to reflect Bone's sepsis definition. The host response was named systemic inflammatory response syndrome (SIRS) and is characterized by criteria including tachycardia (heart rate >90 beats/minute), tachypnea (a respiratory rate >20/minute, hyperventilation, or the need for mechanical ventilation), hypo- or hyperthermia (a core temperature <36.0 °C or >38.0 °C), and leukopenia or leukocytosis (a white blood cell count <4,000/mm^3 or >12,000/mm^3). While SIRS may be induced by several noninfectious impacts such as trauma, burns, major surgery, etc., sepsis was defined as SIRS caused by infection [10]. In this context, severe sepsis was defined as sepsis combined with an acute infection-related organ dysfunction. Organ dysfunctions include septic encephalopathy, acute renal failure, hepatic dysfunction, lactacidosis, pulmonary failure, and coagulopathy. Septic shock occurs if sepsis is accompanied by arterial hypotension unresponsive to fluid resuscitation.

According to the consensus definition of sepsis, the patient fulfills all SIRS criteria. The symptoms and the clinical examination suggest a pneumonia. Hypoxemia and mental deterioration represent at least two acute organ dysfunctions. The patient is likely suffering from severe sepsis caused by community-acquired pneumonia.

1.2 Developing the Diagnosis

The diagnostic workflow for patients with severe sepsis or septic shock is shown in Fig. 1. In general, the basic diagnostic procedure corresponds to the approach necessary to diagnose any infection. After obtaining the patient's history and clinical examination, a focus of infection is suspected. It also needs to be clarified whether infection is accompanied by acute organ dysfunction. If acute organ dysfunction is present, the patient is at a higher risk of an unfavorable outcome, and therapeutic measures have to be initiated in parallel to the diagnostic workup even if the diagnosis of sepsis is most likely not yet confirmed. Such therapeutic measures include

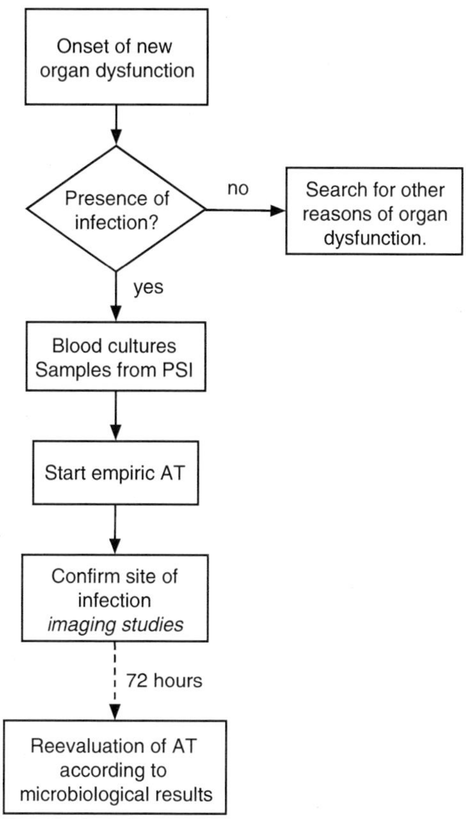

Fig. 1 Basic diagnostic workflow for patients with severe sepsis or septic shock. *PSI* presumed site of infection, *AT* antimicrobial therapy

immediate antimicrobial therapy and vital support depending on the patient's condition such as oxygen insufflation, endotracheal intubation, fluid resuscitation, etc.

1.2.1 Microbiological Workup

It is crucial to diagnose the microbiological nature of the infection in order to assess the adequacy of the empirical antimicrobial therapy. Current guidelines recommend to obtain at least two sets of blood cultures before starting antimicrobial therapy. Additionally, microbiological samples should be collected from the presumed site of infection [8].

1.2.2 Confirming the Site of Infection

The site of infection needs to be confirmed if the site of infection was suggested by the patient's history, symptoms, and clinical examination. Imaging techniques such as ultrasonography, X-ray, or CT scans are used for this purpose. Besides confirming the suspected infection, imaging studies add important information as it can be used to assess the course of infection and rule out whether an interventional or surgical source control is necessary in addition to the antimicrobial therapy. Depending on the patient's condition, it may be justified to perform a complete CT scan of the chest and abdomen including oral and intravenous contrast if the diagnostic workup did not reveal a site of infection [11].

1.2.3 Application of a Sepsis Score

Although the sepsis definition should be easy to implement into clinical diagnosis of this disease, criticism about the low specificity especially of the SIRS definition has been expressed [12, 13]. Furthermore, quality improvement studies showed that sepsis remains often undiagnosed and treatment is initiated too late [14–18]. Scores might be helpful in identifying patients in need for rapid therapy. The Mortality in Emergency Department Sepsis (MEDS) Score (Table 1) has been developed to allow for a risk stratification of sepsis in the emergency department [19]. Recent studies have confirmed that the MEDS Score helps to identify patients with a high risk of death and performs even better than biomarkers such as procalcitonin [20, 21]. However, it is unknown whether application of this score in the clinical practice would improve recognition of sepsis in the emergency department. Other scores such as the Modified Early Warning Score or the Rapid Emergency Medicine Score, which are not specifically designed for sepsis, may similarly predict the deterioration of a patient [22]. Staff training seems to be the more adequate tool to improve recognition of sepsis in the emergency department [23].

1.3 Closing the Case

The patient is diagnosed with severe sepsis caused by community-acquired pneumonia and submitted to the intensive care unit with an empirical antimicrobial therapy consisting of ceftriaxone and clarithromycin. The blood culture and the tracheal aspirate revealed Streptococcus pneumoniae. The antimicrobial treatment was de-escalated to a monotherapy with ceftriaxone at day 3.

Table 1
The mortality in emergency department sepsis (MEDS) score (modified from [19])

Variable	Points
Terminal illness (<30 days expected survival)	6
Tachypnea or hypoxia	3
Septic shock	3
Platelets <150,000/mm^3	3
Bands >5 %	3
Age >65 years	3
Lower respiratory infection	2
Nursing home resident	2
Altered mental status	2

The MEDS Score predicts a 28-day mortality. A score of 12 estimates a 15 % and a score of 15 a 50 % 28-day mortality

This case report showed that the application of sepsis definition and the basic workflow for infectious diseases would lead the physician to the correct diagnosis and adequate therapy. However, diagnosis of severe sepsis or septic shock is often not that easy. As SIRS is a very unspecific host response and also present in many noninfectious diseases, the differentiation of SIRS of an infectious versus a noninfectious origin is a complex task for every physician. Furthermore, blood culture is only positive in 30 % of the patients with sepsis [24], and results are available only within 3 days.

2 A Complex Case of Sepsis

A 64-year-old male patient with an uncomplicated postoperative course after pancreatectomy suffered from ventricular fibrillation on day 5 due to hypokalemia. Return of spontaneous circulation was achieved after 10 min of cardiopulmonary resuscitation. Afterwards, the patient was on mechanical ventilation and needed a moderate dosage of norepinephrine for circulatory support. On the following day, the patient developed fever with a body temperature of 38° C and a leukocytosis of 14 Gpt/l; C-reactive protein was rising. A spontaneous breathing trial to prepare extubation failed despite adequate response when addressed. The patient's condition was starting to deteriorate on day 7 as vasopressor support was increasing, body temperature was 38.5° C, and leukocytosis was 25 Gpt/l.

The patient fulfills SIRS criteria with fever, leukocytosis, and need for mechanical ventilation. On the day after ventricular fibrillation, it is difficult to differentiate a noninfectious origin from an infectious origin of SIRS. Cardiopulmonary resuscitation (CPR) in a patient after surgery may explain the observed host response. On the other hand, the patient may develop a postoperative infectious complication.

As the consensus definition of SIRS and sepsis does not help in solving this typical diagnostic conflict, a group of experts developed the PIRO (Predisposition, Infection, Response, Organ dysfunction) concept for improved characterization and staging of patients with sepsis [25]. Detection of microbial components by polymerase chain reaction (PCR) and biomarkers were named as future tools to describe the conditions *infection* and *response* within the PIRO system. Several biomarkers have been developed to aid the physician in the differentiation from an infectious and noninfectious origin of SIRS, but only few of them are commercially available. PCR might also improve the diagnosis of infection by proof of the underlying pathogen since—in opposite to culture-based methods—the results would be available within 1 working day.

3 Biomarkers

3.1 C-Reactive Protein

C-reactive protein (CRP) is an acute phase protein and is released from the liver after stimulation of IL-6 and other cytokines [26]. Secretion is started 4–6 h after stimulation and peaks at 36 h. CRP can aid in the diagnosis of infection [27]. However, CRP has a slow kinetic after onset of infection, is elevated also in minor infections, and is elevated in many noninfectious causes of inflammation such as trauma, surgery, or rheumatic disorders [28, 29]. A meta-analysis showed a low specificity of 0.67 and a sensitivity of 0.65 to differentiate bacterial from noninfectious causes of infection [30]. Given these facts, CRP has only limited capacities in differentiating noninfectious SIRS from sepsis. This was shown in a group of critically ill patients with SIRS where CRP performed inferior to procalcitonin and sTREM-1 [31]. Nevertheless, CRP may be a good marker to monitor success of antimicrobial therapy as CRP levels decrease when adequate anti-infectious therapy is initiated [32–34].

3.2 Procalcitonin

Procalcitonin (PCT) is the prohormone of calcitonin which is normally produced in the C-cells of the thyroid gland but is only present with <0.1 ng/ml in the blood of healthy humans. Depending on the severity of sepsis, PCT is massively released into the blood within 4–12 h after onset of infection [35–37]. A recent meta-analysis including 3,244 patients from 30 studies estimated a sensitivity of 0.77 and a specificity of 0.79 to discriminate sepsis from noninfectious SIRS [38]. The median discriminating cutoff was 1.1 (interquartile range 0.5–2.0) ng/ml, but this cutoff differed significantly across the studies [38]. Patients with septic shock have the highest PCT levels averaging between 4 and 45 ng/ml [39]. Moderately elevated PCT values around 1 ng/ml may be suggestive of fungal infections [40]. However, the quality and number of available studies do not allow to start empirical antifungal therapy solely based on PCT levels.

Circulating PCT levels decrease with a halftime of about 24 h when the infection is sufficiently treated. Increasing or persistent elevated PCT levels are predictive of an unfavorable outcome [41–43]. This observation was confirmed in several prospective studies where PCT-guided antimicrobial therapy in patients with lower respiratory tract infections resulted in a significant reduction in the duration of antimicrobial therapy without jeopardizing the treatment result [44–46]. It has been suggested that such a concept may also work in the critically ill patient with severe sepsis or septic shock [47, 48], but it has not been proven in large prospective studies. This hypothesis is currently tested in a prospective randomized multicenter study (SISPCT study; Clinical Trials ID: NCT00832039) of which results are expected in 2014. As any other biomarker, PCT can be elevated also in noninfectious diseases such as severe trauma, in surgery [49], after cardiac arrest [50], in patients with medullary thyroid carcinoma [51], and in several other inflammatory stimuli [39].

3.3 Interleukin-6

Interleukin (IL)-6 is the fastest biomarker as it reaches peak levels within 2 h after the infectious stimulus and persists much longer in the bloodstream than TNF and IL-1 [52]. Serum levels of IL-6 are closely related to the severity and outcome of sepsis in patients [53, 54] and decrease in patients where the infection is controlled [55]. However, convincing data from large prospective studies are missing. The data about the capability of IL-6 to discriminate sepsis from SIRS are inconsistent, showing both a good power [56] and a moderate discriminating power [57, 58]. The role of this cytokine as sepsis biomarker remains to be established [59].

3.4 sTREM-1

The triggering receptor expressed on myeloid cells-1 (TREM-1) is upregulated on phagocytes after exposure to bacteria and fungi [60]. Activated phagocytes release the soluble TREM-1 (sTREM-1) into the blood after onset of infection [61]. Higher sTREM-1 levels are predictive of an unfavorable outcome [62]. A recent meta-analysis calculated a sensitivity of 0.79 and a specificity of 0.8 for the diagnosis of sepsis [63]. This would be comparable to procalcitonin, but the number of available studies for sTREM-1 is still low. The role of sTREM-1 in the diagnosis of sepsis remains yet undefined, and larger studies are necessary to clarify this issue.

3.5 Lipopolysaccharide-Binding Protein

Lipopolysaccharide (LPS)-binding protein (LBP) is an acute phase protein that forms a complex with LPS. This complex is of immediate importance for the transcription of cytokines and other proinflammatory mediators [64, 65]. In human serum, LBP is constitutively present at a concentration of 5–10 μg/ml. LBP levels increase in sepsis patients within 24 h [66, 67]. However, the discriminative power of LBP to differentiate sepsis from noninfectious SIRS is poor [66, 67] and is not predictive of outcome [66, 68]. Currently, LBP does not play a role in the diagnosis of sepsis.

4 The Role of PCR-Based Pathogen Detection in Sepsis

Several studies have addressed the performance of PCR in various settings. A recent meta-analysis to compare multiplex PCR with blood culture included 34 studies [69]. The sensitivity for detecting bacteremia and fungemia was 0.75 and specificity 0.92. In general, multiplex PCR has twice as many positive results than a single set of blood cultures [70, 71]. However, the availability of PCR results takes more time than expected when the PCR was applied under clinical conditions. Time to positivity was about 24 h in the clinical setting instead of the suggested 6–8 h [71]. Faster availability of the results would need a 24 h a day and 7 days a week coverage of technicians and equipment.

Several studies suggest a good detection of invasive fungal infections; a meta-analysis reported a sensitivity of 0.95 and a specificity of 0.92 for the PCR-based diagnosis of invasive fungal detection [72]. Furthermore, time to prescription of antifungals was shorter when PCR was available as a diagnostic tool compared to blood culture alone [73]. Prospective randomized studies are missing for critically ill patients, but PCR-based algorithm for amphotericin B application in patients after bone marrow transplantation reduced mortality in a prospective randomized trial [74].

The clinical data of the PCR are promising. However, this technique has several limitations which need to be further investigated or solved before it can be generally applied into the clinical practice. The sensitivity is too low to rule out infection. Despite the high frequency of positive results, more than half of the PCRs remain negative in patients with severe sepsis or septic shock [70, 71]. Multiplex PCR can only detect those pathogens covered by the target list of the assay. Likewise, only specific resistances such as methicillin resistance or vancomycin resistance are available depending on the applied assay. The PCR method is time consuming at the bench, and lack of staff can delay time to positivity significantly beyond the proposed 6–8 h [71]. Even under optimal conditions, the time lag is still too high to consider the PCR result in the initial decision about antimicrobial therapy. It has been therefore suggested that PCR-based pathogen detection can only serve as an add-on to the conventional culture-based methods but cannot replace blood cultures [75].

5 Developing the Diagnosis

The workflow shown in Fig. 1 is the necessary workup for any patient with suspected sepsis. However, it is not suited to differentiate sepsis from any other noninfectious origin of SIRS. An extended flow chart of the diagnostic workup is suggested in Fig. 2. The addition of a biomarker to the clinical diagnosis is recommended both by the PIRO concept as well as in international guidelines [8, 25]. The use of PCT is favored in both publications. Despite the known shortcomings of this biomarker, PCT is to date still the best investigated biomarker under clinical conditions. A PCT value >1 ng/ml in a patient with suspected sepsis should trigger an antimicrobial therapy [76]. On the other hand, it is rather unlikely that a patient with PCT <0.1 ng/ml suffers from severe sepsis. However, a PCT increase may be missed if sepsis started only several hours ago. It is therefore advisable to repeat the PCT measurement after 12–24 h.

The currently available data do not allow for a recommendation to generally apply PCR-based pathogen detection into the diagnosis of severe sepsis or septic shock although the data are promising. If this method is used, a blood sample for PCR-based pathogen detection should be withdrawn together with the initial blood culture by using the same sterile precautions. The PCR result cannot be used for the decision about empirical antimicrobial therapy because of its time to positivity, but a positive PCR result could trigger an early adaption of the empirical antimicrobial therapy. However, such an approach has never been tested in prospective studies. Due to the limitation of this method, de-escalation of antimicrobial therapy solely on a PCR result currently cannot be recommended [75].

6 Closing the Case

The patient fulfilled SIRS criteria the day after cardiopulmonary resuscitation (CPR). The PCT on that day was 2 ng/ml, but it was assessed to be elevated because of the CPR. Clinical examination and chest X-ray did not reveal an infectious focus. The patient remained without antimicrobial therapy. However, extubation failed and the patient developed an acute organ dysfunction the following day. PCT was increased to 8 ng/ml. Septic shock was suspected, and antimicrobial therapy was initiated with meropenem after taking blood cultures and a blood sample for PCR-based pathogen detection. A CT scan revealed an intra-abdominal abscess. The PCR was positive for vancomycin-resistant enterococci (VRE), and linezolid was added to the empirical antimicrobial therapy. The blood culture was negative. The VRE remained unconfirmed in any culture-based techniques,

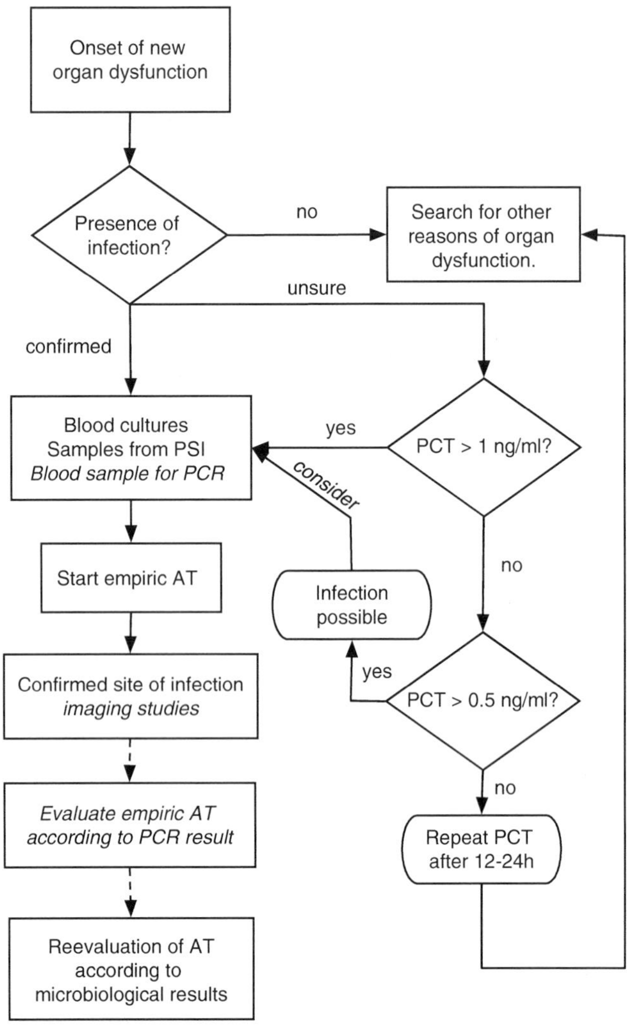

Fig. 2 Advanced diagnostic workflow for patients with severe sepsis or septic shock. This proposed workflow includes PCT as an example of a biomarker for differentiation of noninfectious SIRS from sepsis as well as PCR-based pathogen detection. *PSI* presumed site of infection, *AT* antimicrobial therapy, *PCT* procalcitonin, *PCR* polymerase chain reaction

but the treating physicians decided to continue the linezolid therapy. Serum PCT constantly decreased after surgical source control, and antimicrobial therapy was discontinued after 10 days.

Severe sepsis and septic shock are infectious emergency situations. Initiation of adequate antimicrobial therapy should be initiated as soon as possible when infection-related organ dysfunction occurs. As there is currently no biomarker available which alone allows a rapid and reliable discrimination between sepsis and SIRS without infection, the decision about empirical antimicrobial therapy remains a clinical decision. However, biomarkers can aid and shorten this decision process when taking into account the general shortcomings of biomarkers. PCT is currently the most investigated biomarker for this purpose and the only biomarker which has been integrated into treatment algorithms.

Likewise, PCR-based pathogen detection cannot rule out infection or help in the decision of empirical antimicrobial therapy. However, it can help to reduce the time until the empirical antimicrobial therapy can be assessed for adequacy. This is especially true for invasive *Candida* infections which are difficult to detect in blood cultures. The integration of PCR results into treatment decisions is far less investigated than biomarkers such as procalcitonin. It is therefore currently not possible to assess the impact on the patient or comment on cost-effectiveness.

References

1. Angus DC, Linde-Zwirble WT, Lidicker J, Clermont G, Carcillo J, Pinsky MR (2001) Epidemiology of severe sepsis in the United States: analysis of incidence, outcome, and associated costs of care. Crit Care Med 29(7): 1303–1310
2. Levy MM, Artigas A, Phillips GS, Rhodes A, Beale R, Osborn T, Vincent JL, Townsend S, Lemeshow S, Dellinger RP (2012) Outcomes of the surviving sepsis campaign in intensive care units in the USA and Europe: a prospective cohort study. Lancet Infect Dis 12(12): 919–924
3. Engel C, Brunkhorst FM, Bone HG, Brunkhorst R, Gerlach H, Grond S, Gruendling M, Huhle G, Jaschinski U, John S et al (2007) Epidemiology of sepsis in Germany: results from a national prospective multicenter study. Intensive Care Med 33(4):606–618
4. Harrison DA, Welch CA, Eddleston JM (2006) The epidemiology of severe sepsis in England, Wales and Northern Ireland, 1996 to 2004: secondary analysis of a high quality clinical database, the ICNARC Case Mix Programme Database. Crit Care 10(2):R42
5. van Gestel A, Bakker J, Veraart CP, van Hout BA (2004) Prevalence and incidence of severe sepsis in Dutch intensive care units. Crit Care 8(4):R153–R162
6. Rohde JM, Odden AJ, Bonham C, Kuhn L, Malani PN, Chen LM, Flanders SA, Iwashyna TJ (2013) The epidemiology of acute organ system dysfunction from severe sepsis outside of the intensive care unit. J Hosp Med 8(5): 243–247
7. Hranjec T, Rosenberger LH, Swenson B, Metzger R, Flohr TR, Politano AD, Riccio LM, Popovsky KA, Sawyer RG (2012) Aggressive versus conservative initiation of antimicrobial treatment in critically ill surgical patients with suspected intensive-care-unit-acquired infection: a quasi-experimental, before and after observational cohort study. Lancet Infect Dis 12(10):774–780
8. Dellinger RP, Levy MM, Rhodes A, Annane D, Gerlach H, Opal SM, Sevransky JE, Sprung CL, Douglas IS, Jaeschke R et al (2013) Surviving sepsis campaign: international guidelines for management of severe sepsis and septic shock, 2012. Intensive Care Med 39(2):165–228
9. Bone RC, Fisher CJ Jr, Clemmer TP, Slotman GJ, Metz CA, Balk RA (1989) The methylprednisolone severe sepsis study group: sepsis syndrome—a valid clinical entity. Crit Care Med 17(5):389–393
10. ACCP/SCCM Consensus Conference Committee (1992) Definition for sepsis and organ failure and guidelines for the use of innovative therapies in sepsis. Crit Care Med 20(6): 864–874
11. Marshall JC (2010) Principles of source control in the early management of sepsis. Curr Infect Dis Rep 12(5):345–353
12. Rangel-Frausto MS, Pittet D, Costigan M, Hwang T, Davis CS, Wenzel RP (1995) The natural history of the systemic inflammatory response syndrome (SIRS). A prospective study. JAMA 273(2):117–123
13. Vincent JL (1997) Dear SIRS, I'm sorry to say that I don't like you. Crit Care Med 25(2): 372–374
14. Larche J, Azoulay E, Fieux F, Mesnard L, Moreau D, Thiery G, Darmon M, Le Gall JR, Schlemmer B (2003) Improved survival of critically ill cancer patients with septic shock. Intensive Care Med 29(10):1688–1695

15. Kumar A, Roberts D, Wood KE, Light B, Parrillo JE, Sharma S, Suppes R, Feinstein D, Zanotti S, Taiberg L et al (2006) Duration of hypotension before initiation of effective antimicrobial therapy is the critical determinant of survival in human septic shock. Crit Care Med 34(6):1589–1596
16. Levy MM, Dellinger RP, Townsend SR, Linde-Zwirble WT, Marshall JC, Bion J, Schorr C, Artigas A, Ramsay G, Beale R et al (2010) The surviving sepsis campaign: results of an international guideline-based performance improvement program targeting severe sepsis. Crit Care Med 38(2):367–374
17. Gaieski DF, Mikkelsen ME, Band RA, Pines JM, Massone R, Furia FF, Shofer FS, Goyal M (2010) Impact of time to antibiotics on survival in patients with severe sepsis or septic shock in whom early goal-directed therapy was initiated in the emergency department. Crit Care Med 38(4):1045–1053
18. Ferrer R, Artigas A, Suarez D, Palencia E, Levy MM, Arenzana A, Perez XL, Sirvent JM (2009) Effectiveness of treatments for severe sepsis: a prospective multicenter observational study. Am J Respir Crit Care Med 180(9): 861–866
19. Shapiro NI, Wolfe RE, Moore RB, Smith E, Burdick E, Bates DW (2003) Mortality in Emergency Department Sepsis (MEDS) score: a prospectively derived and validated clinical prediction rule. Crit Care Med 31(3): 670–675
20. Lee CC, Chen SY, Tsai CL, Wu SC, Chiang WC, Wang JL, Sun HY, Chen SC, Chen WJ, Hsueh PR (2008) Prognostic value of mortality in emergency department sepsis score, procalcitonin, and C-reactive protein in patients with sepsis at the emergency department. Shock 29(3):322–327
21. Zhao Y, Li C, Jia Y (2013) Evaluation of the Mortality in Emergency Department Sepsis score combined with procalcitonin in septic patients. Am J Emerg Med 31(7): 1086–1091
22. Ghanem-Zoubi NO, Vardi M, Laor A, Weber G, Bitterman H (2011) Assessment of disease-severity scoring systems for patients with sepsis in general internal medicine departments. Crit Care 15(2):R95
23. Bastani A, Galens S, Rocchini A, Walch R, Shaqiri B, Palomba K, Milewski AM, Falzarano A, Loch D, Anderson W (2012) ED identification of patients with severe sepsis/septic shock decreases mortality in a community hospital. Am J Emerg Med 30(8):1561–1566
24. Calandra T, Cohen J (2005) The international sepsis forum consensus conference on definitions of infection in the intensive care unit. Crit Care Med 33(7):1538–1548
25. Levy MM, Fink MP, Marshall JC, Abraham E, Angus D, Cook D, Cohen J, Opal SM, Vincent JL, Ramsay G (2003) 2001 SCCM/ESICM/ACCP/ATS/SIS international sepsis definitions conference. Crit Care Med 31(4):1250–1256
26. Gabay C, Kushner I (1999) Acute-phase proteins and other systemic responses to inflammation. N Engl J Med 340(6):448–454
27. van Vugt SF, Broekhuizen BD, Lammens C, Zuithoff NP, de Jong PA, Coenen S, Ieven M, Butler CC, Goossens H, Little P et al (2013) Use of serum C reactive protein and procalcitonin concentrations in addition to symptoms and signs to predict pneumonia in patients presenting to primary care with acute cough: diagnostic study. BMJ 346:f2450
28. Eberhard OK, Haubitz M, Brunkhorst FM, Kliem V, Koch KM, Brunkhorst R (1997) Usefulness of procalcitonin for differentiation between activity of systemic autoimmune disease (systemic lupus erythematosus/systemic antineutrophil cytoplasmic antibody-associated vasculitis) and invasive bacterial infection. Arthritis Rheum 40(7):1250–1256
29. Meisner M, Adina H, Schmidt J (2006) Correlation of procalcitonin and C-reactive protein to inflammation, complications, and outcome during the intensive care unit course of multiple-trauma patients. Crit Care 10(1):R1
30. Simon L, Gauvin F, Amre DK, Saint-Louis P, Lacroix J (2004) Serum procalcitonin and C-reactive protein levels as markers of bacterial infection: a systematic review and meta-analysis. Clin Infect Dis 39(2):206–217
31. Su L, Han B, Liu C, Liang L, Jiang Z, Deng J, Yan P, Jia Y, Feng D, Xie L (2012) Value of soluble TREM-1, procalcitonin, and C-reactive protein serum levels as biomarkers for detecting bacteremia among sepsis patients with new fever in intensive care units: a prospective cohort study. BMC Infect Dis 12:157
32. Schmit X, Vincent JL (2008) The time course of blood C-reactive protein concentrations in relation to the response to initial antimicrobial therapy in patients with sepsis. Infection 36(3): 213–219
33. Hoeboer SH, Groeneveld AB (2013) Changes in circulating procalcitonin versus C-reactive protein in predicting evolution of infectious disease in febrile, critically ill patients. PLoS One 8(6):e65564
34. Povoa P, Teixeira-Pinto AM, Carneiro AH (2011) C-reactive protein, an early marker of community-acquired sepsis resolution: a multicenter prospective observational study. Crit Care 15(4):R169

35. Becker K, Müller B, Nylen ES (2001) Calcitonin gene family of peptides. In: Becker K (ed) Principles and practice of endocrinology and metabolism, 3rd edn. J.B. Lippincott Co., Philadelphia, PA, pp 520–534
36. Bloos F, Marshall JC, Dellinger RP, Vincent JL, Gutierrez G, Rivers E, Balk RA, Laterre PF, Angus DC, Reinhart K et al (2011) Multinational, observational study of procalcitonin in ICU patients with pneumonia requiring mechanical ventilation: a multicenter observational study. Crit Care 15:R88. doi:10.1186/cc10087
37. Brunkhorst FM, Heinz U, Forycki ZF (1998) Kinetics of procalcitonin in iatrogenic sepsis. Intensive Care Med 24(8):888–889
38. Wacker C, Prkno A, Brunkhorst FM, Schlattmann P (2013) Procalcitonin as a diagnostic marker for sepsis: a systematic review and meta-analysis. Lancet Infect Dis 13(5):426–435
39. Reinhart K, Meisner M (2011) Biomarkers in the critically ill patient: procalcitonin. Crit Care Clin 27(2):253–263
40. Dou YH, Du JK, Liu HL, Shong XD (2013) The role of procalcitonin in the identification of invasive fungal infection-a systemic review and meta-analysis. Diagn Microbiol Infect Dis 76:464–469
41. Karlsson S, Heikkinen M, Pettila V, Alila S, Vaisanen S, Pulkki K, Kolho E, Ruokonen E, Finnsepsis Study Group (2010) Predictive value of procalcitonin decrease in patients with severe sepsis: a prospective observational study. Crit Care 14(6):R205
42. Charles PE, Tinel C, Barbar S, Aho S, Prin S, Doise JM, Olsson NO, Blettery B, Quenot JP (2009) Procalcitonin kinetics within the first days of sepsis: relationship with the appropriateness of antibiotic therapy and the outcome. Crit Care 13(2):R38
43. Schuetz P, Maurer P, Punjabi V, Desai A, Amin D, Gluck E (2013) Procalcitonin decrease over 72 hours in US critical care units predicts fatal outcome in sepsis patients. Crit Care 17(3):R115
44. Christ-Crain M, Jaccard-Stolz D, Bingisser R, Gencay MM, Huber PR, Tamm M, Muller B (2004) Effect of procalcitonin-guided treatment on antibiotic use and outcome in lower respiratory tract infections: cluster-randomised, single-blinded intervention trial. Lancet 363 (9409):600–607
45. Christ-Crain M, Stolz D, Bingisser R, Muller C, Miedinger D, Huber PR, Zimmerli W, Harbarth S, Tamm M, Muller B (2006) Procalcitonin guidance of antibiotic therapy in community-acquired pneumonia: a randomized trial. Am J Respir Crit Care Med 174(1):84–93
46. Schuetz P, Briel M, Christ-Crain M, Stolz D, Bouadma L, Wolff M, Luyt CE, Chastre J, Tubach F, Kristoffersen KB et al (2012) Procalcitonin to guide initiation and duration of antibiotic treatment in acute respiratory infections: an individual patient data meta-analysis. Clin Infect Dis 55(5):651–662
47. Heyland DK, Johnson AP, Reynolds SC, Muscedere J (2011) Procalcitonin for reduced antibiotic exposure in the critical care setting: a systematic review and an economic evaluation. Crit Care Med 39(7):1792–1799
48. Nobre V, Harbarth S, Graf JD, Rohner P, Pugin J (2008) Use of procalcitonin to shorten antibiotic treatment duration in septic patients: a randomized trial. Am J Respir Crit Care Med 177(5):498–505
49. Sponholz C, Sakr Y, Reinhart K, Brunkhorst F (2006) Diagnostic value and prognostic implications of serum procalcitonin after cardiac surgery: a systematic review of the literature. Crit Care 10(5):R145
50. Schuetz P, Affolter B, Hunziker S, Winterhalder C, Fischer M, Balestra GM, Hunziker P, Marsch S (2010) Serum procalcitonin, C-reactive protein and white blood cell levels following hypothermia after cardiac arrest: a retrospective cohort study. Eur J Clin Invest 40(4):376–381
51. Giovanella L, Verburg FA, Imperiali M, Valabrega S, Trimboli P, Ceriani L (2013) Comparison of serum calcitonin and procalcitonin in detecting medullary thyroid carcinoma among patients with thyroid nodules. Clin Chem Lab Med 51(7):1477–1481
52. Song M, Kellum JA (2005) Interleukin-6. Crit Care Med 33(12 Suppl):S463–S465
53. Miguel-Bayarri V, Casanoves-Laparra EB, Pallas-Beneyto L, Sancho-Chinesta S, Martin-Osorio LF, Tormo-Calandin C, Bautista-Rentero D (2012) Prognostic value of the biomarkers procalcitonin, interleukin-6 and C-reactive protein in severe sepsis. Med Intensiva 36(8):556–562
54. Pettila V, Hynninen M, Takkunen O, Kuusela P, Valtonen M (2002) Predictive value of procalcitonin and interleukin 6 in critically ill patients with suspected sepsis. Intensive Care Med 28(9):1220–1225
55. Tschaikowsky K, Hedwig-Geissing M, Braun GG, Radespiel-Troeger M (2011) Predictive value of procalcitonin, interleukin-6, and C-reactive protein for survival in postoperative patients with severe sepsis. J Crit Care 26(1):54–64

56. Oberhoffer M, Russwurm S, Bredle D, Chatzinicolaou K, Reinhart K (2000) Discriminative power of inflammatory markers for prediction of tumor necrosis factor-alpha and interleukin-6 in ICU patients with systemic inflammatory response syndrome (SIRS) or sepsis at arbitrary time points. Intensive Care Med 26(Suppl 2):S170–S174
57. Harbarth S, Holeckova K, Froidevaux C, Pittet D, Ricou B, Grau GE, Vadas L, Pugin J (2001) Diagnostic value of procalcitonin, interleukin-6, and interleukin-8 in critically ill patients admitted with suspected sepsis. Am J Respir Crit Care Med 164(3):396–402
58. Tsalik EL, Jaggers LB, Glickman SW, Langley RJ, van Velkinburgh JC, Park LP, Fowler VG, Cairns CB, Kingsmore SF, Woods CW (2012) Discriminative value of inflammatory biomarkers for suspected sepsis. J Emerg Med 43(1): 97–106
59. Reinhart K, Bauer M, Riedemann NC, Hartog CS (2012) New approaches to sepsis: molecular diagnostics and biomarkers. Clin Microbiol Rev 25(4):609–634
60. Bouchon A, Facchetti F, Weigand MA, Colonna M (2001) TREM-1 amplifies inflammation and is a crucial mediator of septic shock. Nature 410(6832):1103–1107
61. Gibot S, Kolopp-Sarda MN, Bene MC, Cravoisy A, Levy B, Faure GC, Bollaert PE (2004) Plasma level of a triggering receptor expressed on myeloid cells-1: its diagnostic accuracy in patients with suspected sepsis. Ann Intern Med 141(1):9–15
62. Jeong SJ, Song YG, Kim CO, Kim HW, Ku NS, Han SH, Choi JY, Kim JM (2012) Measurement of plasma sTREM-1 in patients with severe sepsis receiving early goal-directed therapy and evaluation of its usefulness. Shock 37(6):574–578
63. Wu Y, Wang F, Fan X, Bao R, Bo L, Li J, Deng X (2012) Accuracy of plasma sTREM-1 for sepsis diagnosis in systemic inflammatory patients: a systematic review and meta-analysis. Crit Care 16(6):R229
64. Guha M, Mackman N (2001) LPS induction of gene expression in human monocytes. Cell Signal 13(2):85–94
65. Wright SD, Ramos RA, Tobias PS, Ulevitch RJ, Mathison JC (1990) CD14, a receptor for complexes of lipopolysaccharide (LPS) and LPS binding protein. Science 249(4975):1431–1433
66. Sakr Y, Burgett U, Nacul FE, Reinhart K, Brunkhorst F (2008) Lipopolysaccharide binding protein in a surgical intensive care unit: a marker of sepsis? Crit Care Med 36(7):2014–2022
67. Prucha M, Herold I, Zazula R, Dubska L, Dostal M, Hildebrand T, Hyanek J (2003) Significance of lipopolysaccharide-binding protein (an acute phase protein) in monitoring critically ill patients. Crit Care 7(6):R154–R159
68. Tschaikowsky K, Hedwig-Geissing M, Schmidt J, Braun GG (2011) Lipopolysaccharide-binding protein for monitoring of postoperative sepsis: complemental to C-reactive protein or redundant? PLoS One 6(8):e23615
69. Chang SS, Hsieh WH, Liu TS, Lee SH, Wang CH, Chou HC, Yeo YH, Tseng CP, Lee CC (2013) Multiplex PCR system for rapid detection of pathogens in patients with presumed sepsis: a systemic review and meta-analysis. PLoS One 8(5):e62323
70. Bloos F, Hinder F, Becker K, Sachse S, Mekontso Dessap A, Straube E, Cattoir V, Brun-Buisson C, Reinhart K, Peters G et al (2010) A multicenter trial to compare blood culture with polymerase chain reaction in severe human sepsis. Intensive Care Med 36(2): 241–247
71. Bloos F, Sachse S, Kortgen A, Pletz MW, Lehmann M, Straube E, Riedemann NC, Reinhart K, Bauer M (2012) Evaluation of a polymerase chain reaction assay for pathogen detection in septic patients under routine condition: an observational study. PLoS One 7(9): e46003
72. Avni T, Leibovici L, Paul M (2011) PCR diagnosis of invasive candidiasis: systematic review and meta-analysis. J Clin Microbiol 49(2): 665–670
73. Bloos F, Bayer O, Sachse S, Straube E, Reinhart K, Kortgen A (2013) Attributable costs of patients with candidemia and potential implications of polymerase chain reaction-based pathogen detection on antifungal therapy in patients with sepsis. J Crit Care 28(1):2–8
74. Hebart H, Klingspor L, Klingebiel T, Loeffler J, Tollemar J, Ljungman P, Wandt H, Schaefer-Eckart K, Dornbusch HJ, Meisner C et al (2009) A prospective randomized controlled trial comparing PCR-based and empirical treatment with liposomal amphotericin B in patients after allo-SCT. Bone Marrow Transplant 43(7): 553–561
75. Pletz MW, Wellinghausen N, Welte T (2011) Will polymerase chain reaction (PCR)-based diagnostics improve outcome in septic patients? A clinical view. Intensive Care Med 37(7): 1069–1076
76. Bouadma L, Luyt CE, Tubach F, Cracco C, Alvarez A, Schwebel C, Schortgen F, Lasocki S, Veber B, Dehoux M et al (2010) Use of procalcitonin to reduce patients' exposure to antibiotics in intensive care units (PRORATA trial): a multicentre randomised controlled trial. Lancet 375(9713):463–474

Index

Nicasio Mancini (ed.), *Sepsis: Diagnostic Methods and Protocols*, Methods in Molecular Biology, vol. 1237,
DOI 10.1007/978-1-4939-1776-1, © Springer Science+Business Media New York 2015

V

Printed by Printforce, the Netherlands